Ben L. Rapien

Fundamentals of Press Brake Tooling

The Basic Information You Need to Know in Order to Design and Form Good Parts

2nd Edition

HANSER

Hanser Publishers, Munich

Hanser Publications, Cincinnati

The Author:
Ben L. Rapien, 5658 Antoninus Drive, Cincinnati, OH 45238, USA

Distributed by
Carl Hanser Verlag
Vilshofener Straße 10 | 81679 Munich | Germany
info@hanser.de
www.hanser-fachbuch.de
www.hanserpublications.com

Library of Congress Cataloging-in-Publication Data

Rapien, Ben L., 1935-
Fundamentals of press brake tooling : the basic information you need to know in order to design and form good parts / Ben L. Rapien. -- 2nd ed.
p. cm.
ISBN-13: 978-1-56990-474-9 (hardcover)
ISBN-10: 1-56990-474-X (hardcover)
1. Sheet-metal work. 2. Machine-tools. 3. Machine parts. I. Title.
TS250.R35 2010
671.8'23--dc22
2010015242

Bibliografische Information Der Deutschen Bibliothek
Die Deutsche Bibliothek verzeichnet diese Publikation in der Deutschen Nationalbibliografie; detaillierte bibliografische Daten sind im Internet über <http://dnb.d-nb.de> abrufbar.

ISBN 978-3-446-42490-6

Production Management: Steffen Jörg
Coverconcept: Marc Müller-Bremer, Rebranding, München, Germany
Coverdesign: Stephan Rönigk

Table of Contents

Introduction

This book will provide many insights into properly designing parts that can be formed using a press brake. It includes a basic understanding of what must be considered in selecting proper tooling, determining minimum machine requirements, using blank size calculations, and arriving at an acceptable bend sequence. Also included is a great deal of diverse information that will save time, money, and allow parts to be produced to a quality acceptable to the final user.

The author of this book has forty-five years of experience working with press brake tooling. This background includes reviewing customer drawings, designing appropriate tooling, supervising the testing the finished dies (to insure the correct shape could be formed at the tonnage available), plus being a consultant for many companies. This accumulation of expertise has allowed me to help many companies solve the problems they have encountered in making acceptable sheet metal or plate type parts using a press brake.

Over my many years in this field, I have served as clinic chairman in SME sponsored "Sheet Metal Forming" seminars, taught "Basics of Sheet Metal Forming" for 18 years at the University of Cincinnati's College of Engineering, given talks across the United States and Canada with Cincinnati Incorporated "Metal Management" seminars, written many articles for technical magazines including "The Fabricator" and "Welding Design and Fabrication," contributed to the Press Brake forming section of the *TMEH Forming Handbook* published by the SME, and served as a Press Brake Tooling Consultant for many companies.

This book is based on an engineering background. However, the technical information given is simplified to allow most press brake operators to understand the concepts. The information provided may be not what the typical engineer wants to hear, but it is based on "real world" experience based on the actual parameters that have to be considered when parts are formed.

Ben Rapien
Cincinnati, Ohio

Acknowledgements

I wish to express my gratitude to all the people at Cincinnati Incorporated who helped me with the charts and photographs for this book. I also offer special thanks to Mr. Perrin G. March III, Mr. John Kahler, Mr. Dennis Bray, and Mr. Ralph Wellington who spent a great deal of time reviewing the text of the book.

The encouragement of my family, especially my daughter Lynne Averbeck who took my original sketches and brought them to life on the computer, was essential to bringing this book to publication.

With the patience and help of my wife Mary Lynne, who edited my original text, my goal of handing on my 45 years experience in the field of press brake tooling has become a reality.

I dedicate this book to my wife Mary Lynne and my six children.

Chapter 1

Review of the Press Brake

In order to make sure the correct forming machine is selected, we will review press brake fundamentals in this chapter.

The press brake is a basic machine tool used to form a variety of sheet metal and plate parts in many different industries. Parts formed can be used to make: file cabinets, office partitions, boxes, metal doors and door frames, display cases in grocery stores, trucks and truck bodies, safes, tote boxes, corrugated pipe, many different types of radius bends, countertops and sinks, popcorn and soda dispensers, siding, roof decking, playground equipment parts for heavy construction equipment (including forklift trucks, cranes, and bulldozers), plus a plethora of other applications.

A "press brake" can form only straight line bends, it is not a press. A new press brake can be modified to allow some simple "press type" dies to be mounted in the press brake and be used successfully. The supervisor who decides to mount press tooling on a press brake must ensure that: 1) all "press" safety provisions are considered; 2) the press brake loading limits are not exceeded; and 3) the operator has been properly trained.

It is important to understand some basic information about press brakes, especially some of the available options and limitations, in order for you to wisely use the press brake to your best advantage. This chapter provides a foundation that allows someone not familiar with the equipment to understand basic terminology.

A) Basic Press Brake Structure

The press brake frame is made "boxlike" in order to maintain rigidity while it is being used. The design is important in order to provide good, consistent bend angles from part to part. All press brakes have the following features.

1) Bed and Housings

Standard press brake construction starts with the bed plate. This is a heavy plate, 3" to 8" thick depending on the machine size, which is located left to right across the bottom front of the press brake. The top of the plate is normally located 30" to 34" above the floor level.

Longer beds or high tonnage machines often require the bed plate to extend below the floor level into a special rectangular hole in the floor called a "pit." The length of the "bed" limits the length of the bend that can normally be formed. The bed is mounted on two plates attached 90° to the bed. The plates, between 2" and 8" thick, are called the "housings" or "uprights" (Fig. 1-1). The housings are separated by the distance needed to form most of the parts produced by the company buying the press brake. This space between the housings is listed as the "nominal forming length" by most press brake manufacturers. The distance between the housings allows the part to be formed to have relatively long flanges, and assures that the part will not hit the housing plates. The housing plate should have a cutout (called a throat) near the bed top to allow a flange of some depth to be formed to a 90° angle along the full length of the bed plate. In most designs, the standard throat depth may be limited to about 8".

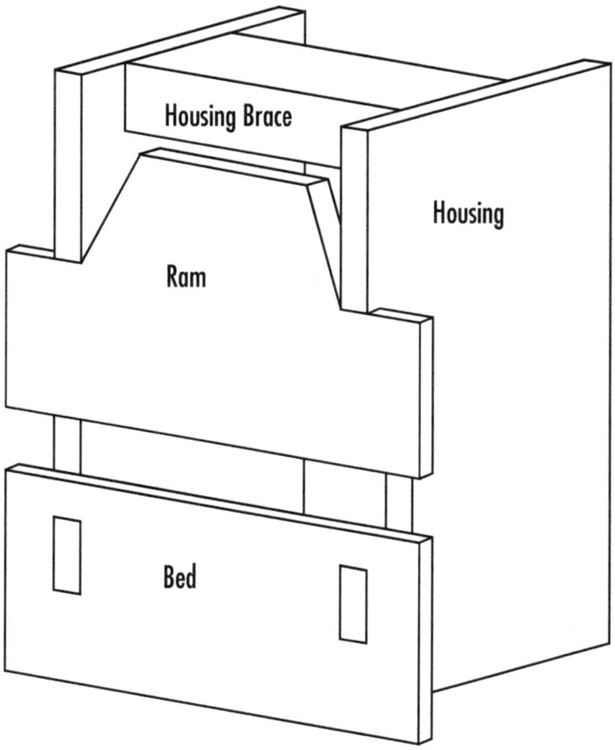

Fig. 1-1. Typical frame design for a down acting press brake.

A few manufacturers actually sell their equipment listing the forming length as the total length of the press brake bed, not the distance between the housings. In many cases, the owners of these press brakes found that some parts could not be formed since the required "part flange" and housing throat were not compatible. Make sure all employees who are assigned work at a specific press brake understand the length and flange limitations.

2) Housing Brace

A spacer plate of various designs is mounted between each housing plate to complete a boxlike design that includes the bed and the two housings. The spacer plate is normally mounted near the top rear of the housings in order to provide proper stiffness to the structure, and to properly hold the housing spacing. This plate is often called the "housing brace."

3) Ram

The last large plate required is called the "ram." It is mounted above the bed top, and is attached to the front of each housing plate. The ram is the same length as the bed plate but it is made from material thinner than the bed in order to provide maximum clearance for part flanges which will move near the front and rear of the ram during the part forming cycle. The ram is designed to be higher than the bed is deep in order to allow the stiffness of the bed and ram to maintain the same deflections under load.

When the ram moves up and down to form a part, it is kept in position by slides and guides that hold the ram to the front of the housings. The slide and guide design is similar to the design of parts found on the bottom of well made dresser drawers. The "hat" or "tee" shaped guide is fastened to the piece of wood separating the drawers, and it does not move. The slide on the drawer is mounted on the bottom of the drawer, normally in the center. The slide is designed to surround the hat shaped or tee shaped guide in order to keep the drawer from moving right to left when it is pulled out and thus becoming stuck in the frame of the dresser. On a press brake, a guide is mounted on each of the housings just above the cutout for the throat area. It also does not move. Attached to the rear of the press brake ram are two ram slides. They are located at each housing and fastened around the guides. They allow the ram to be securely held to the housings, yet allow the ram to be moved up and down easily while maintaining a parallel position to the bed top (without tilting right to left) throughout the forming stroke. Keeping a minimum running clearance between the slides and guides, plus proper lubrication, is required to provide a maintenance free operation, and to establish a consistent "bend line" location.

A unique press brake was designed in Europe during the 1960s, whereby the bed and ram positions are reversed. The ram moves up toward the stationary bed. The design is still available today, but it is restricted to smaller tonnage machines.

B) Types of Press Brakes

1) Mechanical Press Brake

The first press brakes that were originally designed and manufactured in the late 1800s were powered mechanically (Fig. 1-2). A large flywheel, normally mounted on the outside of the right- hand housing, was held on a shaft from which gears, a clutch, and a full length shaft holding two eccentrics (one at each housing) powered the drive of the press brake. When the clutch was engaged it allowed the ram to move up and down a specific distance in order to form a part and provide clearance for its removal. A screw adjustment was devised to adjust the bottom of the stroke location incrementally in order to provide for the different height tooling that would be mounted in the forming area. Two large screws were mounted to each of newly designed "slides." The screw was threaded into a part called a **pitman** that attached the slide assembly to an eccentric which was located near the top of each housing and provided the stroke (the up and down motion) for the ram.

In order to install tooling, the press brake ram had to be moved to the bottom of the stroke. After turning the press brake on, the motor activated the flywheel and allowed it to reach the required number of revolutions per minute (RPM). A mechanical "treadle," attached to a

Fig 1-2. Mechanical press brake. (Photo courtesy of Cincinnati Incorporated)

number of links to activate the clutch was pushed down by foot pressure. If the press brake was operated by an air activated clutch, an electrical foot switch was used to lower the ram. As the ram was lowered from the top of the stroke to the bottom, the operator watched a moving stripe painted on the eccentric until it aligned with a similar fixed painted stripe located on the pitman (Fig. 1-3). When the painted stripes matched so they were in "line," the ram was at the bottom of the stroke. The operator stopped the ram motion at the bottom of the stroke.

The ram position could be incrementally adjusted up or down by the two screws between the pitmans and the slides which were turned by activating the ram elevating motor. The motor was run using push buttons on the right hand side of the ram. The readout of the ram elevating motor was in thousandths of an inch, which allowed the operator to adjust the exact setting of the ram position very carefully in order to create the proper die space needed for the press brake tooling.

The mechanical press brake remained popular into the 1980s. The fact that a mechanical machine could be overloaded and still operate, plus its ability to cycle at a high forming speed were factors in maintaining its popularity. For many years, the mechanical press brake was much easier to set up than the hydraulic press brake, which had entered the marketplace in the late 1940s.

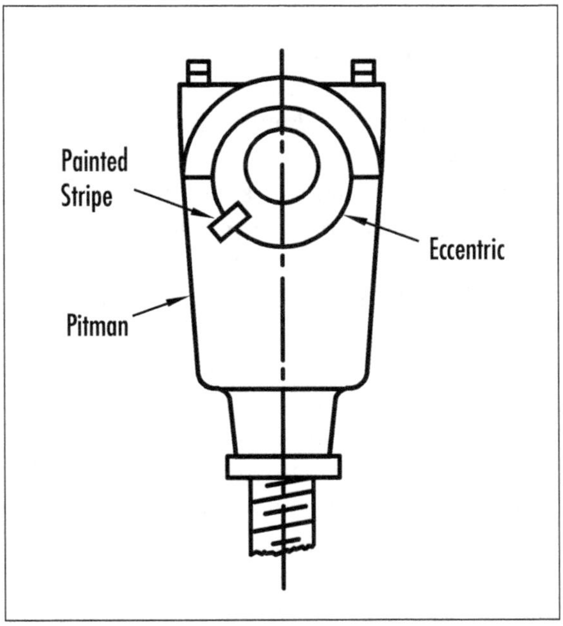

Fig. 1-3. Pitman and eccentric.

2) Hydraulic Press Brake

The hydraulic press brake, developed in the late 1940s, also had a bed, ram, and housing brace. The mechanical flywheel and gears were replaced with hydraulic cylinders. When the hydraulic system was designed, engineers found it difficult to have the cylinders on each housing move up and down together, hold parallelism with the bed top during the stroke, and not drift out of level in between each stroke. Leveling devices using steel tape were made available in 1958. They attached to each of the cylinders and valves and, in addition to improved hydraulic power sources, solved the leveling problem for 10 to 15 years. The hydraulic press brake was considered "slow" when compared to the popular mechanical press brake. It did, however, find a niche in its ability to form heavier steel plate (Fig. 1-4).

Fig. 1-4. Hydraulic press brake. (Photo courtesy of Cincinnati Incorporated)

In the late 1970s, printed circuit boards, encoders, and high-tech electronics were designed into the hydraulic control systems. At the same time, better hydraulic components were becoming available. Newly designed valves, high pressure hoses, and new motors allowed the new generation of hydraulic press brakes to approach the forming speeds of the mechanical press brake.

The appeal of the hydraulic press brake became apparent during the 1970s. The longer stroke length available from cylinders compared to the eccentric limitations of the mechanical design, plus the constant tonnage available throughout the forming stroke, allowed the hydraulic machines to form thicker materials. The fact that a hydraulic press brake could not be "overloaded" was viewed by management as an advantage, since machine damage caused by some overloads on the mechanical press brakes were very costly and time consuming to repair. The ability to change the speed of the stroke, easier methods of setting up tooling, plus the many features introduced through electronics have cemented the hydraulic design as the best available in the current marketplace. The determining factor was that hydraulic press brakes were considerably less expensive than a comparable tonnage mechanical press brake.

One of the initial difficulties with the hydraulic press brake was the inability of the operator to know when the ram was at the bottom of the stroke. By the late 1950s, the machines had "directional lights" which indicated the direction of the ram motion. It wasn't until 1975 that a light was added to the side of the housing that indicated the ram was at the exact bottom of the stroke. Hand cranks adjusted the ram reversal point, but the actual reversal of the ram was done with "limit" switches. The actual reversal point depended on the downward speed of the ram to activate the switch. The ram speed caused a consistent downward overshoot of the limit switch to occur before the ram would actually reverse its direction. The repeatability of the ram, even for the most advanced hydraulic press brakes, was ±0.002".

It wasn't until the 1980s that vibration proof encoders were developed with sufficient reliability to permit them to be incorporated into press brake controls. The advantage was that the ram repeatability could be improved to ±0.0004" at the bottom of the stroke.

Before selecting a hydraulic press brake for your company, check two factors. The machine controls must provide a simple method for the operator to know when the ram is at the bottom of the stroke. It is also important for the ram setting to be easily adjusted so that a proper bending position can quickly be determined. The operator must be able to incrementally set the ram position to compensate for material springback or to easily set up special tooling.

Present hydraulic press brake designs take advantage of many electronic options that have been developed since 1990. Current machines often offer excellent tonnage readouts to monitor the forming loads; programs to assist the operator to determine a safe bend sequence and to recall the tooling and machine setting of previous jobs; the ability to interface with computers in other parts of the plant; and the option to use programs to help determine the proper blank size.

A major design feature incorporated into many of the higher "tech" hydraulic press brakes is the capability of the operator to input the forming angle that is needed into the control. This allows the operator to quickly form different angles from part to part with minimum setup time. It is important to remember that this concept is based on the mathematical travel of the upper die into the lower die cavity, plus any allowance for springback provided in the machine software. All of these programs are based on "air bending" the part (see Chapter 3). If bottoming dies or other type tooling is being used, the option will not work properly.

The hydraulic press brake can be used for many types of special dies. The versatility of some machines may be limited because their design restricts the tooling selection to a few simple vee-shaped dies. Always review with the manufacturer, prior to obtaining a press brake, if the press brake you are selecting can use the tooling needed to form your product. The many different tooling applications discussed in this book provide only a sample of what can be and has been formed with a well designed press brake.

3) Electric Press Brakes

Since the 1990s, a new type of drive has been adapted to provide power to a press brake. An electric motor type drive is incorporated as a power source that allows better control of ram speed and resultant bottom of stroke repeatability (according to manufacturer's literature). This design is in its infancy, and may develop into a viable option in a few years. The main limitations at this time appear to be the difficulty in designing machines with sufficient tonnage capabilities to compete in the general-purpose press brake market. It may be wise to investigate all of the advantages, as well as limitations, prior to purchasing a press brake with this type of drive.

C) Basic Press Brake Features

1) Die Space

The maximum space available for the press brake tooling is the distance measured from the bed top to the bottom of the ram nose when the ram is at the bottom of its stroke and the ram adjustment is at the top (for a mechanical press brake), or when any additional stroke of a hydraulic press brake is not needed. In many cases, a stroke of about 3" to 4" is sufficient to bend the part and have room to remove it from the press brake. There are differences between the mechanical and hydraulic press brakes.

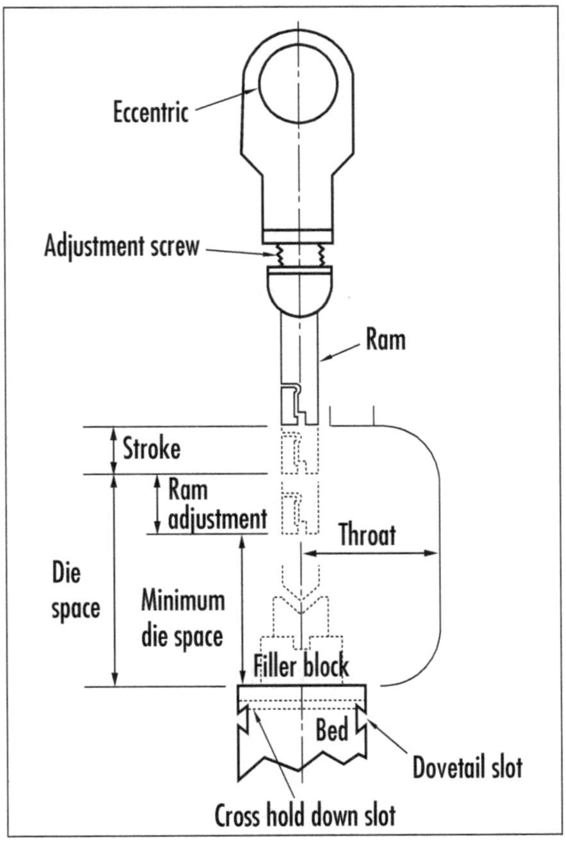

Fig. 1-5. Die space, throat, and stroke of a mechanical press brake.

a) Mechanical Press Brake. On a mechanical press brake, the forming stroke is fixed. Normally, the eccentric is made to provide a 3" ram travel. On larger press brakes, a 4" stroke is supplied as standard. When the ram is cycled to the bottom of the stroke, a maximum of 12" is available on most machines to install the tooling (Fig. 1-5). Since the majority of press brake tooling (upper and lower dies) is only 6" to 7" high, the ram must be adjusted downward to allow the upper die to penetrate into the lower die to form the part. The ram adjustment (screw travel) of the ram is normally limited to a maximum of 6". To ensure that the minimum die space is filled, a bar of steel called a filler block or die holder is placed under the lower die (Fig. 1-6). The filler block not only helps to "fill up" some of the die space, it also makes it convenient for the operator to move the block front to back in order to align the lower die with the upper die.

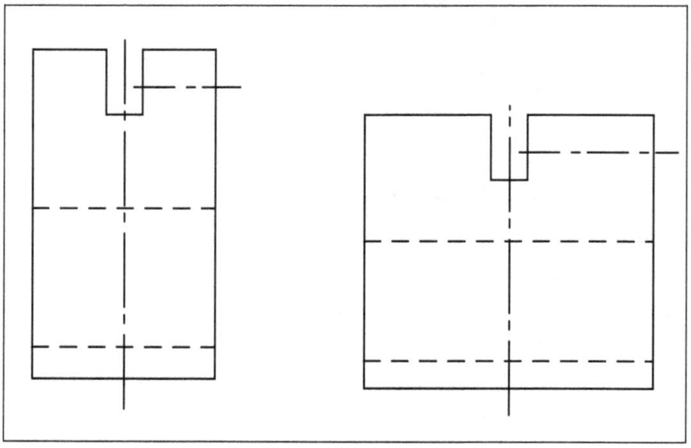

Fig. 1-6. Typical shapes of filler blocks.

b) Hydraulic Press Brake. When the stroke of the cylinder has lifted the ram up to its maximum opening, the distance between the bed top and ram nose is called the **open height** (Fig. 1-7). When the stroke descends its entire distance, the space from the bed top to ram nose is called the **closed height**. The distance of the closed height of many hydraulic press brakes is the same as that found on a similar size mechanical press brake. This allows the same height die holder and forming tools to be used on either type machine. On a hydraulic press brake, the stroke length can be adjusted by the operator to obtain the optimum opening to easily remove the part, yet minimize the time required for the forming stroke.

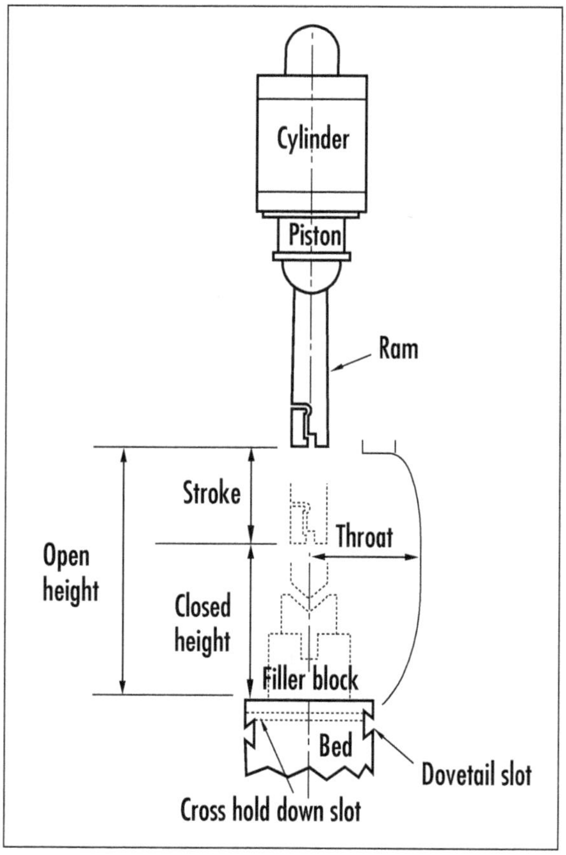

Fig. 1-7. Die space, throat, and stroke of a hydraulic press brake.

2) Throat

The gap or relief cut into each housing (Fig. 1-7) is somewhat different for each size press brake. A proper throat design should allow a flat sheet, as long as the length of the bed, to be formed when placed on top of a full length lower die. The top of the lower die plus filler block is normally about 7" above the bed top. The flange length of the part can be as deep as the nominal depth of the throat. The metal should be able to be formed to a good 90° bend, allowing for any overbend required due to material springback. If the part has previous bends or is formed to an angle sharper than 90°, it could hit the housing in the throat area and cause damage to the part, tooling, or machine.

3) Standardized Method for Fastening the Tooling in the Press Brake

The ram nose is machined to provide a

slot or cutout to locate the top or upper die. A clamping bar, usually made in consistent lengths of 6", 12", or 24" (depending on the manufacturer), is provided to lock the tongue or stem machined on the top of the die in place. The size of most die tongues is 0.5" wide by 0.625" high. The ram nose width is the sum of the flat ram surface in the rear of the tongue slot, plus the 0.5" opening for the die tongue, plus the width of the clamping bar (Fig. 1-8).

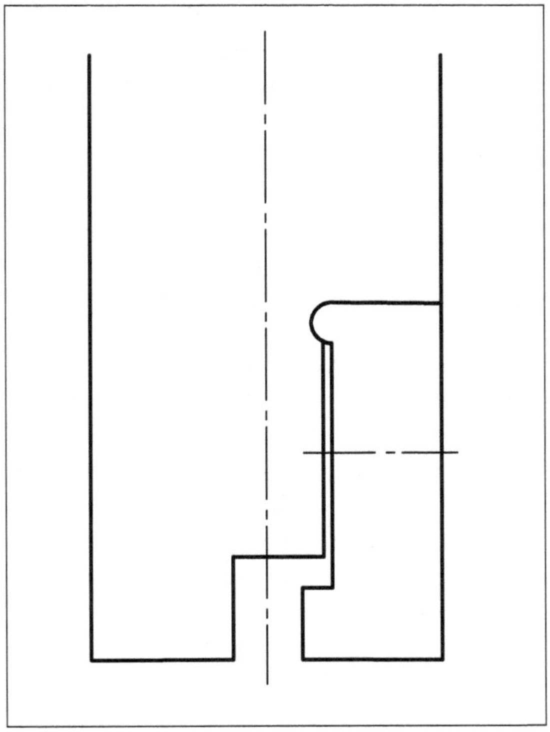

Fig. 1-8. Ram nose and die clamp.

The width of the bed top is always larger than the width of the ram nose. The actual size of the bed top will be about $^1/_2$" thinner than the thickness of the bed plate selected for the size machine being used. This difference results from each side of the bed top being machined down the front and rear anywhere from 4" to 12" to provide a consistent surface for measuring and aligning tooling. Many beds are designed with cross slots (often T slots) to allow the die holder or lower die to be bolted to the bed top. Some bed tops are machined with a full length slot directly under the $^1/_2$" tongue slot in the ram nose. The slot may range from 0.5" wide to as much as 0.75" wide, depending on the manufacturer. This allows the filler block or lower die to be lifted up, using a crowbar, to place shims under the block to compensate for deflection of the bed during the forming cycle or possible die wear.

A few press brakes are designed with set screws along the front and rear of the bed top with the screws reaching into a full length bed slot. This is a cheap but poor design since the bed slot has to be wide enough to allow for some front to rear adjustment. This minimizes the shoulder contact of the lower die with the bed top, which can result in indention of the bed top near the bed slot. Drilled and tapped holes can taper near the top surface of the bed top allowing for potential damage from concentrated die loads at the hole locations. Any wear on the bed top resulting from loading or unloading the lower dies will create a problem with holding good angular consistency.

The ram of a press brake must have some method for **tilting** the ram. It is imperative that the upper and lower dies be parallel from end to end if a good, straight bend is to be made. When a press brake has been used for a long period of time, it is highly probable that some variations will appear in the die space area. Housing stretch, worn tooling or filler blocks, or the need to change the bend angle from one end to the other requires the ability to adjust the ram to hold parallelism. Most press brakes have the ability to tilt the ram from one end to the other, from 0.25" to as much as 1", depending on the press brake size and the manufacturer.

4) Options That May be Added to a New Press Brake

When a press brake is purchased, a variety of options are available from most manufacturers which add flexibility to the machine or limit it to specific type of work.

a) Additional Stroke. If the press brake is a mechanical drive design, the standard 3" stroke is often increased to 4". This allows more room for removing the formed part. This feature is especially useful if previous bends have been made in the rear flange. The disadvantage is that there can be a reduction of forming capacity available for some parts since the maximum tonnage is not available until the stroke is about 0.25" up from the bottom of the stroke. The 3" stroke has the maximum tonnage available about 0.375" up from the bottom of the stroke. If the additional stroke is obtained by simply changing the eccentric size which increases the machine stroke from the standard 3" eccentric to an eccentric that will provide a 4" stroke, the standard 12" die space will be affected. The 4" eccentric will cause the ram to move 0.5" further up and 0.5" further downward. As a result, the normal 12" die space is reduced to 11.5", which may affect some tooling choices. If a new machine is being purchased, the option of having additional stroke, plus changing the housing size to obtain standard die space, should be considered. If your company currently has a mechanical press brake with a 4" stroke, and is rated less than 250 tons of forming capacity, it may be a good idea to confirm the available maximum die space.

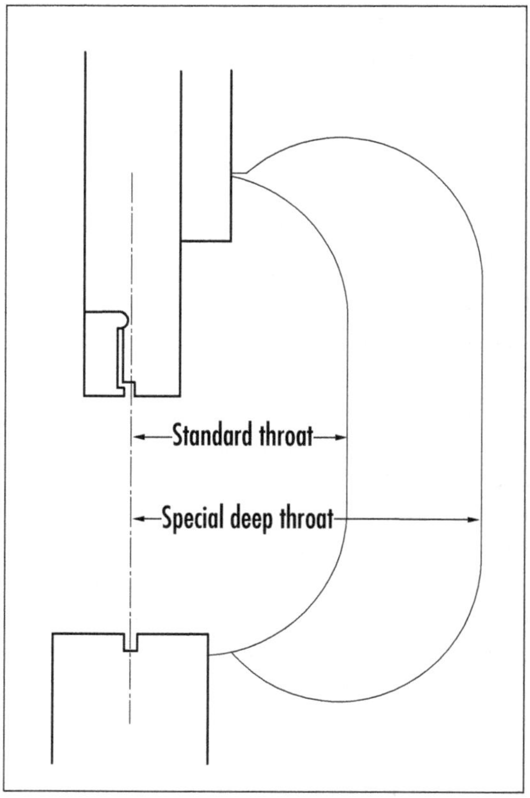

Fig. 1-9. Special deep throat in a press brake housing.

If the machine is a hydraulic design, adding stroke does not affect the tonnage. With hydraulic press brakes, the location of the bottom of the stroke stays the same. The additional stroke increases the open height of the total die space. For example, if the stroke is increased from 12" to 18", the actual increase of stroke is 6" over the standard. The height of the cylinder and housing, however, increases 12". This allows the cylinder to travel up and down the additional distance. For some press brakes, the additional stroke can affect the overall height of the machine.

b) Deeper Throats (Fig. 1- 9). Some parts utilize the full length of the bed and ram, but the back flange interferes with the housings because the standard throat depth is not sufficient. A new machine with a unique throat allows those parts to be formed. Each size press brake does have design limitations as to the maximum depth of throat. Changing the depth of throat will also affect the size of both housings.

It is extremely poor design to suggest only one housing be made with a deeper throat while the other housing be kept standard. The difference in the deflection of both housings under load would make forming a straight bend very difficult. Some new press brakes have encoders on each of the housings, which might minimize this concern. However, adding another variable to the problems of making a good full length 90° bend is considered high risk at this time.

c) Machining the Bed (and ram) for **Angle Brackets.** Some possible tooling applications require the lower die to be larger than the standard design bed top. In order to obtain support for this type of tooling, the bed (and ram if needed) can be modified with special machining and drilling from the top of the bed downwards in the front and rear to provide a heeled step to mount a special 90° **angle bracket** in the front and rear. Depending on the size of the angle bracket selected, the width of the bed top is increased to the width of the standard bed top plus the width of each angle bracket (Fig. 1-10).

In some cases, especially if punching die sets are being used, angle brackets may be required on the ram as well as the bed. Remember that the ram is thinner than the standard bed top. When angle brackets, which are normally made the same size for the bed as well as the ram, are mounted on both bed and ram, the mounting holes will have a different front to rear spacing between the ram and the bed. Often the tooling designer is not aware of the difference, and provides the same mounting pattern for both the top and bottom of the die set.

Angle brackets are very useful but may have to be made in unique lengths. A length of angle bracket mounted on the rear of the bed may have to fit between the support arms of an NC back gage. Lengths of angle brackets also cover up the dovetail slots machined in the front and rear of some bed tops; these are the slots that allow sheet supports or simple gages to be mounted. Prior to purchasing, a review of what angle lengths are needed is mandatory.

d) Permanently Wide Bed (and Ram). Some larger press brakes that form a great deal of heavy plate using wider dies require a wider-than-standard bed top for proper die support.

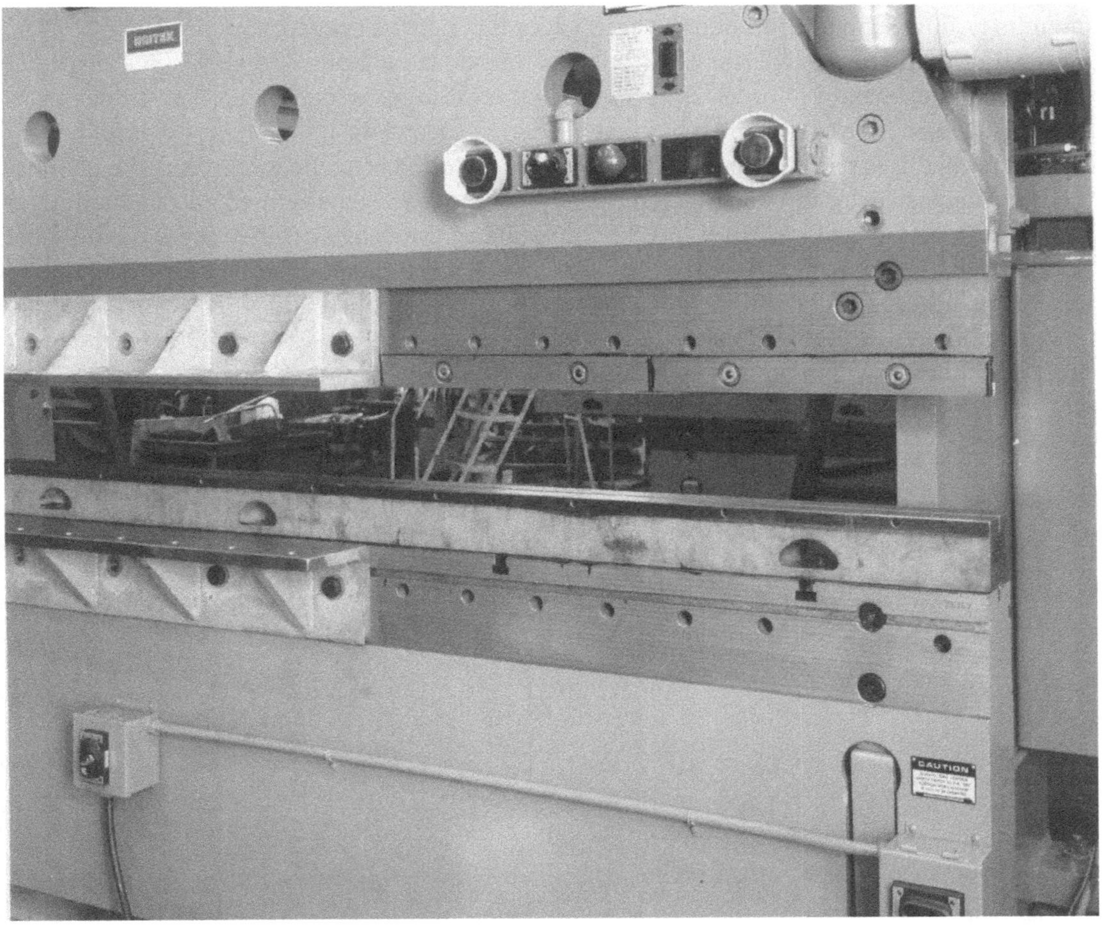

Fig 1-10. Angle brackets mounted on the ram and bed of a press brake.
(Photo courtesy of Cincinnati Incorporated)

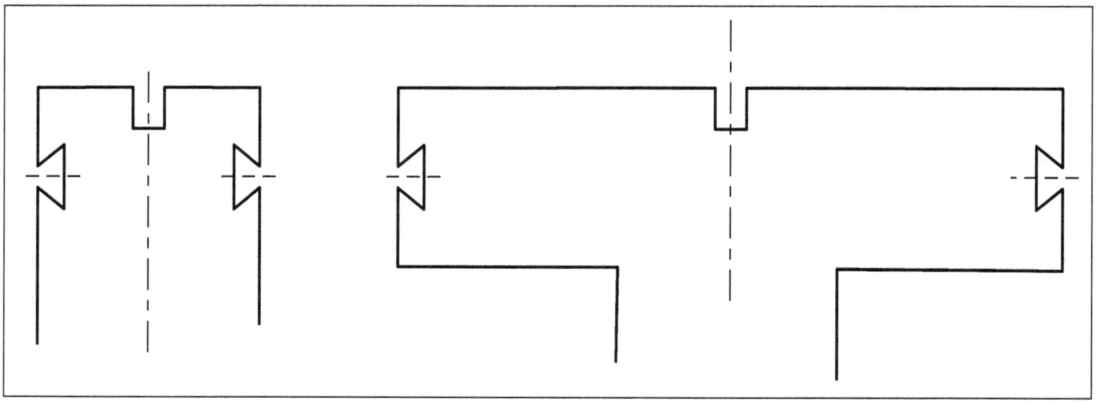

Fig. 1-11. Standard bed top size compared to a special permanently wide bed top.

The best way to provide good die support is to make a permanently wide bed top (Fig. 1-11). The actual width depends on the work that is expected to be done on the machine. If the manufacturer normally machines dovetail slots in the front and rear of the standard bed top, those same slots can be machined in the front and rear of the wider bed. This allows for easy mounting of the support arms and the back gages that may be needed.

e) Additional Die Space. Some work that can be done on a press brake will require more die space than standard in order to fit higher dies used to make certain products. Such parts may include deep boxes or large tubes that need high special dies.

f) Extended Beds and Rams. The bed and ram can often be made longer than the nominal distance suggested by the manufacturer. This may be a less costly method of obtaining a longer die forming surface, but it has the limitation of the flange depth being limited by the throat depth. For some unique parts, the ram extension can be modified to allow the parts to be formed around the end of the ram. This type of part might be the outside cover for a box, or something like the outside body of a washing machine. In all cases, the builder of the press brake will place a maximum tonnage limitation on the extension to ensure that the machine will not be damaged.

Chapter 2

Press Brake Tooling (Normally Called Dies)

Most press brake tooling is made from simple rectangular bars that initially measure 1" to 2.5" wide and 4" to 6.5" high. One end of the bar width will have a tongue or stem that is used to clamp the upper die to the press brake ram or the lower die to the filler block. The other end of the bar is machined with the forming nose radius and angle if the die is an upper die. The top surface of the lower die is machined with a matching **vee-shaped** configuration.

A) Standard Die Tongues

One side of the narrow end of the bar is selected as the mounting side. A 0.5" wide by 0.625" high tongue is machined centered vertically on a theoretical centerline (Fig. 2-1). The flats remaining on each side of the tongue are called the **shoulders**. The shoulders carry the forming load produced when the metal is being formed. This allows the load to be transmitted to the ram. The tongue width must be parallel and held to a maximum tolerance of 0.500"/0.490". The tongue height is held to a $^5/_8$" scale height. The die shoulders must be in the same plane from one side to the other within a 0.005" tolerance.

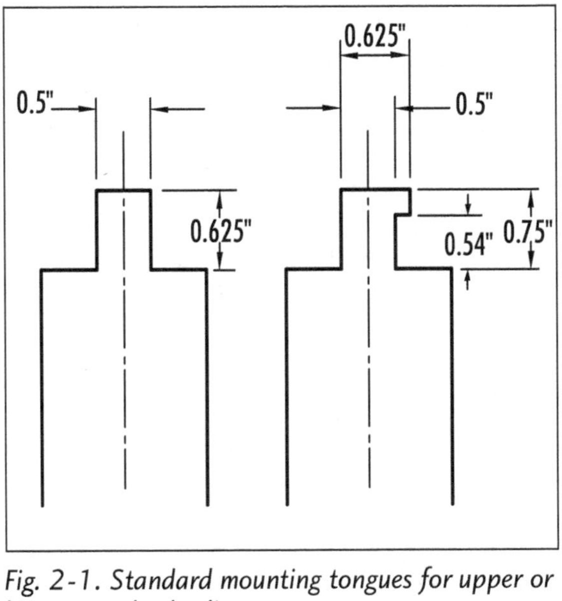

Fig. 2-1. Standard mounting tongues for upper or lower press brake dies.

Some upper dies are made with a special tongue called a hook tongue. This design uses the 0.500" width but a total height of 0.750". About 0.540" up from the shoulders, the front of the tongue is stepped outwards to a 0.625" width. This step fits into a relieved notch machined into the ram die clamps which mount on the front of the ram nose. Sufficient clearance is allowed in the die clamps for easy installation of the die. The hook created by the $^1/_8$" step ensures that abnormally heavy dies, or dies which may be subject to some type of load that would pull the upper die from the ram, will keep the upper die in place. Before purchasing any press brake tooling, make sure the tongue design of the tooling will work with the press brake clamping system that is available.

Press brakes supplied by some European and Japanese manufacturers have a completely different clamping system (Fig. 2-2). Tooling required for that equipment is not interchangeable with a conventional press brake. In some cases, special adaptors can be used. An adaptor requires a certain amount of die space, which may not be available in a standard size machine. Adaptors fastened to the ram will require the lower die to be mounted closer to the bed top. This may create problems with an NC gaging system already mounted on an existing press brake.

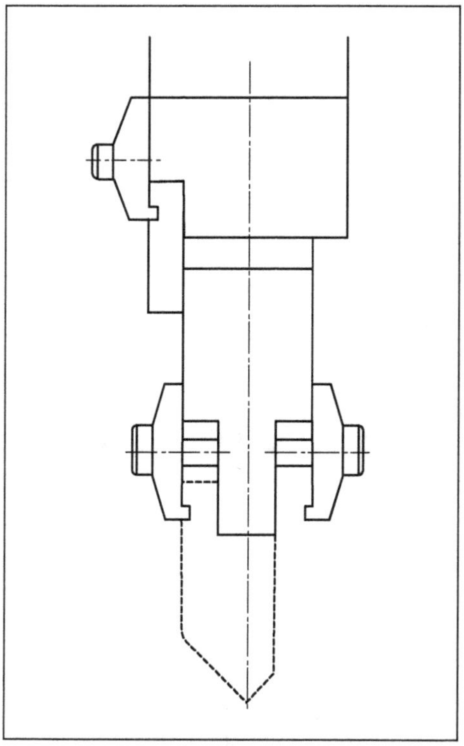

Fig. 2-2. European-style clamping system often used on up acting press brakes.

B) Standard Press Brake Tooling

The upper die can also be called: the male die, punch, knife, or blade. Many of these names are unique to certain industries or different areas of the country. This text will use the term **upper die** to minimize confusion (Fig. 2-3).

To machine the upper die, the end of a bar with an acceptable tongue is marked with a centerline from the center of the tongue to the other end of the bar. When the tongue of the bar is mounted in a special die holder, the other end can be machined to provide a symmetrical angle around the centerline. In most cases, the angle will be 90°, an angle between 80° to 85°, or an acute angle of approximately 30°. The acceptable tolerance for most commercial press brake tooling from the centerline of the tongue to the centerline of the nose is ±0.020", but the variation must stay parallel from one end of the die to the other.

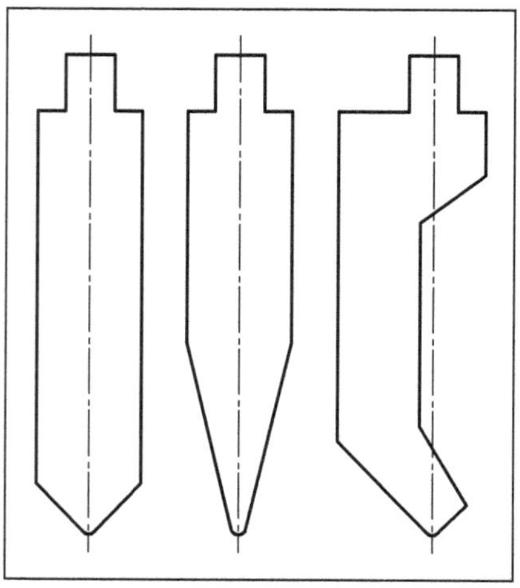

Fig. 2-3. Typical upper dies.

It is important to remember that the angle of a press brake die is the angle that is measured by using a combination square to measure the angle of the upper die. An angle less than 90°—for example 80°—would be defined as the angle of the upper die. An angle greater than 90°, i.e., an obtuse angle—for example 120°—would also be defined as the angle of the upper die.

Many engineering drawings do not show the inside angle of the formed part, but rather show the complementary angle. A 120° obtuse angle is actually shown as a 60° angle (outside). In some cases, the routing also calls for a 60° angle. If this incorrect measurement is not recognized, tooling specifying an incorrect angle may be purchased, or the press brake operator may form the wrong angle.

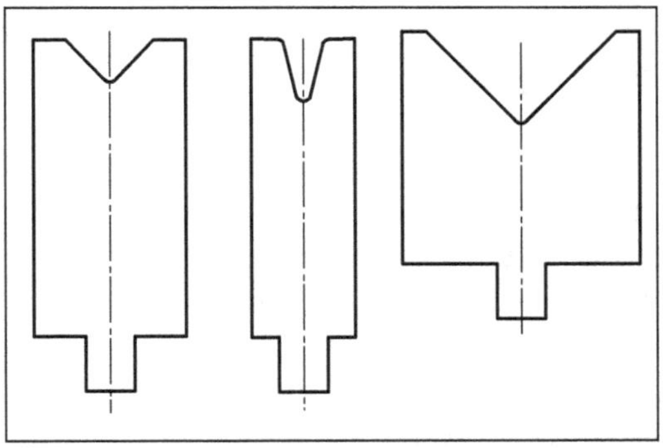

Fig. 2-4. Typical lower dies.

The lower or bottom die (sometimes called the vee or female die) should be machined with the same angle as the upper die. An opening cut into the top surface of the lower die will match the angle of the top die and provide a vee shaped appearance. The tongue of the lower die and centerline tolerances of the vee die opening are the same as the upper die (Fig. 2-4).

Two reasons why the angle of the upper and lower dies should be the same are:

1) when the dies are installed into the machine, the upper die will slide over the lower die without undo effort, and

2) alignment of the lower die with the upper die after both dies are fastened is much easier to visually check; this minimizes setup time.

The die bars must be straight enough that an operator can easily push the upper or lower die in the tongue slot of a standard filler block along its length without the die becoming wedged in the slot. The shoulders of the die bar should be close enough to the surface of the filler block so the operator can, by pressing down with reasonable hand pressure, force any gap between the shoulders and filler block to be closed. Some special dies or very long lengths (12' or longer) may not be able to be held quite as straight. Care must always be taken when loading and unloading the dies from die racks. Longer dies must clear the die rack supports prior to being lifted to keep the bar from being kinked. A **kink** can never be completely straightened and will cause continuing setup problems. Therefore, the saying "straight as a die" now has meaning.

Although over 95% of press brake tooling will be simple "vee" dies designed to form only one straight line bend in the part, tooling of many unique shapes is being used to form special bends or to improve productivity. Some of these special applications will be discussed in later chapters.

C) Installing the Press Brake Dies

Most manufacturers will provide instructions describing the best way to load the tooling into their press brakes. The normal method to install simple vee dies is to have the filler block or die holder located (ref. Fig. 1-6), but not bolted down, on the bed top. The ram is adjusted to the height of the lower die plus the height of the upper die that will be required when the two dies are located in the machine. A simple scale or tape measure can be used to check this distance. This is an approximate distance and should be on the high side.

The lower die is removed from the storage racks. If it is very short and can be safely lifted by the operator, the die can be carried to the left hand side of the bed top. Starting with the end that will be on the right hand side after installation, and with the desired side toward the front, the die is located with the tongue in the slot of the filler block (ref. Fig. 1-5). The operator pushes the lower die along the top of the filler block until the trailing end of the bar is positioned near its left hand edge. If the lower die is too heavy, a crane assist may be needed.

The upper die is then removed from storage, by hand or crane depending on its weight. The nose of the upper die is located in the vee opening of the lower die. The tongue of the upper die is slipped into the tongue slot of the ram. The die clamps must be open. It is normally easier to slide the upper die into the machine if the upper die is initially positioned slightly lower than the vee opening. The end of the upper die closest to the vee opening must be tilted upward allowing the nose of the upper die to slide into the left hand end of the lower die. This permits the upper die tongue to be moved back and forth to be aligned with the ram slot. As the die becomes engaged into the vee opening and the ram slot sufficiently to ensure the upper die will not fall out when released, the operator must make a judgment visually. When the upper die is pushed further onto the lower die and the upper die starts to become parallel (it is now tilted upwards slightly), the ram setting must allow the die to be installed with the tongue remaining in the ram slot and sufficient room available for the upper die to be slid across the rest of the vee die length to complete the installation. After any needed ram adjustment has been made, the operator can continue to push the upper die into the machine until both ends of the upper and lower dies (assuming both dies are the same length) match. If the dies are not the full length of the press brake, the operator will start to push both dies together, causing both the upper and lower dies to be located symmetrically between the housings.

After the dies are properly located, the operator tightens the die clamps on the ram nose securing the tongue of the upper die so that it can be lifted up out of the vee die opening without falling out of the machine. The operator then raises the ram slightly, lifting the upper die away from the vee opening. The set screws in the front of the die holder are tightened to clamp the tongue of the lower die against the straight, machined surface opposite the set screws. Looking at the dies from the end of the bed and ram, the operator can see if the upper and lower dies are in reasonable alignment. This would mean the centerlines of the upper and lower dies seem to be closer than $^1/_{32}$". If serious misalignment is present, the lower die must be moved forward or backward on the bed top so that an acceptable alignment is obtained. The operator then slowly lowers the ram and visually watches the upper die being forced upward, causing the shoulders of the upper die to seat against the ram nose. The ram movement is then stopped, and the operator retightens the die clamps, which may not have been tightened enough to hold the upper die securely to the ram.

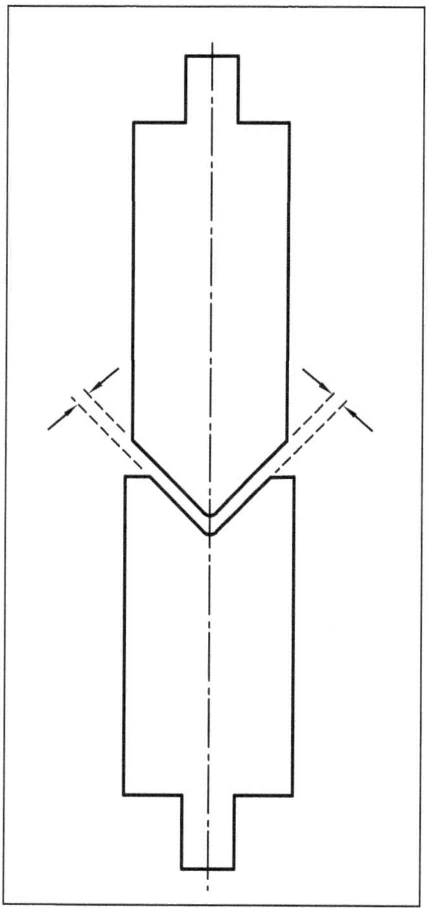

Fig. 2-5. Equal clearance required between the nose of the upper and sides of the vee opening in the lower die.

When the die clamps are tightened, the operator should only use the T- wrench that has been supplied by the press brake manufacturer. Some setup personnel feel that all bolts must be tightened using an extra pipe that makes the "T" of the T- wrench supplied longer, allowing a greater force to be used for clamping the die. This actually will elongate the clamping bolts and, over time, causes them to fail.

After the dies are clamped in position, the ram is

Fig. 2-6. Die clamp and bolts used to fasten the upper die in the press brake ram.
(Photo courtesy of Cincinnati Incorporated)

lowered to **seat** the shoulders of the upper die. The operator must take extreme care that not too much machine tonnage is used. Excess tonnage may cause the upper die to sink into the ram nose or to damage a die that had special cutouts and has a limited tonnage rating. Some new hydraulic press brakes have a setup mode that limits the machine to about 10% of its maximum capacity during the setup function. This helps minimize any problems unless very short die lengths are being use (see Chapter 5, section A).

The ram is then raised slightly, and final alignment of the dies can be completed. The sloping sides of the upper die nose and the sides of the vee die should be within 0.0015" of the front or rear gap of each other at both ends of the die (Fig. 2-5). The lower die is adjusted to obtain alignment. At that time, the bolts that clamp the filler block to the bed top are tightened, completing the setup (Fig. 2-6).

It must also be pointed out that final die alignment cannot be done safely by simply hitting the dies "hard." There were many press brake operators that thought that if a set of vee dies were hit harder, the lower die and filler block would move front or back to obtain acceptable alignment. Obviously, if the upper die strikes a misaligned lower die, both dies will tilt slightly under the pressure. Increasing the pressure on the lower die by lowering the ram elevation will not move the lower die and the filler block in any direction on the bed top. Besides not making sense from an engineering point of view, in practice, all the "hard" hit did was cause the upper and lower die to rock either forward or backward during each forming stroke. The quality of the bend is thus affected, as well as abnormal wear and tear on the tooling and press brake. Always follow the manufacturer's suggestions if they vary in any way from the above procedure.

D) Standard Die Steel

The majority of press brake tooling is made from standard higher carbon steel bars available from many steel companies. The most common is a basic alloy steel bar called 4140, identified by a four-digit numbering system was established by the Society of Automotive Engineers (SAE) and is used today throughout industry (Fig. 2-7). A similar system was developed by the American Iron and Steel Institute (AISI). The last two digits specify the approximate carbon content in hundredths of 1%, while the first two digits specify the type of steel and the percentage of alloying element. The alloy is required if the bar is to be hardened. If the last two digits of the identifying number show that the carbon content is between 30 and 50 points (or hundredths of 1%), the steel will need some type of alloying elements to be added to allow the bar to be hardened to provide reasonable die life. Some companies use only a high carbon steel bar made of a carbon content of 0.80 to 0.90 points of carbon. No alloying elements are required. The higher carbon content is sufficient to allow the steel to be hardened without any need for alloys. This type of bar will provide additional die life when compared to alloy-type bars of the same hardness.

The Society of Automotive Engineers (SAE) has set up a simple standard for identifying the chemical composition of many types of common steel. The numbering system is similar to the system estabilished by the American Iron and Steel Institute (AISI). The SAE standard uses a four digit identifying number. The first number indicates the general type of steel. The second number identifies the main type of alloy used in the steel. The last two numbers indicate the average carbon content in hundredths of 1%. The identifying numbers that seem to apply to the steel used in press brake work are:

Carbon Steel..1xxx
  Plain Carbon Steel................................10xx

Molybdenum Steels..................................4xxx
  Chromium 41xx
  Chrominum-Nickel 43xx
  Nickel...46xx

Fig. 2-7. SAE numbering system to identify different types of steel.

The competitive problems steel companies in the United Stated are experiencing have limited the availability of many types of steel and bar sizes that were readily available only a few years ago. Some companies that want to design their own tooling may find that they will have to pay a premium price since many bar sizes (even those listed in steel company catalogs) are no longer available. The maximum "square" size limitation on bars of steel used for press brake tooling is 6" × 6". For other maximum sizes, the limitation is 36 square inches of cross section in the bar. The steel companies also place a minimum weight requirement if bars of steel are purchased from the mill. When press brake tooling suppliers select their die steel cross sections, they will only purchase those sizes that have the potential for higher usage.

Dies made from 4140, 1080, or 1090 type steels are heat treated to a hardness range of 250 to 305 Brinell. (Note: Some bars made from 4140 are sold under trade names in order to have a marketing advantage.) All heat treated bars will have some hardness range. As the bar is cooled after being heat treated to obtain a required hardness, the outside of the bar will cool first. The inside of the bar will cool more slowly. The faster the bar is cooled, the higher the hardness. Since narrow bars will cool faster than square bars, the common narrow bar sizes used for many press brake dies will be at the higher end of the Brinell hardness range given above. Square bar hardness will likely be in the center to low end of the range.

Over 90% of the simple, commercial type tooling sold is made from the steel described above, and provides acceptable die life without further hardening. When wear does occur, the worn surfaces of the bars can be remachined to provide "like new" nose radii or vee die openings for future bending operations. Some types of metals and sharp edges caused from punched holes, notches, or burnt surfaces will cause abnormal wear after short usage.

E) Flame Hardening Standard Press Brake Dies

In order to obtain longer die life, the nose radius of the upper die or the corner radius of the vee opening can be flame hardened. In many cases, the forming cycles on a standard upper or lower die can be increased by a factor of four compared to the normal number of hits that would be expected with the standard 250 to 305 Brinell hardness of conventional die steel. The big limitations are that the bar cannot be reworked to obtain a "like new" working surface, and the bar straightness will not be as consistent as the standard hardness die steel bar.

Flame hardening press brake tooling is very specialized and probably should be done under the supervision of the company that supplied the tooling. The main problem is ensuring that the flame hardened bar stays straight enough to be used without cracking into several pieces. Another issue is keeping the nose radius from developing cracks. Heat caused by the flame hardening will cause the bar to distort. Since the die steel bar (previously heat treated) now has the nose radius or vee die corners flame hardened to a hardness of over 50 on the Rockwell "C" scale, the straightening operation requires not only experience but also some trial and error. A flame hardened die is seldom as straight as a normal heat treated die bar. The minimum nose radius that should be flame hardened is $^{1}/_{32}$". Sharper radii are subject to **heat checks**, which are small cracks that, if used during the forming process, will often cause small sections of the sharp radius to break off.

F) Press Brake Dies Requiring Tool Steel Inserts

Die steel used for normal press brake work is not tool steel. Tool steel is a different type of steel made by specialty steel companies. This type of steel is used only in short lengths after it is heat treated to a hardness of over 56 Rockwell C . The steel will crack if bent too much through deflection of the bed and ram. Many press dies use tool steel for forming inserts, punches and dies, or notching inserts. Shear blades for cutoff shears are made of tool steel. Round-shaped tool steel knives are used for slitting. Squaring shears are designed to keep the long tool steel knives from deflecting while cutting material, thereby eliminating any cracking problems. Tool steel inserts are only used in press brake tools when unique materials such as wire or wire cloth are to be formed. Some dies that require a great deal of unit force—such as closing a hinge shape in heavy material—may need tool steel inserts only in certain parts of the die.

The hardened tool steel die sections used in press brake tooling are normally made in lengths between 8" and 12". Longer sections, 24" or longer, have a history of cracking after limited use. This can lead to a great deal of expense in replacing the sections, as well as lost production time, since the die cannot be used until it is repaired. All tool steel die sections must be machined to shape prior to heat treatment. The high hardness of tool steel after it has been hardened prohibits any conventional machining and limits any shape modifications of the inserts to grinding. Special grinding wheels are very expensive. Normally they are not an option if only a few parts are needed.

When steel is heat treated, some part distortion should be expected. Longer machined bars that are heat treated will have some bow, camber, possible twist, and some slight growth or elongation when the material cools down after it is removed from the heat treat furnace. If the correct type of tool steel is selected for the press brake die inserts, the short inserts will be able to be used without extensive additional work. Prior to designing the tooling, the die designer should do the following.

1) Select an air hardening tool steel. Some tool steels require an oil quench after they are removed from the furnace. The quench will cause added distortion to the insert. Air hardening steels will minimize any potential distortion.

2) Confirm the need for short die sections with the company doing the heat treatment.

3) Possibly drill and tap a hole in the end of each insert so that the insert can be hung vertically while in the furnace. This will also reduce any tendency of the part to distort.

Steel is considered high tensile when the carbon content reaches 0.6 % carbon content or higher. Tool steels will almost always have a carbon content of over 1%. Different types of tool steel have many different alloying elements added to provide different types of wear characteristics. For this discussion, chrome is the alloy that is used in the air hardening tool steel selected for press brake tooling. The chrome content can vary for the different grades of tool steels from 1% to as much as 12%. The grade that seems to provide the best results for press brake tooling inserts is the 12%, or AISI D2 tool steel. The common reference is high carbon-high chrome steel.

Since 1970, a type of press brake die called **European-Style Tooling** has been available to form simple vee bend shapes. Because of its high cost and special tongue design, it is still classified as a separate type of press brake tooling. Most of the material used to make this category of press brake die seems to be tool steel of some type (see Chapter 21).

Chapter 3

The Basic 90° Bend

Press brake bending falls into two basic categories with several compromise options. The first is the foundation for all press brake work and is called **air bending**. The second type is called **bottom bending**.

A) Air Bending

Air bending is defined as three points of contact with the part to form a straight line angle (Fig. 3-1). The nose of the top or upper die forces the part to be formed into the vee shaped lower die. The included angle machined on both the upper and lower die must not allow any contact with the part except the nose of the upper die and corners of the vee opening in the lower die. When the upper die has penetrated deep enough into the lower die to produce the required angle (this is at the bottom of the forming stroke), the upper die is returned to the top of the stroke releasing the now formed part. When the part is released, the two legs of the newly formed part will spring back somewhat until the stresses in the formed part are balanced. If the material is simple cold rolled steel, it is common for the metal to open 2° to 4° from the angle actually made during the forming stroke.

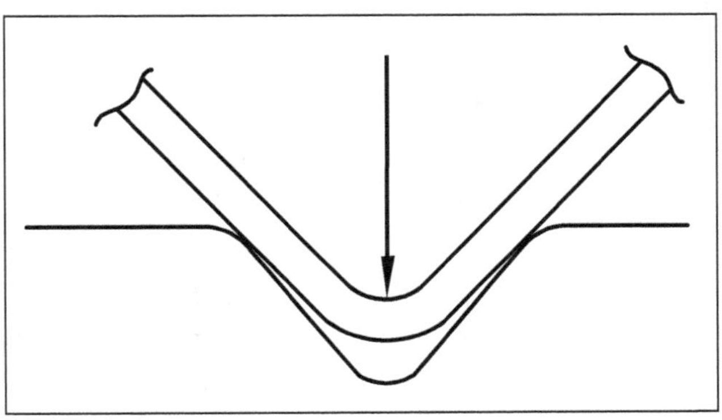

Fig. 3-1. Air bend showing three points of contact with the press brake tooling.

The greater majority of press brake forming is making a simple 90° vee bend in a part. To allow for springback, the angle cut on the upper and lower dies will be machined to an angle less than 90°, normally between 75° and 85°. This allows the part to have only three points of contact with the tooling and no contact with the other surfaces.

The nose radius of the upper die should be equal to, or slighter less than, the metal thickness that is being formed. The sharper the nose radius, the greater the die wear. Special nose radii are often required for aluminum, high tensile material, or exotic materials.

There are two simple **rules of thumb** that have been used for years to choose tooling that will give the most consistent and accurate air bend when forming mild steel. The recommended vee die openings found on air bend tonnage charts are based on these methods.

The first rule, developed in the 1920s to determine the best vee die opening, is to multiply the material thickness by 8 and round the answer to the nearest simple fraction. For example, 16 gauge mild steel has a nominal thickness of 0.060". Multiply 0.060" × 8, and the answer is 0.48". To select the proper vee opening, the answer is rounded up to 0.5".

Press brake operators also found that when forming mild steel, the inside radius in the bent material was a function of the vee die opening. Although the inside radius is a parabolic shape rather than a true radius, it is common practice to measure this arc with a simple radius gage that closely fits the formed part. Therefore, the second rule is that the expected inside radius is 0.156 ($^5/_{32}$) times the vee die opening being used. If the vee die opening is greater than 12 times the vee opening, it becomes apparent that the inside radius is actually elliptical, and any dimensional radius called for on a drawing is an estimate. If an attempt is made to form a part using a vee opening less than 6 times the material thickness, the inside radius will not be a radius since the material will try to form a theoretical inside radius of less than one metal thickness—which is impractical to air bend.

Based on the above rules, a 0.5" vee opening (calculated for 16 gauge) × 0.156 will equal an approximate 0.075" inside radius. Note that the rule, which applies mostly to mild steel material, does not refer to the material thickness being used. If the first example of 16 gauge mild steel recommends that a 0.5" vee opening be selected, the resultant 0.075" inside radius will be slightly larger than the 0.060" material thickness. If 18 (0.048) gauge mild steel was formed using the same 0.5" vee die opening, a similar 0.075" inside radius would be formed into the thinner material. If 14 (0.075) gauge mild steel was formed over the same die, the resultant inside radius would be very close to metal thickness. Therefore, for most of the

Material Thickness of Standard Carbon Steels

Standard Gauge Number	Weight Pounds/Sq Ft.	Inch Equivalent	Tolerance Range
10	5.625	0.1345	0.1419-0.1271
11	5.000	0.1196	0.1270-0.1121
12	4.375	0.1046	0.1120-0.0972
14	3.125	0.0747	0.0821-0.0710
16	2.500	0.0598	0.0635-0.0568
18	2.000	0.0478	0.0508-0.0449
20	1.500	0.0359	0.0388-0.0344
22	1.250	0.0299	0.0313-0.0284

Fig. 3-2. Common steel gauge thicknesses used in most press brake work.

common gauge thicknesses normally used for press brake forming, a vee die opening of 6 times the metal thickness rounded to the next simple fraction will produce an inside radius close to one metal thickness. Consult the next section (B) describing forming tolerances to understand why the eight times metal thickness vee die opening remains the recommended and most used vee opening selection. See the chart of different gauges of mild steel showing the nominal thickness plus the possible tolerance range (Fig. 3-2).

It is also interesting to note that each gauge thickness has a weight in "pounds per square foot" (lb/ft^2) that is a simple number. For example, 16 gauge is listed at 2.500 lb/ft^2. The "gauge" system for steel was established in the late 1880s to allow the steel companies to regulate their production. The width of the steel being rolled could be set, and the length of material rolled over a specific time period could be measured. To determine the weight per square foot, the thickness had to be determined. The steel industry devised a gauge system to facilitate calculation of the tonnage of the steel being processed. Refer to Fig. 3-2 which illustrates the comparative lb/ft^2 versus material thickness for the more popular gauges used in press brake work. The current gauge thickness of steel was standardized as a federal law passed by the U. S. Congress on March 3, 1893. The **gauge** system law is based on a steel density of 489.6 pounds per cubic foot (lb/ft^3).

B) Air Bend Forming Tolerances (Angular Only)

Since mild steel may not be consistent from piece to piece, coil to coil, or heat to heat, angular variations must be expected. The material could change in chemistry, which affects the tensile and yield strength. The rolling of the material during the manufacturing process may cause thickness variations that affect the angular consistency.

Other variations result from worn tooling, press brakes that do not consistently repeat at the bottom of the stroke, or poor setup by the operator or setup person. Most of the angular variation encountered will be found to be material variations. If the press brake is properly maintained, it should repeat to the bottom of the stroke each time within an acceptable tolerance. Worn tooling, once it has been set up and shimmed to produce an acceptable part, does not change from part to part. If the operator is locating the part properly, and assisting the part upwards during the forming stroke as required, the part tolerance should not be affected. It should be noted that if a formed part is removed from the press brake with a correctly formed angle, and then dropped on the floor or thrown into a container, the formed angle can open up and be out of tolerance.

If only the standard gauge tolerances are considered, a simple sketch, showing a drawing of a part having some thickness that is formed into a 90° angle, can be used to determine tolerances. The part sketch should show an inside and outside radius of the part. The sketch should include three marks: one mark to show where the top die contacts the part on the inside of the bend, and two marks on the outside of the material to show where the part would contact the vee die corner radii.

The sketch illustrates a part of nominal gauge thickness as it would look at the bottom of the forming stroke with the appropriate tooling contact. Fig. 3-3 illustrates (by use of dotted lines) possible material variations within a gauge range. If the material is thicker, the outside surface is pushed further down into the vee die cavity, resulting in an angle overbend. If the material is thinner than nominal, the outside surface does not penetrate into the vee die sufficiently to make the proper angle. Thus the angle remains open. Since only the material thickness was changed, it becomes clearly evident that material variations will cause angular variations when using simple air bend dies. If the material thickness becomes thicker than the material

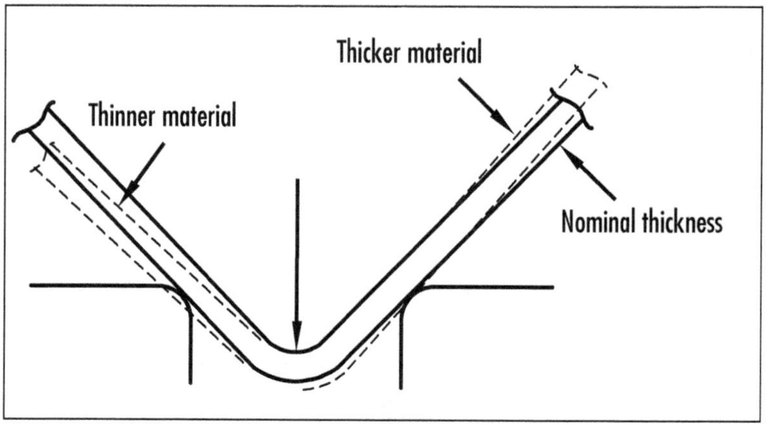

Fig. 3-3. Potential angular variation that can be caused by material thickness tolerances.

used for the original setup, an over bend angle can be expected. If the material thickness is thinner than the material used for the original setup, the bend angle will be open.

Each gauge of material can be carefully sketched using a magnified scale, or using computer graphics that could measure angular variations that would not only show a 90° bend but also show their thicker and thinner tolerances as described above. It would be found that the average angular variation for gauge material would be about ±2°.

Practical experience has shown that a normal stack of material supplied to a press brake will not have the entire range of tolerance allowed on the tolerance chart. Some material variations can be anticipated, since to produce a coil of steel, in order to keep the strip tracking in a straight line, the center of the sheet is made slightly thicker than each edge. When the coil is cut or blanked to the material dimensions needed to make a particular part, some thickness difference will occur. How much, or in what direction, will not be known unless each part is measured and marked prior to making the required bends. In almost all cases, this is impractical both from a cost and time viewpoint.

Experience in working with sheet metal has proven that material variations in sheets of mild steel up to 10 gauge thick and as long as 10' will cause an actual angular variation of ±0.75° when air bending. Additional variation should be expected from the initial test part, which seemed to be acceptable, but may have had variation due to machine deflection, die wear, or machine repeatability. In sheet metal (10 gauge or thinner), surface hardness caused by the rolling operation in the manufacturing process, and chemistry changes in the material, all add some possibilities for variations. Because of the many other factors that must be considered, an additional ±0.75° must be added to the tolerance range. The total tolerance range is the addition of tolerances that are expected from probable material variations, plus the variations caused by all of the other unknown factors just listed. A realistic tolerance that should be considered when air bending 10 gauge or thinner mild steel up to 10' long is ±1.5°.

For plate, an additional degree is required, since the material variations are much greater. Tolerance for air bending material 7 gauge and thicker will be ±2.5° up to 1/2" thick plate. Heavier materials are often formed to an improved tolerance by using more than one stroke of the ram, and it is important to remember that any discussion of tolerance is based on using the recommended upper and lower dies.

To hold a consistent bend requires a vee die opening that allows the legs of the part to penetrate down into the vee die sufficiently to allow each leg or flange to have a **flat** distance of 2.5 metal thicknesses past the outside radius of the part before contact with the corners of the vee die. The flat is needed to provide control of the bend angle. The recommended "8 times metal thickness" vee die opening provides a good flat to allow consistent parts to be formed within the tolerance range discussed. A smaller vee opening (e.g., 6 times metal thickness vee opening) will actually form a slightly smaller inside radius, but the flat from the outside radius to the contact with the vee die corners will also be reduced. This reduction of the flat surface results in additional angular variations in the part. A larger vee die opening will provide a greater flat, but also increases the size of the inside radius. The larger radius will result in more springback when the forming pressure is released, introducing more potential part variation.

The practical tolerance for air bending sheet metal up to 10 gauge thick, and 10' long, is ±1.5°. This variation is often felt to be more than can be accepted but, as with all tolerances, the maximum range possible does not normally occur in one part. A standard statistical bell-shaped curve should reflect the actual bend variations. This means the greater majority of parts will be formed with much less variation. Most production runs require only a few parts of each shape to be formed. With the availability of high-tech, computer access press brakes, air bending is regaining its popularity, which had dropped somewhat from the 1960s through the 1980s.

C) Forming with Bottoming Dies

To obtain better angular consistency, or to compensate for repeatability or deflection problems of the press brake, a forming method called **bottoming** may be selected (Fig. 3-4). Bottoming often creates problems for the press brake operator. The forming method has four different definitions depending on the tooling design and how it is used during the forming cycle. Any simple straight line forming where the formed part touches the sloping "vee" section, in addition to the corners of the vee opening, is no longer an air bend. It must be classified as some type of bottoming die because completion of the bend will require more force than would be required to make a similar air bend.

1) True Bottoming

The upper and lower dies are machined so that the forming surfaces have the same angle as the angle of the part that is to be formed. If a 90° angle is required, the upper and lower die surfaces are machined to a 90° angle symmetrical around the centerline. The radius of the tip or nose of the upper die is machined with a one metal thickness radius, or to the closest simple fraction. The tooling for machining radii is often limited to specific fractions, and then converted to corresponding decimal dimensions. It is common practice, since most bottoming work is preformed using materials 14 gauge or thinner, to select die bars of the same width for the upper and lower dies.

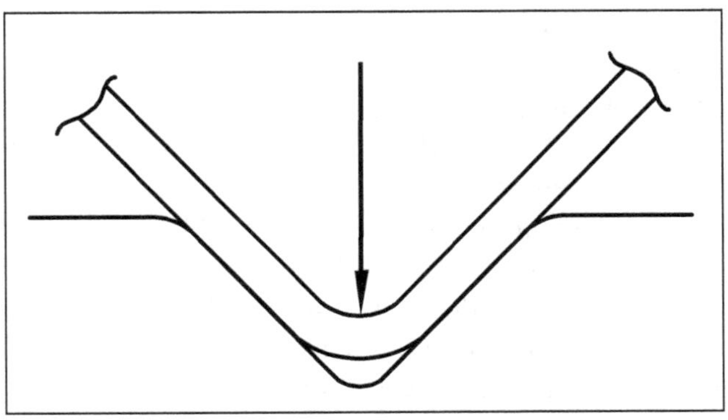

Fig. 3-4. Bottoming illustrating the material contacting the sides of the vee die opening.

Often the vee opening selected is the same 8 times metal thickness vee die opening recommended for an air bend die. Some operators, however, are more comfortable with the vee die opening being 6 times metal thickness. This opening causes the material to initially form to an inside radius of approximately one metal thickness. When material is formed, either using the air bend method or with bottoming type tools, as the part is forced into the vee opening, an inside radius is formed into the metal. Although called a radius, it is actually some type of "parabolic" shape. This is very important to know since it helps to explain what happens to the legs of the part during a forming cycle using bottoming dies.

During the forming cycle, several functions occur that can affect the quality of the final angle. The nose radius of the upper die is machined with a true radius. The inside radius formed on the inside of the part is an elliptical shape due to the part being air bent as it travels into the die cavity. The elliptical shape will be slightly larger than the radius machined on the die. When the outside legs of the part strike the sloping sides of the vee die opening, several conditions may result. Depending on the position of the top die at the bottom of the stroke, and the amount of force or tonnage striking the part, the operator may find, as shown in Fig. 3-5, one of the following.

Stage 1) The inside radius of the part will follow the 0.156 times the vee opening rule, as in air bending.

Stage 2) If the stroke pushed the part down to the bottom of the vee die using only the force required to air bend the part, the formed angle would spring open, probably 2° to 4°, when the upper die returns to the top of the stroke.

Stage 3) If the forming stroke had been lowered slightly so that the tonnage at the bottom of the stroke built up to about 1.5 to 2 times the normal air bend tonnage, then the pressure was released as the ram returned to the top of the stroke, the resultant angle will be **overbent** by several degrees. The overbent angle will be very consistent in tolerance but will not be the desired final angle.

Stage 4) If the bottom of the stroke ram setting is increased so that the tonnage at the bottom of the stroke builds up to 3 to 5 times the tonnage required for a simple air bend, the corners of the upper die will force the overbent legs of the part back to the desired angle, normally 90°.

The obvious question is: "Why does the part overbend to an angle less than 90° when the die angle apparently should limit the flange motion?" The answer is fairly simple. Take one hand and hold it up in front of you. Keep your four fingers together and open your thumb to form an angle between your thumb and forefinger. Notice the large elliptical shape that your skin makes between the thumb and forefinger. Take the forefinger of the other hand and start to press it down into the center of the elliptical area between the thumb and forefinger. Immediately, your thumb and forefinger will start to move together, reducing the size of the original angle you had made. The same phenomenon occurs when a bottoming operation is used. The upper die radius is a true radius. The shape formed in the material when it is pushed down into the vee die is somewhat elliptical. At the bottom of the stroke, as tonnage is built up, the part will overbend just like your fingers did. The flanges will overbend until they touch the corners of the top die. If the pressure is released at that time, the flanges may spring back. If the part were struck hard enough that the area contacted by the upper die exceeded the yield point of the material, springback would be eliminated. If released from the forming pressure at that time, the part may still be in an overbent condition. It will remain there until the upper die is set lower to allow the corners of the upper die to wedge the flanges open to an acceptable 90° angle. This requires a great deal of tonnage. The sharper the nose radius of the upper, the greater the amount of overbending.

True bottoming will produce a good consistent angle and an inside radius of one metal thickness. However, as pointed out, the forming tonnage required will be 3 to 5 times the

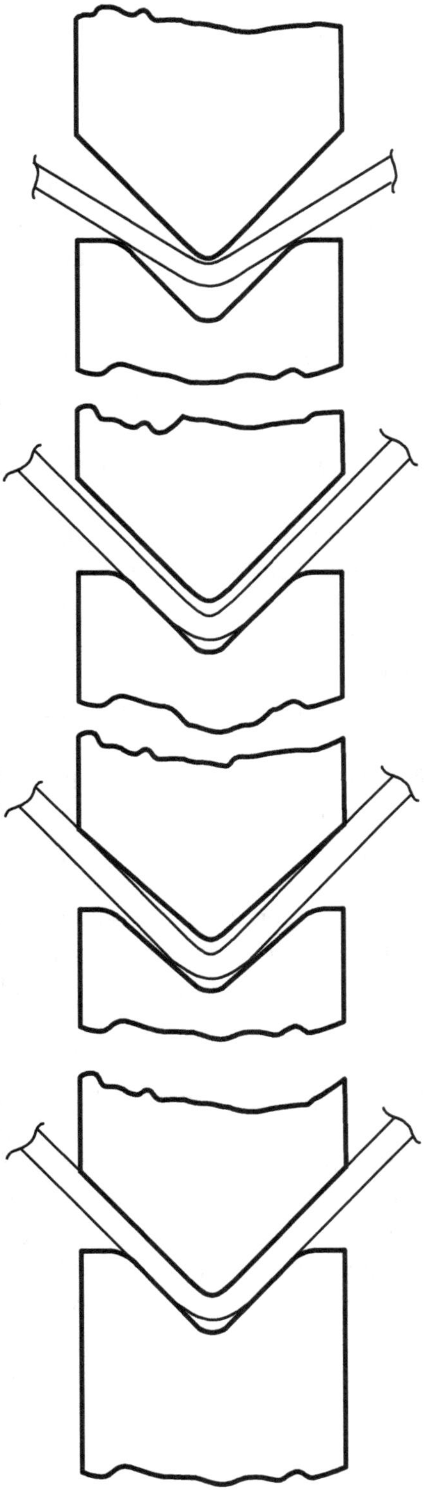

Stage 1 — Part forms the same as if is being "air bent." The inside radius being formed will follow the .156× the vee opening rule.

Stage 2 — Part contacts the bottom of the vee opening and the forming tonnage builds up to slightly more than the normal air bend tonnage. If the ram is reversed at this point, the part will spring open to an angle greater than the die angle.

Stage 3 — The ram travels farther down into the lower dies and the forming tonnage is increased. The material overbends slightly until the inside legs of the part touch the corners of the upper die. If the pressure is released at this point, the part may spring back to an angle close to the desired angle or it may stay overbent. If it remains overbent, the angle will be very consistent but it will simply be the wrong shape.

Stage 4 — The ram is lowered to allow the corners of the upper die to push the flanges of the formed angle back to the angle of the upper and lower die. At the same time, the inside radius reforms the inside radius that has been formed in the part to the same radius that is machined on the nose of the upper die.

Fig. 3-5. Stages that occur when bottoming a simple 90 degree bend.

tonnage needed to form the same angle by using the air bend method. Since the forming tonnage becomes so high, often requiring a much larger press brake, most bottoming work is limited to 14 gauge or thinner material. All parts, prior to selecting the forming process, should be reviewed to determine if sufficient tonnage is available to properly form the part.

2) Bottoming with Springback

A skilled press brake operator may often be able to form a variety of parts using the overbending function that occurs in a bottoming forming cycle as described previously (Fig. 3-6). The operator must carefully adjust the forming cycle stroke to allow the angle to overbend, but not be "set." When the ram moves back to the top of the stroke, the formed angle will spring back to the required shape. This method only requires about 1.5 times the normal air bend tonnage, and may provide an angular accuracy slightly better than air bend tolerances. The disadvantage is that, if the part is hit too hard, the angle will stay overbent. Then, only bottoming tonnage will allow the upper die to push the legs back to 90°.

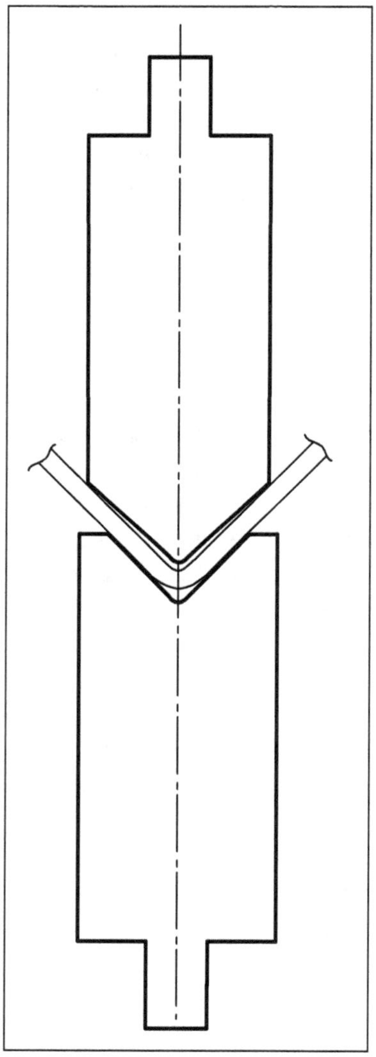

Fig. 3-6. Part movement while overbending the flanges when "bottoming with springback" is the forming method chosen.

This forming method requires a great deal of operator skill to obtain good parts consistently (ref. Fig. 3-5, Stages 2 and 3). Many users of small tonnage press brakes attempt to use this method, even using sharp nose upper dies, in an effort to form their parts. Often the operator will rehit overbent parts several times in an effort to square the legs of a 90° bend angle.

If **bottoming with springback** forming is done with an upper die that has a nose radius smaller than metal thickness, the upper die will produce a crease or groove in the inside surface of the radius. This crease will occur when the top die contacts the material and pressure is built up to start the bending of the material into the vee opening. Some people will mistake this crease as a sharp inside radius. The actual part shape is the normal inside radius with a crease in the center.

There are a number of companies selling what is called "high precision" press brake tooling (often associated with the European style tooling discussed in Chapter 21) that promotes 88° angles on their dies. This falls into the "bottoming with springback" concept. This type of die is not designed to work with "programmable angle" press brake options available in many new high-tech machines, since they are programmed to work only with true air bend dies. The 88° dies do not fall into this category since they require that the material actually touch the sides of the lower die to reduce some of the springback.

3) Coining

Some designers of parts believe that the inside radius of the part should be smaller than metal thickness. The only way this can be done is to force a small radius on the upper die (smaller than one metal thickness) into the inside radius that has been formed into the metal during the air bend portion of the forming stroke. The sharp nose radius on the upper die pushes down into the part at the bottom of the stroke and reforms the inside into a smaller radius. When solid metal is displaced or changed in shape, it is like the flat surfaces of a metal disc being reformed into a new shape, such as a penny, dime, or nickel. In this case, the displacement of the metal creates the new desired part, which is called a coin. When the upper die displaces the metal in the inside radius of the part, the forming method is called **coining**. The force required to displace the metal of the inside radius of a part to a $^1/_2$ metal inside radius will range from 5 to 10 times the tonnage required to air bend that material using the recommended vee die opening (Fig. 3-7).

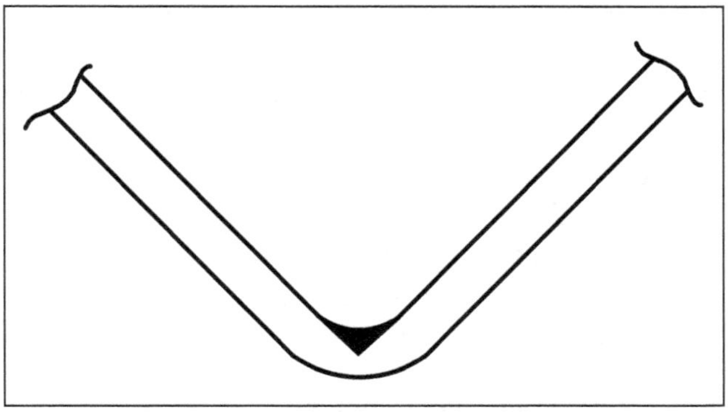

Fig. 3-7. Material displacement caused by coining the inside radius of a 90 degree bend.

There is a mistaken belief that a sharper inside radius made by coining will result in a smaller outside radius. This thinking can be disproved on the drawing board. A part, using the gauge thickness in question, should be drawn to a magnified scale showing the material at a typical 90° angle. The inside radius should be drawn to the same estimated radius that would be formed if the recommended vee die had been used. A line along the inside of each flange should be extended to illustrate a sharp, or 0", inside radius. The small area now shown by the two straight lines at 90° and the curved line of the inside radius illustrates the amount of material that would be displaced if a sharp corner was actually made in the part.

The displaced material can only dissipate into the outside radius. If the small amount of material in the sharp inside corner is measured and incorporated into the outside radius of the part, the actual outside radius might be several thousandths of an inch smaller than originally formed. Tests formed by The Cincinnati Shaper Company in the 1960s found that hitting parts in 16 gauge and 10 gauge mild steel up to 100 tons per foot (100 tons/ft) only changed the outside radius of the formed part 0.008". The resultant tonnage also caused the part shape to backbend from excess pressure at each corner of the vee die opening, providing a totally unacceptable formed final angle.

4) Bottoming Using Angles Other Than 90°

For many parts, there is a need for bottoming type accuracy, but the press brake does not have the available tonnage to form the part with true bottoming dies. The tonnage needed to bring the part to a consistent "overbent" position is only about 1.5 to 2 times the charted air bend tonnage for that gauge of mild steel. Once the part reaches a set overbent angle, the angle along the length of the bend line will be very consistent. If the part is one that will be repeatedly formed, it may be a good idea to have a special set of vee dies cut with an angle greater than 90°. This will allow the material to be somewhat "bottomed" at the lower tonnage. Instead of forming to an unwanted overbent angle of 88°, if the dies were machined to an angle of 92°, the formed part overbends 2°, resulting in the desired 90° bend.

Some materials will spring back unless hit at a tonnage greater than the available press brake capacity. This is often true when stainless is to be formed. Stainless is often formed using bottoming dies, resulting in springback to an angle 2° to 3° greater than desired after the pressure is released. When inspected, the angle will be very consistent along the bend line. If the die is made having an 87° or 88° included angle, instead of 90°, the operator will be able to make an acceptable 90° bend angle using the bottoming with springback concept.

The dies that have been cut to a special angle are not general purpose dies. The operator must learn to use them in order to obtain good angles. They will solve a tonnage limitation problem and provide good consistency. They will demand that the tons/ft tonnage needed for the longest part must also be held if shorter lengths of the same part must also be made. If the 92° dies used to correct the part "overbend" problem for long parts were used with shorter length parts, but were formed at a tonnage normally needed for true bottoming, the resultant part angle would probably have a 92° (or whatever angle that was machined on the die) angle along the bend line. The same logic would prevail if a short piece of stainless was truly bottomed using the 88° dies—the final angle might be the 88° machined on the dies. This method is a good reminder that hydraulic press brakes have tonnage limitations. They cannot be overloaded. When a mechanical press brake was used, the operator often thought: "if the angle is not correct, hit it harder!" This logic caused many overloads, along with high repair bills.

5) Bottoming Tolerances

True bottoming or coining tolerances will cut the normal tolerances expected from air bending in half. Instead of the ±1.5°specified for air bending 10 gauge and thinner up to 10' long using the recommended vee die opening, a bottoming (or if the material is coined) tolerance of ±0.75° variation can be attained. To hold tighter tolerances, a great deal of operator inspection will be required with time allowed to measure and rehit some of the bends. The optimum tolerance is ±0.5°. If enough time is spent on each part, and if the material specifications are closely held, some parts have been held to the equivalent of machining tolerances. If this is required, allow sufficient time for a great deal of handwork by a skilled operator, since this will approach "craftsman"-type work.

"Bottoming with springback" tolerances will vary between air bend and bottoming tolerances. Due to the many possible die and material combinations, an acceptable tolerance range that can be expected in a typical production run cannot be provided.

Chapter 4
Air Bend Tonnage

The method of calculating the pressure or the force required to air bend mild steel is based on engineering "beam" formulas. (A typical beam formula might be used to determine the maximum loading on a span between two bridge piers.) Several variations of the formula, plus comparisons with actual measured forming tonnages used to form mild steel, have produced the air bend tonnage chart being used in industry today. The first known chart was published by The Cincinnati Shaper Company (now Cincinnati Incorporated) in 1939. The charts are designed to show the bending force required to bend a specific material thickness (gauge) of mild steel using a specific vee die opening. The force requirement shown on the chart is shown in tons/foot or tons/meter. Important items to know when using the air bend tonnage chart are discussed below.

A) Yield and Tensile Strength

The maximum **tensile strength** is the force required to actually cause a piece of metal to fail or break apart. If a strip of metal was clamped at both ends, and some method of pulling each end in an opposite direction was activated, two types of material strength could be measured (Fig. 4-1). As soon as the strip starts to lengthen, the force being used has caused the strip to exceed its yield point. Yield point is when the material starts to lengthen without any additional increase in the applied force. If the pressure is released just after the material starts

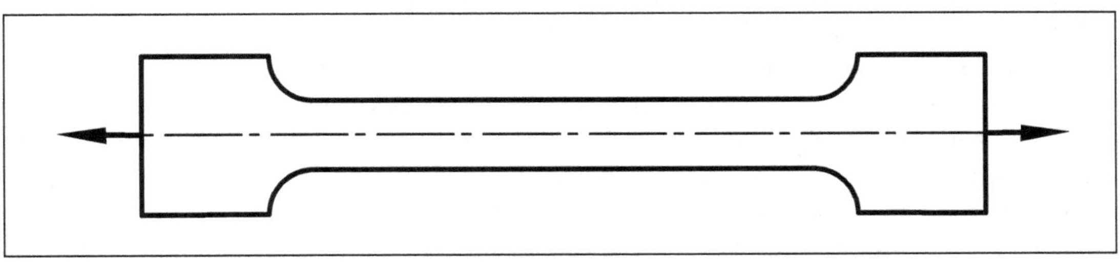

Fig. 4-1. Typical strip or round bar used in a "tensile strength" test.

to lengthen, but the strip keeps its new length without changing its cross section, the material has not exceeded its yield strength. If force being applied causes the material to "thin out" in cross section or width, the yield strength of the material has been exceeded. If the pulling force continues to lengthen the strip so that the material not only becomes longer but the cross section of the strip also thins out and the material actually is pulled apart, the ultimate "tensile strength" of the metal has then been exceeded.

B) Mild Steel

Mild steel is defined as having a maximum tensile strength of 60,000 pounds per square inch (PSI) and maximum yield strength of 40,000 PSI.

The basis for this definition is that all of the press brake load calculations done in the 1920s and 1930s were based on the strength of the steel available at that time. Since the first **air bend tonnage** chart has been widely circulated, and since mild steel currently uses the same parameters, these **air bend tonnage charts** still make up the foundation of calculating press brake forming tonnages today.

C) Tonnage Calculations

The engineering formula used to estimate the "air bend" forming tonnage ignored any friction or effect of the metal being pulled down into the vee die. It is important that the top corners of the vee opening are machined to at least a one metal thickness or greater radius to allow the material to easily move down into the vee cavity. A vee opening with too sharp of a corner radius may cause the bends to be inconsistent, and the forming tonnage will be affected.

D) Tonnage Variations

Calculated tonnage figures should be expected to vary because specific figures are used in an engineering formula, and the steel being used will vary in thickness as well as tensile and yield strength. The actual air bend tonnages may vary as much as ±10% from a published chart. When only mechanical press brakes were being used, these variations were often ignored. The assumption was that the machine could be overloaded slightly and still form a part that calculated close to the maximum available tonnage. In addition, there was no reliable method available to measure the actual load on the mechanical press brake while it was being used in a production mode. Since the introduction of the hydraulic press brake, which is a machine that cannot be overloaded, the knowledge of the actual forming tonnage required has become much more important.

E) Tonnage Estimate Compared to Press Brake Tonnage

Do not calculate a charted air bend forming tonnage to the maximum available tonnage of a hydraulic press brake. As described in the above paragraph, material variations will cause the actual tonnage to vary from part to part. To allow for this known variation, the maximum air bend (charted) forming tonnage calculation should be limited to 85% of the maximum machine tonnage rating. If machine size limits a particular part from being formed, a larger vee die opening might be considered to reduce the air bending tonnage. The disadvantage would be that the inside radius will become larger (see the discussion on selecting the correct vee die opening in Chapter 3) and the blank size could change.

All air bend tonnage charts (Fig. 4-2 and Fig. 4-3)—developed to calculate the proper air bend tonnage—use the tensile and yield strength of the material of mild steel. The steel must have the ability to form to an inside radius of one metal thickness, and even have the inside radius "coined" to a one-half metal thickness without part failure. Several questions must then be answered. Can mild steel be air bent to allow an inside radius of less than metal thickness to be formed? The answer is NO! Remember that the inside radius is a function of the vee die opening. If the vee opening is made at 6 times the metal thickness, an inside radius approaching the one metal thickness will be formed. The angular accuracy will also become more inconsistent.

If a vee die opening less than 6 times metal thickness is used, the metal will try to form to an inside radius less than metal thickness. Therefore, it will not produce a radius, but rather a shape with an indention caused by the nose of the upper die sinking into the metal. In addition, four other results can be expected.

1) Very heavy marking will occur on the outside of the part where it contacted the vee opening.

2) The bend angle will flare open due the extreme pressure from the outside corners of the vee opening creasing the material.

3) The loss of any acceptable flat past the outside radius of the part can also be expected, which will minimize consistent angularity.

4) An increase in the forming tonnage from the effect of the outside indentions or grooves in the part flanges (restricting part movement into the vee die cavity) cannot be calculated.

The option to obtain a smaller inside true radius is "coining," which will require a great amount of additional forming tonnage. As pointed out previously, many parts are formed using a sharp nose radius on the upper die which gives a small groove in the inside radius. This can give an illusion of a sharper inside radius and may be acceptable for many parts.

Remember: Purchased dies of 90° are not air bend dies unless they are being used to bend an angle greater than 90 ° (e.g., 95°). The forming tonnage is NOT the tonnage that is listed on an air bend tonnage chart. This point is emphasized since some companies selling press brake tooling are providing 90° angles on the majority of their press brake dies. They have standard air bend tonnage charts in their tooling catalogs, but do not explain that the forming tonnage will exceed the chart tonnage if 90° bends are made.

Information on forming materials other than mild steel can be found in Chapter 6.

Tons Pressure Required per Linear Foot to Bend Mild Streel with Air Bend Dies

Thickness of Metal		Width of Vee Die Opening																						
Gauge	Inches	1/4"	5/16"	3/8"	7/16"	1/2"	5/8"	3/4"	7/8"	1"	1 1/8"	1 1/4"	1 1/2"	2"	2 1/2"	3"	3 1/2"	4"	5"	6"	7"	8"	10"	12"
20	0.036	3.1	**2.3**	1.7	1.4	1.1	...	...	...	...	...	...	...	...	...	...	...	...	...	...	...	...	...	...
18	0.048	5.3	4.0	**3.0**	2.5	2.2	1.7	1.3	...	...	...	...	...	...	...	...	...	...	...	...	...	...	...	...
16	0.060	...	...	5.6	4.5	**3.8**	2.8	2.2	1.8	1.5	...	...	...	...	...	...	...	...	...	...	...	...	...	...
14	0.075	...	...	...	7.6	6.3	**4.7**	3.5	3.0	2.5	2.1	1.8	...	...	...	...	...	...	...	...	...	...	...	...
12	0.105	...	...	...	...	...	9.7	8.0	**6.5**	5.6	4.6	4.1	3.2	...	...	...	...	...	...	...	...	...	...	...
11	0.120	...	...	...	...	...	...	11.1	9.0	**7.5**	6.3	5.5	4.4	2.9	...	...	...	...	...	...	...	...	...	...
10	0.135	...	...	...	...	...	...	...	11.9	9.9	**8.5**	7.3	5.8	4.0	...	...	...	...	...	...	...	...	...	...
3/16	0.188	...	...	...	...	...	...	...	...	...	16.4	14.3	**11.2**	7.5	5.7	4.4	...	...	...	...	...	...	...	...
1/4	0.250	...	...	...	...	...	...	...	...	...	...	29.5	22.7	**15.4**	11.4	9.0	7.4	6.1	...	...	...	...	...	...
5/16	0.313	...	...	...	...	...	...	...	...	...	...	...	39.8	27.0	**19.7**	15.3	12.7	10.5	7.7	...	...	...	...	...
3/8	0.375	...	...	...	...	...	...	...	...	...	...	...	...	42.3	30.9	**24.0**	19.6	16.3	12.3	9.5	...	...	...	...
7/16	0.438	...	...	...	...	...	...	...	...	...	...	...	...	...	45.8	35.4	**28.6**	24.4	17.3	14.8	11.2	...	...	...
1/2	0.500	...	...	...	...	...	...	...	...	...	...	...	...	...	...	48.8	39.7	**33.3**	**24.6**	19.4	15.9	13.1	...	...
5/8	0.625	...	...	...	...	...	...	...	...	...	...	...	...	...	...	...	...	58.3	43.1	**33.3**	27.4	23.3	16.9	...
3/4	0.750	...	...	...	...	...	...	...	...	...	...	...	...	...	...	...	...	...	...	53.5	43.6	**36.5**	27.1	21.0
7/8	0.875	...	...	...	...	...	...	...	...	...	...	...	...	...	...	...	...	...	...	...	64.6	52.9	**39.7**	31.6
1	1.000	...	...	...	...	...	...	...	...	...	...	...	...	...	...	...	...	...	...	...	91.2	76.2	**56.3**	44.2

The tonnages highlighted are for Vee Die openings eight times the thickness of the metal. These normally are used for average right angle bending work, as they give an inside radius of the bend approximately equal to the thickness of the metal. Different radii of bends require different tonnages as shown in this example.

Three bends in 1/4" x 10' mild steel

1. Standard inside radius from standard die opening.

   1/4" over 2" die requires 15.4 T/ft.
   15.4 T/ft. x 10 ft. = 154 tons.
   Inside R is approximately 5/16"

2. Smaller inside radius from smaller die opening.

   1/4" over 1 1/2" die requires 22.7 T/ft.
   22.7 T/ft. x 10 ft. = 227 tons.
   Inside R is approximately 15/64"

3. Larger inside radius from larger die opening.

   1/4" over 3" die requires 9.0 T/ft.
   9.0 T/ft. x 10 ft. = 90 tons.
   Inside R is approximately 15/32"

NOTE: The three bends differ in inside radius, in tonnage required, and in size press brake required.

Fig. 4-2. Air bend tonnage chart providing the number of tons/foot to air bend mild steel.
(Courtesy of Cincinnati Incorporated)

Metric Tons Required per Linear Meter to Bend Mild Streel with Air Bend Dies

(approximate tensile strength of 42 kg/mm2)

VEE DIE OPENING mm.	4	6	8	10	12	16	20	24	28	32	36	40	48	64	80	96	128	160	200	250	300
INSIDE RADIUS mm.	0.7	1	1.3	1.6	1.9	2.5	3.2	3.8	4.4	5	5.7	6.3	7.5	10	12.5	15	20	25	31.3	39	47
MIN. FLANGE LENGHT mm.	2.8	4.2	5.7	7.1	8.5	11.3	14.1	17	20	23	26	29	34	46	57	68	91	113	142	177	212
METAL THICKNESS - mm.																					
1	20	11.6	**8.1**	5.7	3.8	...	...	...	...	...	...	...	...	...	...	...	...	...	...	...	...
1.5	...	...	19.8	14.9	**12.1**	7.9	6.0	4.8	...	...	...	...	...	...	...	...	...	...	...	...	...
2	...	...	...	28.9	22.8	**15.3**	11.1	9.3	7.5	6.0	...	...	...	...	...	...	...	...	...	...	...
2.5	...	...	...	...	37.7	25.1	**19.3**	15.5	12.4	10.4	9.4	...	...	...	...	...	...	...	...	...	...
3	...	...	...	...	...	38.9	29.8	**23.3**	18.8	15.8	13.6	12.2	...	...	...	...	...	...	...	...	...
3.5	...	...	...	...	...	...	42.7	33.1	**26.5**	22.2	18.9	16.5	12.8	...	...	...	...	...	...	...	...
4	...	...	...	...	...	...	...	44.3	35.2	**29.5**	25.0	22.1	17.2	...	...	...	...	...	...	...	...
4.5	...	...	...	...	...	...	...	...	44.7	37.4	**32.0**	28.0	22.1	...	...	...	...	...	...	...	...
5	...	...	...	...	...	...	...	...	...	46.2	39.8	**34.9**	27.9	19.3	...	...	...	...	...	...	...
6	...	...	...	...	...	...	...	...	...	...	66	55	**44**	30	...	...	...	...	...	...	...
8	...	...	...	...	...	...	...	...	...	...	...	114	88	**59**	45	35	...	...	...	...	...
10	...	...	...	...	...	...	...	...	...	...	...	...	150	103	**76**	60	40	...	...	...	...
12	...	...	...	...	...	...	...	...	...	...	...	...	...	162	119	**94**	62	48	...	...	...
16	...	...	...	...	...	...	...	...	...	...	...	...	...	...	248	195	**129**	96	73	55	...
20	...	...	...	...	...	...	...	...	...	...	...	...	...	...	...	339	229	**171**	126	94	...
25	...	...	...	...	...	...	...	...	...	...	...	...	...	...	...	...	407	305	**224**	166	131

The tonnage highlighted are for Vee Die openings eight times the thickness of the metal. These normally are used for average right angle bending work, as they give an inside radius of the bend approximately equal to the thickness of the metal. Different radii of bends require different tonnages as shown in this example:

Three bends in 6 mm x 3 mm mild steel

1. **Standard inside radius from standard die opening.**

   6 mm over 48 mm die requires 44 T/(metric) / M. 44 T/M x 3 M. =132 T (metric)

   Inside R is approximately 7.5 mm.

2. **Standard inside radius from standard die opening.**

   6 mm over 36 mm die requires 66 T/(metric) / M. 66 T/M x 3 M. =198 T (metric)

   Inside R is approximately 5.7 mm.

3. **Larger inside radius from larger die opening.**

   6 mm over 64 mm die requires 30 T/(metric) / M. 30 T/M x 3 M. =90 T (metric)

   Inside R is approximately 19 mm.

NOTE: The three bends differ in inside radius, in tonnage required, and in size press brake required.

Fig. 4-3. Air bend tonnage chart providing the force in metric tons per linear meter to air bend mild steel. (Courtesy of Cincinnati Incorporated)

Chapter 5

Fundamental Information

Press brakes have been used for over 125 years, yet many of the practical concerns are seldom put into writing. These include: preventing the equipment from being damaged, problems resulting from improper blank preparation, and the normal limitations that should be expected when forming steel and simple design limitations. The following sections highlight some of the more common concerns.

A) How to Protect the Ram Nose and Filler Block from Damage

If a hammer is used to hit a piece of soft wood, the flat surface of the hammer will sink into the wood until the wood is compressed enough that its compressive strength equals the downward force of the hammer. The result of the hammer blow will exceed the yield point of the wood (which is less than the compressive strength), causing an indention or new shape to be made on the surface contacted by the hammer.

Steel also has a yield point. Normally it is somewhat greater than 30,000 pounds per square inch (PSI). If the yield point is exceeded, a new shape will be made. To be reasonably sure accidental damage will not happen to the surface of the ram nose, or to the top surface of the filler block or die holder, the pressure applied to those surfaces should not exceed a 30,000 PSI load. If the 30,000 PSI figure is divided by 2000 pounds per ton, a figure of 15 tons per square inch (tons/in.^{2}) can be used for easy calculations.

The tongue of the upper or lower die is only used for clamping the tools in place. They are normally not designed to carry the forming load. The shoulders of the die, in the front and rear of the die tongue, must be made in the same plane in order to uniformly carry the forming load. If the shoulders do not have sufficient area to keep the transfer of forming pressure from the die shoulders to the ram or filler block to less than 15 tons/in.^{2}, there is a great risk of the die shoulders sinking into the ram nose or filler block, thereby causing permanent damage. A

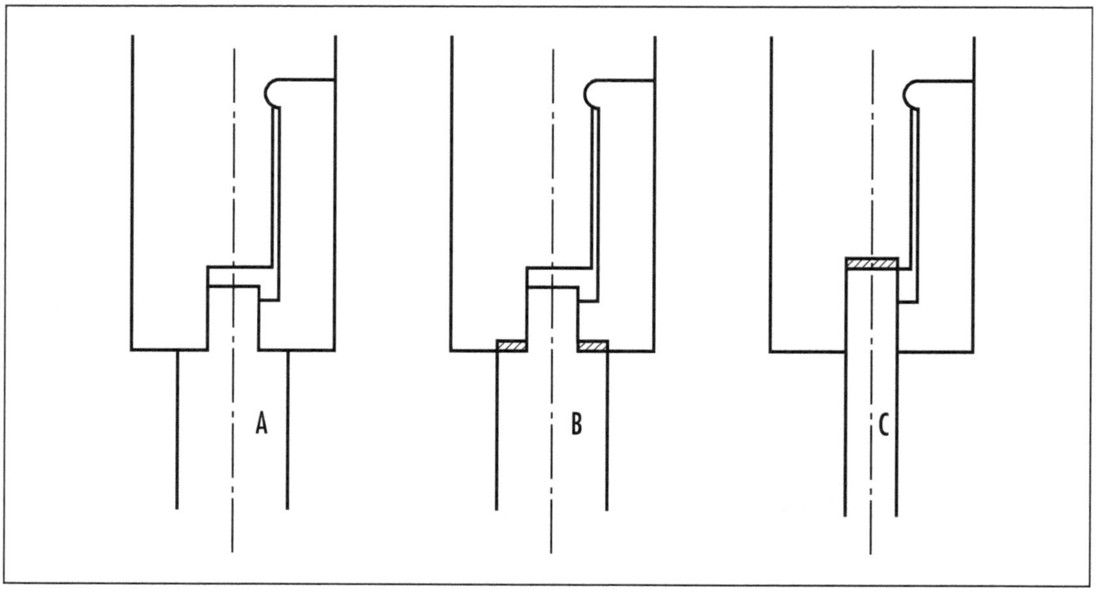

Fig. 5-1. Standard upper die (A) shoulders shown carrying a safe forming load. Distortion that will occur to a ram nose if the upper die (B) is overloaded causing the die shoulders to "sink" into the nose of the ram. Distortion caused by a $^1/_2$" stem die (C) sinking into the ram nose due to an overload.

ram nose that is not smooth, but has damage from dies having been sunk into its surface, will cause abnormal bend line variations, because the bend angle will change where there is no solid backing under the die shoulders (Fig. 5-1).

The calculation to determine a safe load for the ram nose is to take the number of square inches per foot that the shoulders of the upper die actually touch the ram nose, and then multiply that figure by 15 tons/in.^{2}. For example, if the upper die bar is 1.5" wide, and if the standard tongue is 0.5" wide centered on the centerline, a 0.5" shoulder should be available in the front and in the rear of the bar. Measuring front to back, the load carrying shoulders would total 1". For each foot of die length (12"), the area able to carry the load would be 1" × 12" or 12 in.^{2}/ft. Multiplying this area by 15 tons/in.^{2} shows that the safe load calculates to be 180 tons/foot (tons/ft). Two design assumptions are made.

1) The design of the upper die is strong enough that it will not fail in an area that may have been relieved (i.e., if a gooseneck die is being used) to clear previous bends in the part

2) The vee die strength is sufficient that it will not split open.

If the forming load is less than the safe load calculation, the shoulders of the die will not sink into the ram nose. This problem is seldom encountered when the air bend method of forming has been selected. If bottoming or dies that coin the metal are used, the required forming tonnage to obtain a good bend will increase greatly over normal air bend tonnages. If a part made from 16 gauge thick mild steel is formed by air bending, the required tonnage is about 3.8 tons/ft. If this same part is bottomed, in order to improve the angular accuracy, the tonnage, as discussed previously, might increase by a factor of three to five times the air bend forming tonnage. The forming range could then fall into a range of 11.4 to 19 tons/ft. If the part was to be coined to obtain a smaller than metal thickness inside radius, the tonnage would increase to a range of 19 to 38 tons/ft. Why is there such a range? When the press brake is initially set up to make the part, a test part is made to determine the required ram setting at the bottom of the stroke. As the various parts are formed, the ram setting may or may not be changed. The blanks often change in thickness (within the tolerance range for the gauge of metal being formed). This means that thicker parts will be hit harder and thinner parts hit less hard. To obtain the best consistency during a production run, the operator often sets the ram at the higher tonnage setting. This improves production since less time has to be spent trying to inspect each bend.

Many dies used for forming thinner materials are made from narrower bars of steel. If the bar thickness is reduced to 1" thick, the shoulder area is reduced to about 6 in.^{2}/ft. The reason is that a 1" bar only has a 0.25" shoulder in the front and the rear, with an allowance of 0.5" for the tongue width. The 0.5" of available shoulder front to back times 12" per foot equals the 6 in.^{2}/ft used for calculating purposes. The calculated safe tonnage is a maximum of 6 in.^{2}/ft × 15 tons/in.^{2}, or 90 tons/ft. Many times, a part requires the die to be cut up into short lengths in order to fit around notches, holes, or other flanges. If the die design chosen is the coining type, and if the ram setting is not adjusted when forming long lengths and short lengths, it is very possible that the small length will be hit abnormally hard if the ram setting is not raised slightly. When longer lengths are formed, the bed and ram deflect more than when a shorter part is made. The ram setting will then be set lower in order for the upper die to penetrate into the lower die to obtain the pressure needed to form the part while accommodating the machine deflection caused by the high forming tonnage. If the shorter flange length is made at the same ram setting that is required for the longer bend length, the shorter length will be hit at a much higher tons/ft load than needed since the machine deflection is much less.

The current thinking in some companies is to use upper dies that are cut into a number of short sections in order to make it easier to change die lengths or setups as different shapes and lengths are made during each shift. If a part has to be rotated during the forming process to form sides that have different length bends, it is very possible that the ram setting will not be changed from cycle to cycle. If a short length flange is bent using a short die section that does not have a die width to provide sufficient shoulders to carry the abnormally heavy forming load, the die shoulders will sink into the ram nose just like the hammer on the piece of wood. The damage done to the ram will be permanent. The sinking into the ram nose will affect all future setups, creating part quality problems until the ram is repaired.

Remember, someone must have the responsibility and authority to check and approve the forming procedure to ensure the press brake is properly set up and will not be damaged. In almost every case, a short die section has been the problem.

B) Springback

When a piece of metal is bent, the inside surface of the part is in compression; the outside surface is in tension. A theoretical line, described as the neutral axis, forms some type of elliptical shape inside the material thickness. The neutral axis is that location within the material thickness that is neither in compression nor in tension. In a flat sheet, a line in the center of the material thickness is called the neutral axis. The length of the neutral axis does not change and is used for measuring the part width. The closer the neutral axis stays to the center of the metal in the arc area that has been bent, the more likely that springback can be expected in the formed angle when the forming stroke is completed.

Most people learned the concept of springback in the first or second grade. Springback can be simply shown by taking a sheet of 8.5" × 11" paper and rolling it into a large tube shape. When the paper is released, it will return to its original flat condition. The reason is that the yield point of the paper has not been exceeded. If the same piece of paper is rolled into a

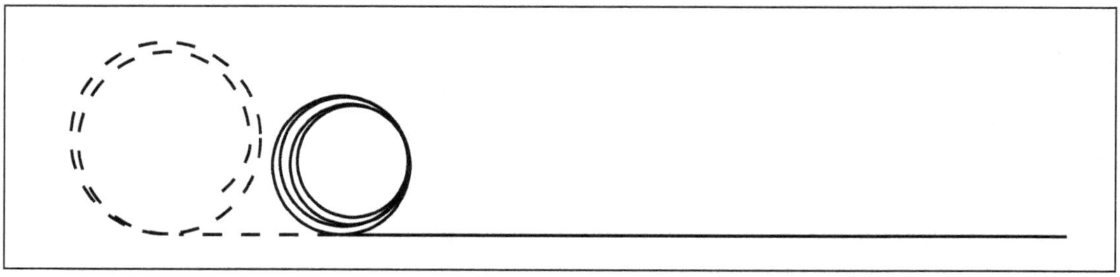

Fig. 5-2. Springback occurs when a simple sheet of paper is rolled into a small tube shape, and then unrolls into a larger tube shape when released.

smaller tube shape, when the paper is released it will open up into some type of circular shape. However, the arc or radius will be much larger than the radius that may have been measured while the paper was rolled up. If the paper is rolled again into a much smaller tube shape, when the pressure is released it will now open up into a smaller tube shape, but will still be slightly larger than the tube diameter that was last rolled. This opening up of a radius after forming is called springback. A larger radius can be expected to have greater springback. The smaller the inside radius, the less springback should be expected (Fig. 5-2).

The same concept applies when forming metal. As the metal is forced down into a vee die opening, the yield point of the steel is exceeded. As the metal travels down into the die cavity, the neutral axis of the metal leaves its theoretical location in the center of the material thickness. It starts to form an elliptical arc (Fig. 5-3). Some of the area in the material located on the inside of the formed arc, above the neutral axis, will be in compression prior to the forming pressure being released. Some of bend area is "set," since it has had its yield point exceeded. Some of the material on the outside of the neutral axis in the formed area will be in tension. When the forming pressure is released, the area of the bend that has been "set" does not move. The area of the bend in a state of compression or tension will try to balance the stresses. This will cause the part angle to open outwards or spring back (Fig. 5-4).

Normal expected springback that occurs when forming mild steel using the recommended vee die opening will be 2° to 4°. If stainless steel (type 304) is formed to a 90° angle (again using the recommended vee die opening), the springback will increase to 4° to 6°. Aluminum can vary from 0° to over 20° of springback depending on the hardness, type, or heat treatment of the material. Exotic materials should be tested prior to finalizing a die selection due to the wide range of springback allowances that may be required.

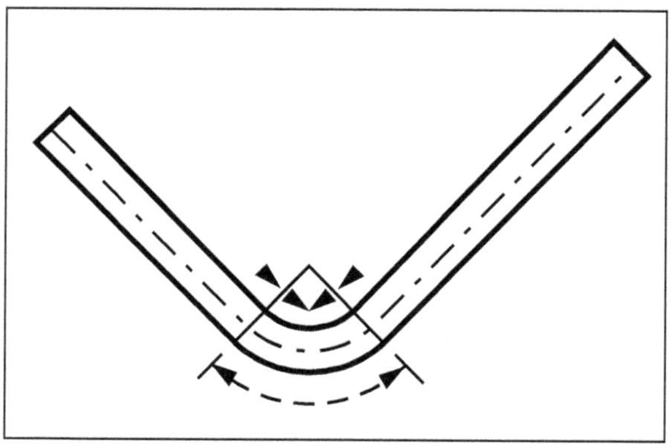

Fig. 5-3. The theoretical "neutral axis" is shown in the center of the straight legs of a formed part. The neutral axis in the bend area creates a parabolic shape, not a true radius.

Note: If the springback is greater than the allowance provided in the die, the operator often attempts to overcome the springback by bottoming out the upper and lower dies. The resultant tonnage will far exceed the normal air bend tonnage calculations. If the press brake is mechanical, it may be overloaded. If the press brake is a hydraulic, the ram may simply reverse and not form an acceptable bend.

C) Bow Along the Bend Line

What can be done if the 90° bend that was formed has a "bow" along the bend line (Fig. 5-5)? This distortion has caused many hours of lost time in the fabricating area. The operator forms a part of some length (normally in excess of 8'). The part is not unusual and in many cases has been made many times. When the part is checked for angularity, the part angle is found to be more than acceptable and within all tolerances. The problem is that when the part is laid on the floor, it does not appear to be straight. The part is then placed on a good smooth surface, and tilted so that each flange is up at the same angle. If the angle is a 90° bend, each leg would be sloping upward at a 45° angle. It can easily be seen that the part touches the flat surface at each end, but the center of the part is a noticeable distance up from the base surface. Why? This problem was never here before.

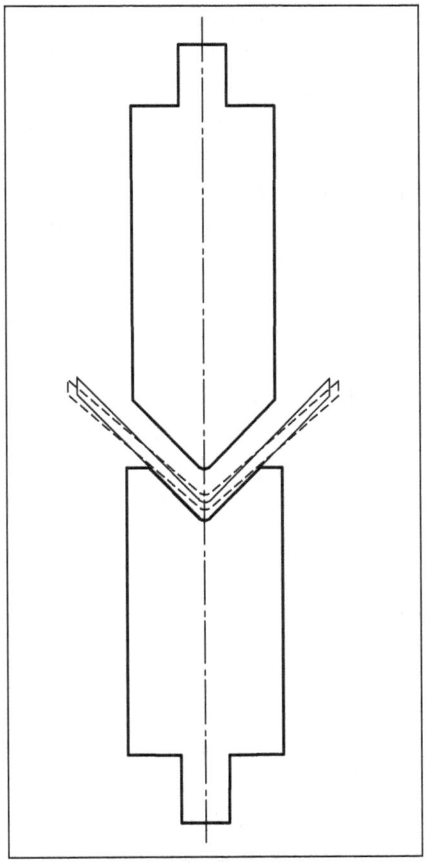

Fig. 5-4. Springback in a part after the part has been formed by air bending, and the upper releases it.

In an attempt to correct the problem, someone will suggest the dies should be shimmed in such a way that the "bow" will be pushed out. Often a lengthy discussion will take place trying to decide what type of straightening operation can be developed to salvage the part. The purchasing department is called to determine what is wrong with the steel. The answer is fairly simple in most cases. First, it must be recognized that the formed angle is correct. All a press brake can do is push the material uniformly down into the lower die in order to make a consistent, acceptable angle. As pointed out above, that is what happened to the part. Any shimming done after an acceptable angle has been made in the part will change the upper die penetration into the lower die and change the resultant angle.

It is known that when a piece of metal is bent, the part will actually grow or lengthen along the bend line. This occurs because the inside of the bend radius is in compression, and the outside is in tension. Consider the following example.

When we were in grade school, most of us had a pink beveled rubber eraser. By holding it

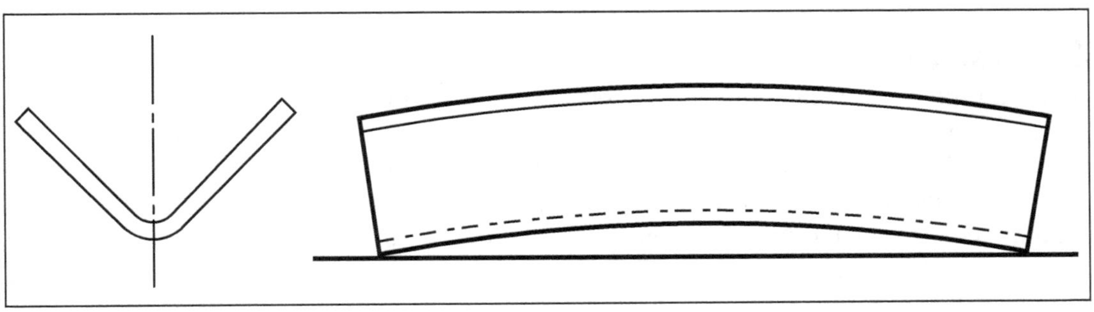

Fig. 5-5. Bow along the "bend line" occurring after the angle has been formed.

at each end, we could bend it into some type of arc or angle. As we held the eraser, we could see that the outside was in tension. The top of the eraser was peaked on each side, and when looking at it from the end, the outside radius was not a true radius but somewhat flattened or stretched. The inside radius was obviously in compression. The inside radius was not a true radius, and the compression had actually caused the rubber to "squash" out to each side much like a marshmallow being compressed. We could see, if we looked at the thickness of the eraser in the radius area, that the eraser was wider on the inside while probably slightly shorter at the top where each "peak" was formed. The same thing happens when we bend a piece of steel. The inside of the metal actually becomes slightly longer along the inside of the bend line. It grows in length in the area of the bend.

If the part that the operator is trying to form has a "hard" surface along the thickness of the part, the hard edge cannot "grow" or lengthen during the forming process. In almost all cases, this is not a problem. However, if the bend line is near the hard edge of the material (such as a short flange), the material in the area of the bend radius will grow in length, but the hard edge will not move. When the bend is made, the angle will be acceptable, but the part will have a noticeable **bow** along the bend line.

Where does the hard edge come from? In most cases, if the material is sheet metal, it comes from dull cutting knives. Dull shear knives do not provide a sharp cut, rather the material will almost be torn apart, thereby causing the edge of the material to have a definite burr plus an abnormal amount of penetration burnish. Sometimes the entire sheared edge becomes hard. Dull slitter knives can cause the same problem. In some parts, dull "hole" punches (where the holes are near the bend line) or dull notching dies (where the notches are at the end of the part near the bend lines) have caused the same problem.

In formed plate, the problem can occur but is more noticeable. Plate is often cut by flame cutting, or by a method that may cause the edge to become hardened. Some laser cutting can cause the edge of the material to become harder. Rolled plate that has not had a trim cut will also cause a bow to form when a simple bend is made near the "rolled" edge.

To quickly determine if the bow is caused by a hardened material edge, take two similar blanks prepared in the same manner. One of the blanks should be formed without any additional operations being done to the material. The second blank should have the edge (material thickness) closest to the bend line, ground with a simple hand grinder to actually remove the hard surface that was caused by the cutting operation. The second blank can then be formed. If a hardened edge, as described previously, is the problem, the second part can easily be seen to contain much less "bow" than the first part. After the problem is definitely identified, the solution may be as simple as changing the edge on the shear knife or sharpening the slitter knives. In the case of plate, an additional grinding operation may be required to ensure that scrap parts will not be made when only a few parts have to be formed.

D) Cracking

Hairline cracks are often found on the outside radius of formed plate (Fig. 5-6). The cracks provide a graphic picture of what happens when the maximum tensile strength of steel is exceeded. The only way cracking can be eliminated is to form the parts with a larger (inside and outside) radius. In most cases this can be done easily. If mild steel is being formed, select a larger vee die opening for future parts. A larger die opening will often provide the inside radius that will resolve the problem. Remember that the blank size will also change. If the material is higher tensile steel, a larger nose radius may be needed on the upper die.

Other problems, caused by the part design, may also have to be evaluated. A short flange

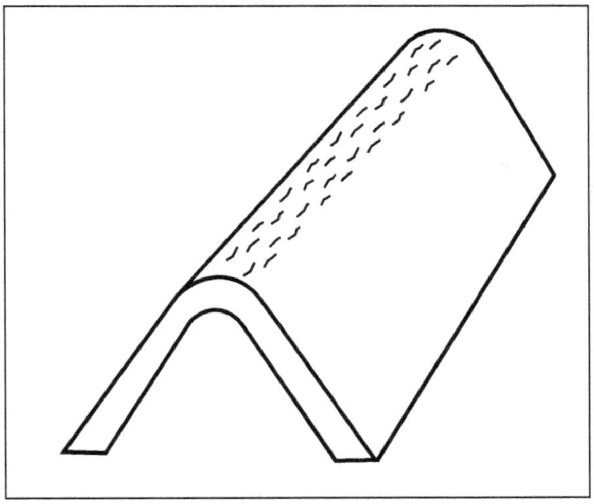

Fig. 5-6. Cracking on the outside radius after the part has been formed.

requirement may require the smaller vee die opening. For some parts, the short flange does not have a cracking problem; however, when the part is indexed forward to form a second, longer leg in the part, cracking may occur in that bend. When the first bend is made, the short flange, not being affected by the weight of the bulk of the material in the front of the press brake, will slip down into the vee die cavity since it has minimum tendency to be affected by friction. When the second bend is attempted, the added weight (now added to the back of the part) can cause more friction as the rear flange is pulled into the die cavity. The natural tendency of the material to sag downwards toward the ground while it is being formed, both in the front and rear of the die opening , is now sufficient to cause enough additional resistance that the material can have its tensile strength exceeded. Cracking may now occur.

In many cases, the cracking will be inconsistent and not occur in every part. Since the part design may require a "borderline" die selection, a careful study of the problem should be made. In some cases, the only thing that can be done is to lengthen the leg of the first bend, form the part using a larger vee die opening, and cut off the excess material that had been added to form the first bend after the forming operation is competed. For some parts, the problem has been solved by carefully lifting the part so that the long section of plate extending out in the front of the press brake is lifted upwards with a crane prior to making a bend. The crane can be incrementally adjusted to keep up with the material as it is formed upwards during the forming cycle.

E) Shimming

When the press brake is used to form any part, the bed and ram are designed to withstand the pressure. The load required to form the part, however, will cause the bed and ram to deflect. Most forming on the press brake is done with the part centered equally between the two housings. This is important because any deflection between the bed top and ram nose will be symmetrical. The greater the required forming load, the greater the bed and ram deflection.

In order to form a good angle along the length of the bend line, the top die must penetrate uniformly into the lower die. If the top die penetrates at each end of the bend line sufficiently to form a good 90° bend, but the center portion of the upper die does not penetrate into the lower die the same distance, it is because the forming tonnage causes the bed and ram to deflect. The center of the part will form an open angle. The resultant shape of the part is the same as if your hands were cupped together with the tips of the fingers and the heel of each hand touching, but the center of the hands were forming an open or "canoe shape." (A simple straight line bend, where no compensation for the deflection of the bed and ram is allowed, may form an angular shape normally called "canoe shaped.")

In some cases, when bottoming dies are used, the resultant bend line could provide an overbent shape (less than 90°) in the center of the part, while the ends have been formed to 90°. This means that the entire length of the bend line was not formed with a consistent, bottoming tonnage (refer to Chapter 3-C-1). The bed and ram deflected during the forming stroke before sufficient tonnage was applied in the middle of the bend line, allowing the overbent flanges to be "pushed" back to 90°.

After some experience with setting up the standard vee type dies, the operator or setup person will learn how to install shims to correct the effect of deflection. Shimming must be done with material that is not compressible under a normal forming load. Brass or bronze shim stock, often used for maintenance work, will not do a good job since most of that type of material has a low compressive strength. It will tend to mushroom outwards under load, just like a marshmallow will when compressed under a simple load.

The most common and unwanted "shim" is dirt. Almost every shop has airborne, oiled dirt particles in its atmosphere. That helps explain why metals don't rust inside the building. It is important that, prior to installing any die into the press brake, the shoulders of each die should be wiped (normally using a shop rag) to remove the majority of dirt that seems to be attracted to those surfaces. Wiping the die shoulders also provides a simple way to ensure that any nicks or bulges in those critical surfaces, caused when the tooling was stored, can be discovered and removed.

A good shim stock is called "blue tempered spring steel." This material, although somewhat expensive, can be obtained in a variety of thicknesses that allow the shimming to be "fine tuned." This shim stock can be purchased in increments from 0.0015" to 0.015". Most users only keep a few thicknesses in stock. This allows most shimming to be close, but could still allow some small angular variations to remain along the part bend line. Another common shim stock is paper. Paper is virtually incompressible under most press brake forming loads. It is not unusual to find common "time cards" being used to shim the press brake setup. Ordinary paper, such as Kraft paper (like a brown paper bag used to carry groceries), can be used successfully.

Most shimming is actually done only under the filler block as opposed to under the vee die shoulders. The bolts used to hold the filler block to the bed top are removed to allow the die holder to be raised. Normally, this is done by using a crowbar; by inserting it at one end of the full length rectangular slot in the bed and lifting that end up, a piece of wood can be inserted between the bed top and the bottom of the end of the filler block. Two identical shim packages are placed on the bed top. One is located along the front length of the filler block, and the other is located at the rear. This provides a space for the hold-down bolts. The filler block is again lifted up with the crowbar, the wood block removed, and the filler block returned to its original position. The bolts are then used to fasten the filler block loosely to the bed top. (Refer to the section on die alignment before fully tightening the bolts—Chapter 2-C and Chapter 24-B and C.)

Normal shim packages using shim stock are made up by estimating how thick the stack of shims should be to compensate for the deflection occurring during the forming cycle. If the distance between the housings is a nominal 10' length (longer or shorter, depending on the machine being used), the stack of shims would be made up of three or four different lengths with the longest about three quarters of the part length. For this example, assume a 10' part is to be formed. If possible, it is desirable to use the same thickness shim stock. If a total stack height of 0.015" of shim is required, the shim package could be made up using one length of 0.005" shim stock about 7' long; one piece of 0.005" thick stock, 5' long, placed on top of the seven foot piece; and finally a 2' to 3' length of 0.005" thick stock laid on top of the first two

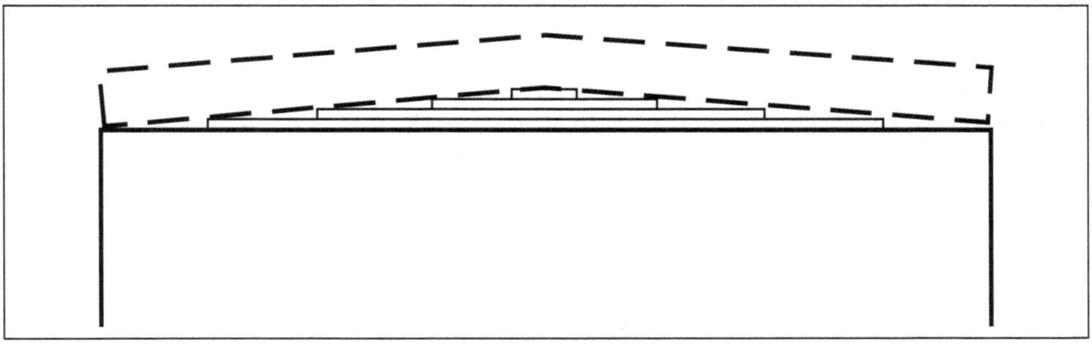

Fig. 5-7. The recommended method of shimming the lower die under the filler block to compensate for bed and ram deflection caused by the forming load. Shimming ensures uniform penetration of the upper die into the lower die along the bend line.

lengths. This method gives support under the filler block so that when the load occurs, the filler block cannot be deflected more than 0.005", which minimizes any noticeable change in the formed angle (Fig. 5-7).

In some cases, where "time cards" are used to create a shim package, the results are not as good as when shim stock is used. With time cards, a similar stacking method is used. Different stacks of cards, ranging from only a few cards measuring several thousandths of an inch thick to a stack of cards having the same 0.015" thickness used in the above example, are located in a symmetrical pattern around the front center of the bed. By lifting the filler block as described previously, the cards can be slid under the filler block. The cards are positioned so that they are supporting the filler block in the same way bridge piers support a bridge. A potential problem might develop as the forming load is built up. The filler block may deflect downward in between the individual stacks of cards since there is no support to limit the deflection until it touches the bed top. A shim stack limits the deflection to the next level of shims. A wavy bend line will often result. The change in angularity can be measured to where each stack of cardboard shims was placed (Fig. 5-8). When an acceptable shim package has been developed, it is important for the operator to record the information. If used for a popular part, simply maintain the developed "shim" package as part of the setup.

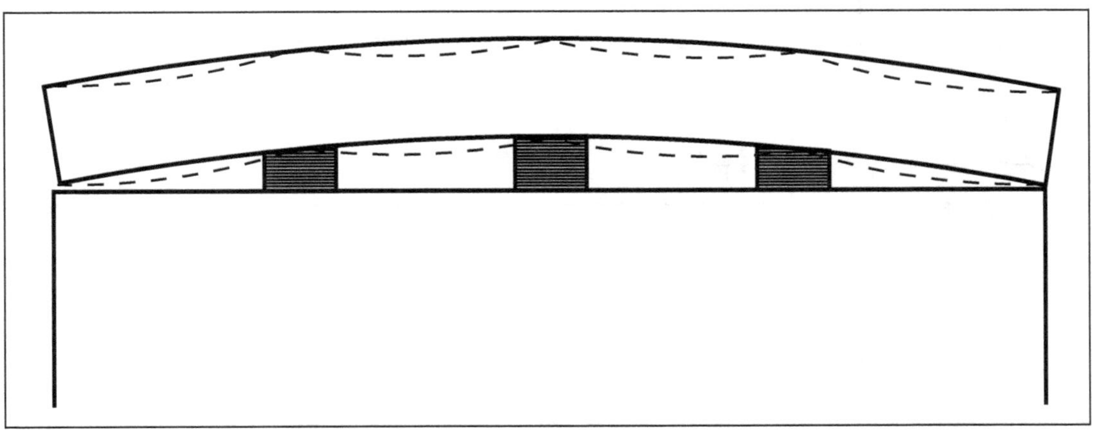

Fig. 5-8. Shimming that is done with time cards will create a "bridge support" design which can allow some die deflection to occur during the forming load between the stacks of shims. This will create a wavy bend line.

If the ram of the press brake has been damaged previously by the ram nose having been indented through the overloading of short die lengths, standard shimming methods may not solve the problem. The best method is to determine where the distortions are located. This can be done by removing the filler block and lowering the ram nose to a location near the bottom of the stroke. Set up a dial indicator so that the measuring finger reads 0.000 under one housing. When moved from a location under one housing to the same location under the other housing, the indicator should still read 0.000. This ensures that the bed and ram nose are as parallel as possible. The indicator finger should not be located on the die clamps, but on the back shoulder of the ram about $^1/_8$" past the die slot opening. Starting at one end of the bed, move the dial indicator along the bed top toward the other housing (Fig. 5-9). Each noticeable variation and location inspected should be recorded. Any serious indentions should be marked on the front of the ram with a permanent marker to identify the location and length of the problem area. Since, in most cases, the ram nose will not be remachined to eliminate the indentions, a careful review may determine the best of three choices. A permanent filler shim can be fastened into the hole. In some cases, a wider die can be used to allow the shoulders to span the indention and provide adequate support to eliminate any die deflection in that area. The last option is to accept the resultant angular variation since it may not be enough to cause a problem.

A few manufacturers actually produce their equipment so that the bed and ram are closer together in the center of the press brake rather than under the housings. Cincinnati Incorporated, a major supplier of press brakes, calls this a "microcrowning" feature. This intentionally deformed bed or ram (or both) minimizes the need for shimming. For most users having this feature, the actual shimming time is reduced by two-thirds.

After many cycles and numerous parts have been produced, especially if the forming tonnages approached maximum capacity on a regular basis, it is likely that the bed and ram actually take a new "set." A permanent hollow between the bed and ram can develop. A new

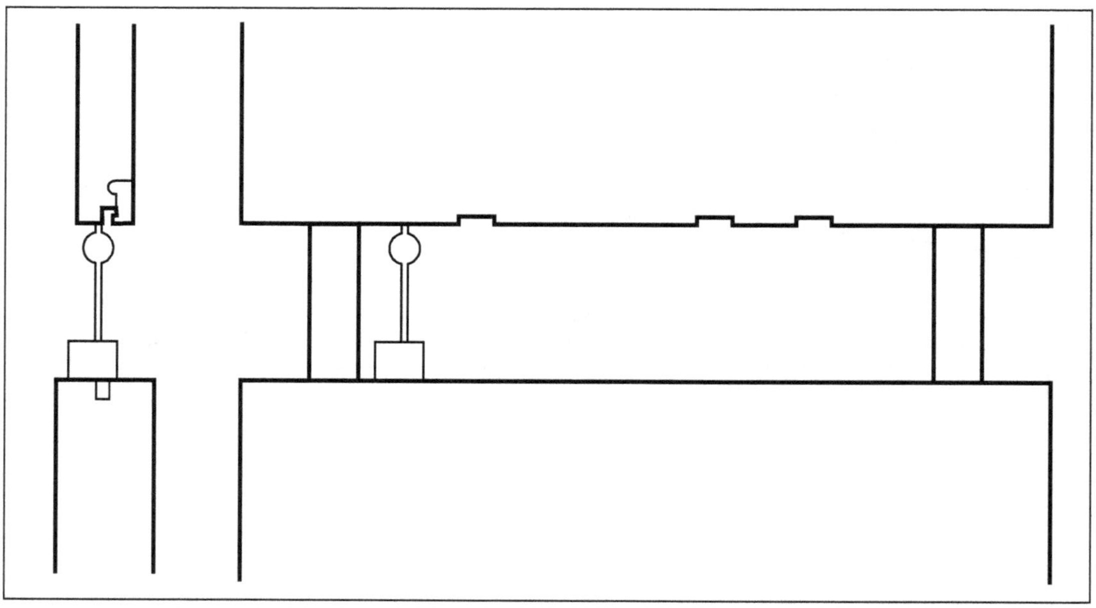

Fig. 5-9. A dial indicator is shown resting on the bed top with the fingers contacting the back of the ram nose near the tongue slot. By moving the indicator slowly from one housing to the other, a pattern of wear or distortion in the ram nose can be determined.

press brake should be supplied with a bed and ram that are either parallel to each other or actually closer together in the center than at the housings. If, after years of use, a hollow has developed, by using the same dial indicator procedure described previously, a permanent shim package can be kept under the filler block. This will provide a bed and ram relationship similar to a new machine.

F) Tonnage Buildup

When making a simple 90° air bend, the actual load required to make the bend builds up very quickly. The reason is that, in order to push the metal down into the lower die, the yield point of the steel must be exceeded. The forming load will reach approximately 85% of the calculated "maximum air bend tonnage" required to form the 90° angle when the metal has only been bent 20° total (10° per side, front and back). The maximum air bend tonnage is reached when the part is bent a total of 40°. The load might drop slightly as the part is forced further down into the die to complete the bend angle (Fig. 5-10). This information provides answers to two questions. It explains why only one tonnage is given to calculate the forming load. The same tonnage is used if the formed angle has been bent to a 150° included angle or to a 30° included angle. It also explains why there are different forming limitations between mechanical and hydraulic press brakes having the same apparent tonnage ratings.

A mechanical press brake, if it was built prior to 1958, was only sold with a tonnage rating called the "midstroke" tonnage capacity. In 1958, a dual rating appeared on new machines

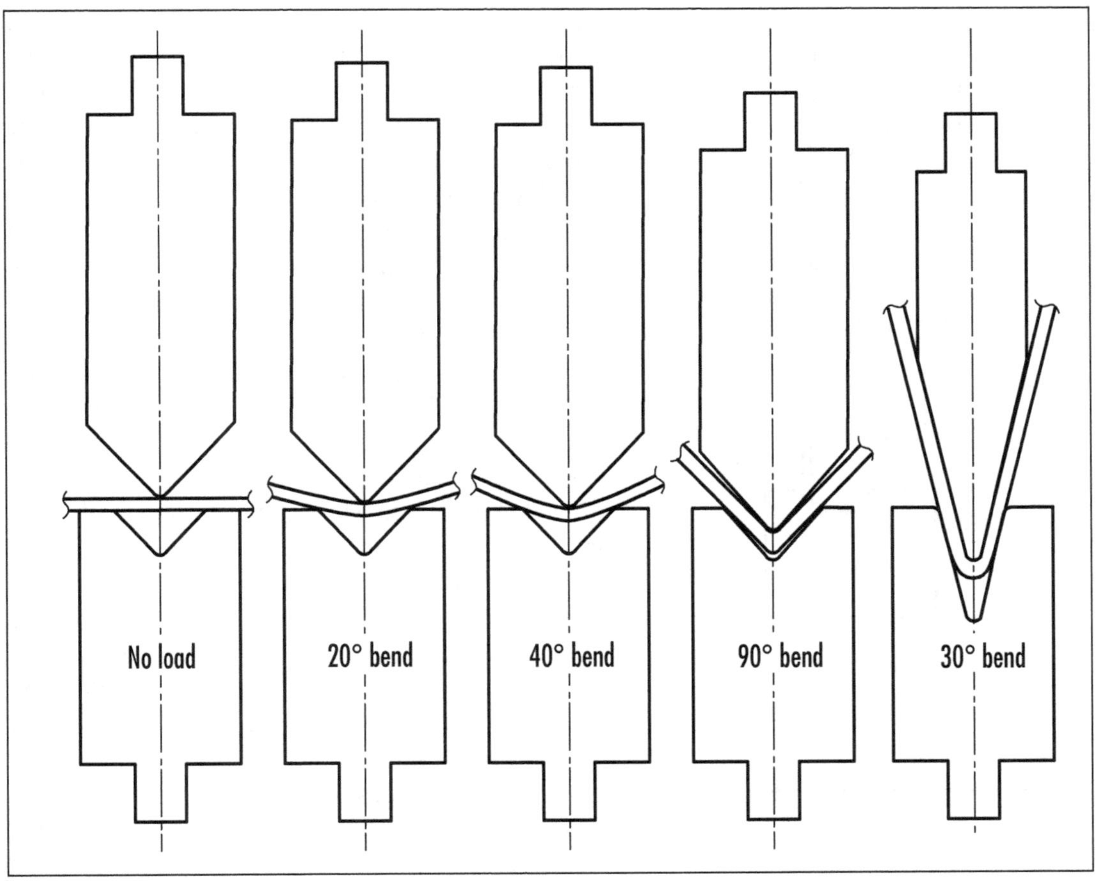

Fig. 5-10. Tonnage buildup.

identifying the midstroke capacity, but also listing a tonnage called "near bottom of stroke" tonnage. This concept was developed from the following facts. As the eccentric rotates the ram to the bottom of the forming stroke, the distance of the eccentric centerline is constantly changing with relation to the forming centerline. This creates a moment arm, which reduces the available forming tonnage. By the time the ram nose is about 0.375" from the bottom of the stroke, the moment arm differential is minimized. This allows a higher "near bottom of stroke" tonnage to become available. Depending on the size of the vee opening, the distance of the forming stroke required will affect how high up on the forming stroke the upper die will contact the part. Since the forming load is near the actual chart value of the air bend tonnage chart when the part is only formed a minimal amount, larger vee die openings (which require a longer forming stroke) restrict the length of a part that can be formed using a mechanical press brake.

A hydraulic press brake provides constant available tonnage at any point in the forming stroke. The length of travel required does not matter as long as the machine tonnage limitations are not exceeded. With constant tonnage throughout the stroke, the position of the maximum forming tonnage on the stroke does not matter. As a result, hydraulic press brakes can often be used to form plate and other shapes that may require long, high tonnage forming strokes.

G) Eccentric Loading of a Press Brake

A press brake is designed to carry a maximum load along the total length of the bed and ram. Each housing or upright is made to carry one-half of the total load. As a result, all press brake loading is expected to be balanced around the centerline of the bed and ram as seen from the front of the press brake. This is important since the bed and ram deflect under load, and the load in the center keeps the theoretical deflection symmetrical. If a very heavy concentrated load is to be formed in the center of the bed and ram, it is wise to contact the press brake manufacturer to confirm that the load will not cause damage to the machine. An example would be a full capacity load two feet long or shorter.

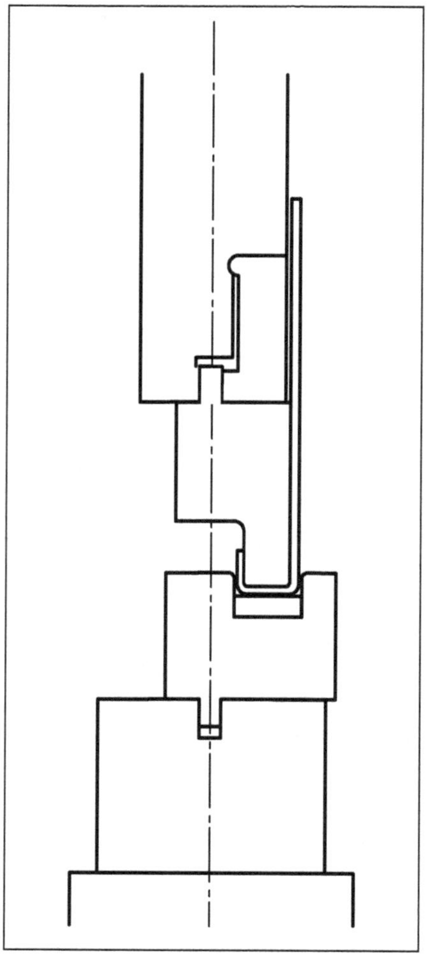

Fig. 5-11. Eccentric load caused by a forming load.

Years ago, it became common practice for operators who needed to make short lengths or small parts using a mechanical press brake to do all of their forming under the right hand housing. This practice occurred because all of the controls for adjusting the ram elevation and activating the stroke were on the right side of the machine. As a result, the right hand housing was overloaded and often stretched, if not actually cracked. Shimming a bed and ram to obtain good parts, if one housing was open more than the other, required a great deal of time and skill. With the introduction of the hydraulic press brake, the controls were often movable. Each cylinder would only produce one-half of the specified capacity. If the operator would attempt to form a part under one housing that exceeded one-half of the total machine capacity, the safety limitations on the machine would signal an overload, and the ram would reverse without forming the part. As a result, many operators who had originally run mechanical press brakes resisted hydraulic press brakes because they felt they were unable to do the jobs they did on their old machines. The basic rule of all press brakes still holds today. For best results, the forming load should be centered between the housings so that the load will stay evenly distributed.

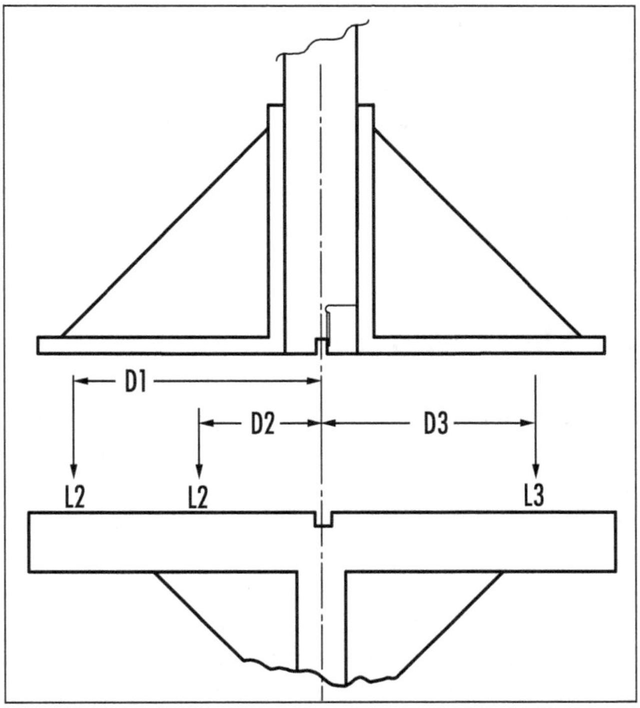

Fig. 5-12. Eccentric load caused by a forming load.

A second design concern must also be kept in mind. The press brake is designed to have the load on the machine kept symmetrical around the centerline of the bed and ram, looking at the end of the press brake front to back. The theoretical centerline of the upper and lower dies are made to carry the forming load in such a way that minimal loading is placed on the slides and guides holding the ram in place. Rather, the load is carried by the parts that provide motion to the ram, and then to the housings (Fig. 5-11 and Fig. 5-12).

In practice, many exceptions must be considered. Special dies such as hemming dies, dies used to form hinges, offset dies, and radius tooling, will all cause some "front-to-back" loading. During the 1940s, many press brakes were made with angle brackets or with permanently wide beds and rams to allow punching to be done. The die sets used for punching often required strange hole configurations to be punched. In many cases, the front to back loads could not be perfectly balanced. After cases of unintentional damage to the slide and guide areas of the ram and housings, specific front to back off-center loading limitations were placed on each size press brake.

Every press brake manufacturer has limitations on front to back eccentric loading. That is the force causing the ram of the machine to be pulled to the front or rear, causing an unacceptable eccentric force to be loaded on the slides and guides of the machine. This same problem occurs when the proverbial "junk" drawer many of us have at home is overloaded and sags downward as it is pulled out. The slide on the bottom of the drawer is then stressed, sometimes causing pieces of plastic to break or screws to loosen. This is the same type of possible damage that can be caused to the slides or guides on an eccentrically overloaded press brake. The manufacturer's limitations are often measured as the number of "inch tons" of eccentric loading allowed before serious damage to the machine can be expected. If that type of damage does occur, it is normal for the manufacturer not to cover (under warranty) the damage to the slide and guide area since the eccentric load limitations contained normal engineering safety factors.

In order to provide a higher inch ton capability, a design known as a **longer slide and guide** option is available, which allows the amount of front-to-back capability to be increased. Any eccentric loading must still be close to symmetrical around the press brake centerline, since each slide and guide can only carry one-half of the total load. Before using a press brake tooling setup that has a known front-to-back eccentric load, check with the original press brake manufacturer to confirm that the machine can do the work without damage. Some manufacturers provide a chart listing their press brake eccentric load limitations in their manual.

H) Short Flanges

Many parts are designed with very small flanges. To form the minimum flange length when making a simple 90° bend, lay the unformed blank over the recommended vee die opening so that the edge of the flange reaches past the **lead-in** radius in the back of the vee opening; also, the part must be supported by at least one metal thickness of material on the rear shoulder of the die. For some parts, a slightly smaller vee die opening could be used, possibly as small as a 6 times the metal thickness vee die opening.

Although the part drawing may specify an outside flange length dimension, the gage setting for the bend provides an inside dimension to the flange length. The minimum flange length is determined as follows. The inside dimension of the part will be one-half of the vee die opening, plus the distance required for the lead-in radius of the lower die, plus one metal thickness or slightly more if the material is very thin (Fig. 5-13). The outside dimension must have an additional metal thickness added. Remember that using a smaller vee die opening will cause the angular accuracy of some bends to be affected, as was pointed out in Chapter 4.

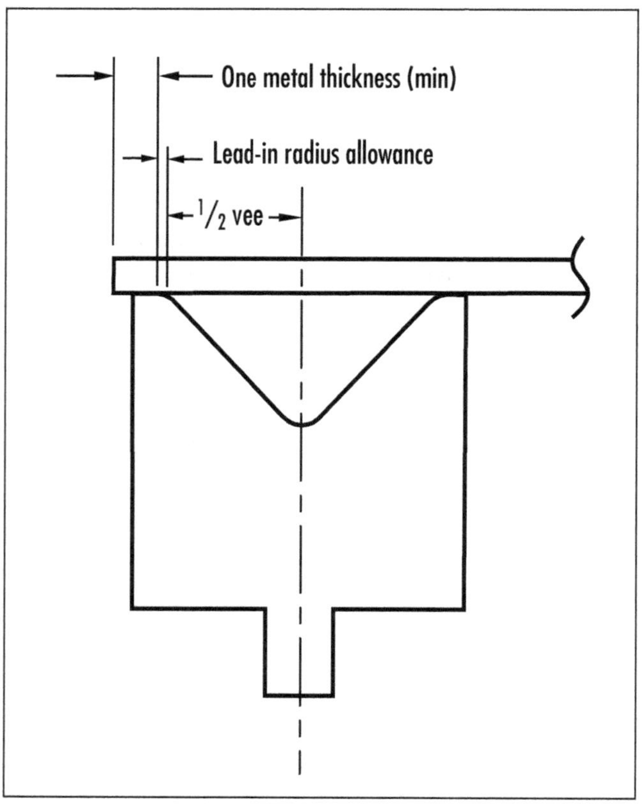

Fig. 5-13. Forming a Short Flange.

For materials thinner than 14 gauge, smaller short flanges have been made by making a special gage ledge in the rear slope of the vee die (Fig. 5-14). This might work for short lengths of some parts, but it is best to consult a press brake tooling supplier to obtain their opinion as to what problems—with regard to forming accuracy, tonnage required, and machine limitations (if any)—must be evaluated.

I) Bed Reference

All mechanical press brakes, and all older hydraulic press brakes, had their method of reaching the bottom of the stroke and having the stroke return to the top controlled by a design that was located on the top of the housing throat. All mechanical press brakes have both eccentrics, which provide the stroke of the ram, located near the top of the housings. The original hydraulic press brakes were designed to allow the hydraulic cylinders to simply bottom out and reverse on some type of adjustable stop, again located near the top of the housing.

During the forming load, all press brake throats will gap or open up depending on the amount of load required. It is exactly like watching a "C" clamp gap as the screw of the clamp is tightened to hold two parts together. If the method of reversal is on top of the throat area of the press brake housing, the opening of the throat area will cause the upper die to not penetrate into the lower die sufficiently to make a good bend unless the operator has compensated for the gapping by setting the ram reversal point lower to allow for the forming tonnage. The forming tonnage will cause the top portion of the housing to move upward. Under load, the lower portion of the housing will appear to move down, but it can't because it is mounted to the floor. As a result, the gapping at the throat will cause the ram reversal at the bottom of the stroke to not have a consistent relationship with the bed top, causing potential problems with the upper die penetrating consistently into the lower vee die.

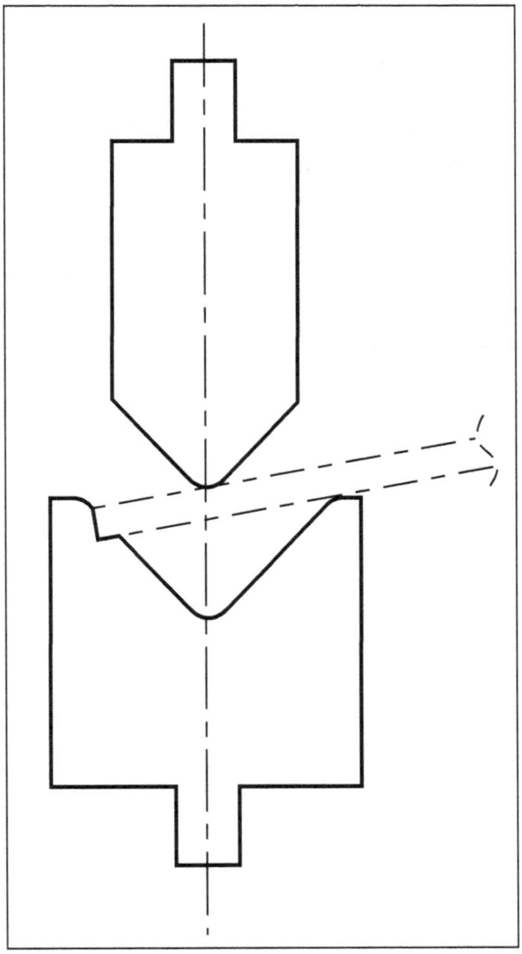

Fig. 5-14. Lower die cut with a special gage ledge in order to obtain a shorter flange. This method is only recommended for material thinner than 14 gauge, unless special backups are added to the dies.

If the part has only one specific bend length, the operator can adjust the ram reversal point to make good bends without changing the reversal point from bend to bend. The problem arises when a part, such as a panel, has short bends at each end and a longer bend line in the front and back of the part (four bends total). If the part is made using air bend dies, the long bends will require a different tonnage than the short end bends, which in turn will cause a different gapping of the throat as the bend length changes. In order to maintain the same ram penetration of the upper die into the lower die and obtain a consistent angle, the operator would be required to change the ram setting for each length of bend line. Since this is impractical in a production setting, most users selected bottoming dies, usually requiring a higher tonnage machine, to form the parts. This meant that the ram setting was set for the longest bend line, and the shorter bends were hit abnormally hard.

In 1958, a standard feature called "bed referencing" was developed for hydraulic press brakes by The Cincinnati Shaper Company. The design placed the adjustable stop that was used to activate the limit switch attached to the ram (which signaled the ram reversal point) below the throat area. As a result, any gapping of the throat was not cause for a change in the upper die penetration into the lower die. While the gapping occurs, the ram simply continues downward until it touches the bottom stop attached to the bed. This "new" design was the reason that air bending has become popular again, and has allowed many parts to be formed using much smaller tonnage hydraulic press brakes.

Not everyone has accepted the change back to "air bending," since it means a slightly different inside radius in the formed angle and the forming tolerances are greater. As hydraulic press brakes became more sophisticated—with the addition of encoders, computers, and software; the ability to use a lower tonnage; and faster press brakes—management has been forced to reevaluate their forming processes to determine what is the best method for making their product.

Chapter 6
Forming Other Types of Material

Mild steel is the basic reference used for forming parts using a press brake. Sheet metal can be purchased as cold rolled steel or as hot rolled steel. The forming tonnage is the same for either type material. Some operators say that hot rolled sheet metal has a grain that can affect the bend angle if the part is struck using the same ram setting. Since most thin, hot rolled sheet metal is actually processed by a "pickling and oiling" (to remove any scale) operation, the visual appearance of a piece of cold rolled or hot rolled pickled and oiled steel is often similar. Occasionally, the hot rolled piece will appear to have a blue cast. After the material is cut up, and flanges formed on all four sides, most operators cannot tell what type of material is being used. Experience has proven that any forming differences between the two materials is insignificant and can be compensated for by the press brake operator.

Plate, only available in a hot rolled condition, does have grain from the rolling operation. A simple angle often has to be formed in several strokes since the material variations from part to part can affect the final angle if the ram setting is not constantly adjusted. Management must allow sufficient time, when plate is being formed, to give the operator time to measure, rehit, and remeasure the angle on each bend if proper dimensions are expected to be held (refer to Chapter 19).

The variations found when forming high tensile steel, aluminum, or some exotic metals are discussed in this chapter.

A) High Tensile Steel

It was pointed out in Chapter 4 that to successfully bend a part, the material must be pushed down into the lower die far enough to exceed its yield strength, not its tensile strength. The same parameters still exist when higher tensile steel is formed. The main difference is that the tensile and yield relationship is not the same with high tensile steel as compared to mild steel. This can be seen in the chart on page 54:

	Mild Steel	A-36	High Tensile Low Yield	High Tensile Medium Yield	High Tensile High Yield
Tensile Strength	60,000 PSI	80,000 PSI	90,000 PSI	80,000 PSI (min.)	100,000 PSI (min.)
Yield Strength	40,000 PSI	36,000 PSI (min.)	35,000 PSI	60,000 PSI (min.)	80,000 PSI (min.)

The figures shown are considered nominal. The published figures will often list a specific tensile strength such as "80,000 PSI (min)." But the question is: what is the maximum? The answer was only found after many discussions with the United States Steel Company, Bethlehem Steel Company, and many steel service centers around the United States. The consensus was that, if a "minimum" figure was shown, the steel strength could increase as much as 15,000 PSI. The only one to disagree is a well known steel company on the West Coast. Anyone who purchases A-36 or high tensile steel on the West Coast would be wise to check the actual tensile and yield strength of their material to avoid machine capacity problems. Some companies have encountered tonnage problems using a hydraulic press brake when the actual tensile or yield strengths exceeded the 15,000 PSI minimum range described.

By reviewing the mild steel figures, it can be seen that there is a 50% difference between the tensile and yield strengths. The thinner gauges of mild steel seldom have a cracking problem when using the recommended vee dies, or even using bottoming type tooling. Mild steel plate may encounter cracking at each end of the bend line on the outside radius. Causes may depend on the vee die opening selected, the quality of the plate, or the grain direction. Most press brake tonnage charts now recommend that the vee die opening be increased to 10 times the plate thickness in order to minimize the possibility of the part having a tendency to crack. The big objection to the increased die opening is that the inside radius of the part will become larger, and the flange length of the part will have to be made longer in order to span the larger vee die opening.

The difference between the tensile strength and the yield strength of the material labeled "high tensile/low yield" is significant. It should seldom "crack." Most of the material in this category is stainless steel, and it is rare for a piece of stainless to crack. The exception is when a stainless steel part has been formed incorrectly and, in an attempt to salvage the part, it is straightened and reformed. When a piece of stainless steel is formed, the area affected by the forming has the yield strength increased from the nominal 35,000 PSI to about 70,000 PSI. As a result, the window between the tensile and yield strength is very small—approximately a 30% difference—which can cause cracking if the part is reformed.

The materials that fall into the "high tensile/medium yield" strength area have a 33% differential between their tensile strength and yield strength. A popular material in this category is called "Corten." The last category shown on the chart, "high tensile/high yield," allows only a 20% difference between the nominal tensile strength and yield strengths. A common material, T-1 steel, falls into this category. Materials with the trade name "Corten" and "T-1" are very popular and readily available from most steel service centers. Refer to the "Bending Factors Chart" (Fig. 6-1) to determine the recommended inside radius required by many high tensile steels. They are categorized by ASTM numbers to allow different steel companies to use their trade names while still having the actual steel easily cross referenced by users.

When bending mild steel, the inside radius is a function of the vee die opening (refer to

Chapter 3-A). It has been found that when high tensile steel is formed—stainless steel being an exception—the material will try to "stick" to the radius of the upper die. The die and material act in a way similar to taking your thumb and pushing it down on a marshmallow. The marshmallow will form around your thumb. In the case of the high tensile steel, if the material forms around a radius that is too small, the outside fibers of the steel will exceed the tensile strength, and failure will occur. In order to prevent this, the steel company supplying a special type of steel will publish a minimum inside radius that can be formed with the grade and thickness of that material. If there is a question about what inside radius can be formed without failure, it is best to contact the supplier and obtain the information in writing prior to conducting any part forming. Normally, if material failure does occur, and the minimum forming radius published by the steel company was not used, the steel supplier will not assume any responsibility for the results.

In recent years, many new grades of high tensile steels have become available. Many of them offer the ability to form the metal to a smaller than normal inside radius without failure. Prior to designing parts with this material, it is recommended that some sample material of the required grade and thickness be obtained and formed to determine if the inside radius, tonnage, and quality are acceptable.

The increase in forming tonnage required to form higher tensile steels is a relationship between: 1) the tensile strength and yield strength of mild steel, 2) the actual tensile and yield strength of the high tensile steel, and 3) any effect a larger required inside radius may have on the forming load.

The first step is to add the nominal tensile and yield strengths of mild steel as shown on the Bending Factor chart. This would be 60,000 PSI plus 40,000 PSI, to equal 100,000 PSI. The nominal tensile and yield strengths of the higher strength are then added. If the high tensile-medium yield steel is selected, the resultant figure is 80,000 PSI (min.) tensile strength plus 60,000 PSI (min.) yield strength. Since, in many specifications, these figures are minimums, one must add in the possible increases that could occur. To calculate, add the 15,000 PSI additional for both the tensile strength and for the yield strength , or 30,000 PSI total. Adding all of the figures above will total 170,000 PSI. In this example, any effect of a larger radius required to eliminate cracking will be ignored. Taking the 170,000 figure and dividing it by the mild steel figure of 100,000, a bending factor of 1.7 is determined. This means that the increase in forming tonnage could be as much as 1.7 times the tonnage shown on the air bend tonnage chart.

The Bending Factors Chart developed by Cincinnati Incorporated (Fig. 6-1) provides not only the ASTM material type information, but also the "bending factors" and required inside radius for many different types of higher strength steels. For specific jobs, it is important that you have brochures and forming information available from your suppliers so that, in the event of a forming problem, you can discuss the problem using the vendor's literature references.

B) Aluminum

There are so many different grades of aluminum available that it is impractical to provide advice for all of them. General ideas will be given to assist the user in deciding what type of aluminum may be best for a specific application. The type of aluminum chosen will be identified by a four-digit number followed by a hyphen and a letter, or a four-digit number followed by a hyphen plus a letter and number. An example might be 3003-0. The "3003" identifies the type of aluminum; the "0" identifies the heat treatment or hardness of the material. The type of aluminum that is chosen is often limited by what is available from the local supplier. Since

BENDING FACTORS CHART

The tensile strength and/or yield strength of many steels are specified as minimum values with no limit on the maximum values. The factors in this chart allow for an increase of 15,000 p.s.i. above the specified minimum values. The factors must be increased for steels exceeding this 15,000 p.s.i. allowance. The actual physical properties and chemical analysis of a steel may meet more than one specification and/or grade within a specification. In this case, the factors for the specification and/or grade with the highest mechanical properties must be used.

ASTM SPECIFICATION			TENSILE STRENGTH K.S.I.	YIELD STRENGTH K.S.I.	MATERIAL THICKNESS INCHES	NOMINAL VEE DIE OPENING	NOMINAL UPPER DIE RADIUS	FORMABILITY FACTOR
NO.	TYPE	GRADE						
A36	Structural Steel		58-80	36-51	.180-.500 .500-1.000	8T 10T	1T 1.5T	1.3
				51-65	.180-.500 .500-1.000	10T 12T	1.5T 2T	1.6
A131	Structural Steel For Ships	ALL	58-71	34 Min.	.180-.500 .500-1.000	8T 10T	1T 1.5T	1.2
A242	High-Strength Low-Alloy Structural Steel		70 Min.	50 Min.	.180-.250 .250-.500	12T 12T	2T 3T	1.7
A283	Low and Intermediate Tensile Strength Carbon Steel Plates of Structural Quality	A	45-55	24 Min.	.180-.500 .500-1.000	8T 10T	1T 1.5T	1.0
		B	50-60	27 Min.	.180-.500 .500-1.000	8T 10T	1T 1.5T	1.0
		C	55-65	30 Min.	.180-.500 .500-1.000	8T 10T	1T 1.5T	1.1
		D	60-72	33 Min.	.180-.500 .500-1.000	8T 10T	1T 1.5T	1.2
A284	Low and Intermediate Tensile Strength Carbon-Silicon Steel Plates for Machine Parts and General Construction	C	60 Min.	30 Min.	.180-.500 .500-1.000	8T 10T	1T 1.5T	1.2
		D	60 Min.	33 Min.	.180-.500 .500-1.000	8T 10T	1T 1.5T	1.2
285	Low and Intermediate Tensile Strength Carbon Steel Plates for Pressure Vessels	A	45-65	24 Min.	.180-.500 .500-1.000	8T 10T	1T 1.5T	1.0
		B	50-70	27 Min.	.180-.500 .500-1.000	8T 10T	1T 1.5T	1.1
		C	55-75	30 Min.	.180-.500 .500-1.000	8T 10T	1T 1.5T	1.2
A299	Carbon Manganese Silicon Steel Plates for Pressure Vessels		75-95	42 Min.	.180-1.000	12T	1.5T	1.5
A366	Steel, Carbon, Cold Rolled Sheet Commercial Quality		Not Specified	Not Specified	ALL	8T	1T	1.0
A441	High-Strength Low Alloy Structural Manganese Vanadium Steel		70 Min.	50 Min.	.180-.500 .250-.500	12T 12T	2T 3T	1.7
A442	Carbon Steel Plates with Improved Transition Properties for Pressure Vessels	55	55-75	30 Min.	.180-.500 .500-1.000	8T 10T	1T 1.5T	1.2
		60	60-80	32 Min.	.180-.500 .500-1.000	8T 10T	1T 1.5T	1.3
A514	High-Yield Strength, Quenched and Tempered Alloy Steel Plate Suitable for Welding		110-130	100 Min.	.180-1.000 Over 1.000	16T 16T	2T 3T	3.1
A515	Carbon Steel Plates for Intermediate and Higher Temperature Service for Pressure Vessels	55	55-75	30 Min.	.180-.500 .500-1.000	8T 10T	1T 1.5T	1.2
		60	60-80	32 Min.	.180-.500 .500-1.000	8T 10T	1T 1.5T	1.3
		65	65-85	35 Min.	.180-.500 .500-1.000	10T 12T	1.5T 2T	1.45
		70	70-90	38 Min.	.180-.500 .500-1.000	12T 14T	2T 2T	1.6
A516	Carbon Steel Plates for Moderate and Lower Temperature Service for Pressure Vessels	55	55-75	30 Min.	.180-.500 .500-1.000	8T 10T	1T 1.5T	1.2
		60	60-80	32 Min.	.180-.500 .500-1.000	8T 10T	1T 1.5T	1.3
		65	65-85	35 Min.	.180-.500 .500-1.000	10T 12T	1.5T 2T	1.45
		70	70-90	38 Min.	.180-.500 .500-1.000	12T 14T	2T 2T	1.6
A517	High Strength, Alloy Steel Quenched and Tempered Plates for Pressure Vessels		115-135	100 Min.	.180-1.000 Over 1.000	16T 16T	2T 3T	3.1
A537	Heat Treated, Carbon Manganese-Silicon Steel Plate for Pressure Vessels	Class I	70-90	50 Min.	.180-1.250 Over 1.250	14T 16T	3T 4T	1.9
		Class II	80-100	60 Min.	.180-1.250 Over 1.250	16T 16T	4T 5T	2.0
A569	Hot-Rolled Carbon Steel Sheets and Strip Commercial Quality		Not Specified	Not Specified	0-.500	8T	1T	1.0

Fig. 6-1. Bending factors chart. (Courtesy of Cincinnati Incorporated)

BENDING FACTORS CHART

ASTM SPECIFICATION NO.	TYPE	GRADE	TENSILE STRENGTH K.S.I.	YIELD STRENGTH K.S.I.	MATERIAL THICKNESS INCHES	NOMINAL VEE DIE OPENING	NOMINAL UPPER DIE RADIUS	FORMABILITY FACTOR
A570	Hot-Rolled Carbon Steel Sheets and Strip Structural Quality	A	45 Min.	25 Min.	.0-.500	8T	1T	1.0
		B	49 Min.	30 Min.	.0-.500	8T	1T	1.1
		C	52 Min.	33 Min.	.0-.500	10T	1.5T	1.15
		D	55 Min.	40 Min.	.0-.500	12T	2T	1.25
		E	58 Min.	42 Min.	.0-.500	12T	2.5T	1.3
A572	High-Strength Low-Alloy Columbium-Vanadium Steels of Structural Quality	42	60 Min.	42 Min.	.180-.250 .250-.500	10T 12T	1.5T 2T	1.4
		50	65 Min.	50 Min.	.180-.250 .250-.500	12T 12T	2T 2.5T	1.6
		60	75 Min.	60 Min.	.180-.250 .250-.500	14T 14T	3T 3.5T	1.8
		65	80 Min.	65 Min.	.180-.250 .250-.500	14T 16T	3.5T 4T	1.9
A588	High-Strength Low-Alloy Structural Steel with 50,000 PSI Minimum Yield Point to 4" Thick	All	70 Min.	50 Min.	.180-.250 .250-.500	10T 12T	2T 3T	1.7
A606	Steel Sheet and Strip, Hot Rolled and Cold Rolled, High-Strength, Low-Alloy with Improved Corrosion Resistance	H.R.	70 Min.	50 Min.	0-.062 .062-.250 .250-.500	8T 10T 12T	1T 2T 3T	1.7
		C.R. Anneal. Norm.	65 Min.	45 Min.	0-.062 Over .062	8T 10T	1T 2T	1.6
A607	Steel Sheet and Strip, Hot Rolled and Cold Rolled, High-Strength, Low-Alloy Columbium and/or Vanadium	45	60 Min.	45 Min.	0-.180 .180-.230	8T 10T	1T 1.5T	1.5
		50	65 Min.	50 Min.	0-.180 .180-.230	10T 12T	1.5T 2T	1.6
		55	70 Min.	55 Min.	0-.180 .180-.230	12T 12T	2T 2.5T	1.7
		60	75 Min.	60 Min.	0-.180 .180-.230	12T 14T	2.5T 3T	1.8
		65	80 Min.	65 Min.	0-.180 .180-.230	14T 14T	3T 3.5T	1.9
		70	85 Min.	70 Min.	0-.180 .180-.230	16T 16T	4T 5T	2.0
A633	Normalized High-Strength Low-Alloy Structural Steel	A-B	63-83	42 Min.	0-1.00	12T	2T	1.6
		C-D	70-90	50 Min.	0-1.00	12T	2T	1.7
		E	80-10	60 Min.	0-1.00	12T	2T	2.0
A666	Hot Rolled Structural Steel, High-Strength Low-Alloy Plate with Improved Formability	50	60 Min.	50 Min.	0-.250 .250-.750	8T 10T	1T 1.5T	1.5
		60	70 Min.	60 Min.	0-.250 .250-.750	8T 10T	1T 1.5T	1.6
		70	80 Min.	70 Min.	0-.250 .250-.750	8T 10T	1T 1.5T	1.8
		80	90 Min.	80 Min.	0-.250 .250-.750	8T 10T	1T 1.5T	2.1
A715	Hot Rolled, High-Strength, Low-Alloy Steel and Strip with Improved Formability	50	60 Min.	50 Min.	0-.230	10T	1.5T	1.5
		60	70 Min.	60 Min.	0-.180 Over .180	8T 10T	1.5T 2T	1.6
		70	80 Min.	70 Min.	0-.180 Over .180	10T 10T	2T 2.5T	1.8
		80	90 Min.	80 Min.	0-.180 Over .180	10T 12T	2T 2.5T	2.0

MATERIAL	TYPE	TENSILE STRENGTH K.S.I.	YIELD STRENGTH K.S.I.	MATERIAL THICKNESS INCHES	NOMINAL VEE DIE OPENING	NOMINAL UPPER DIE RADIUS	FORMABILITY FACTOR
Stainless Steel	302 304	80-90	30-40	0-.500 .500-1.000	8T 10T	1T 1.5T	1.25
Aluminum	3004-H34	35 Typical	21 Typical	0-.125 .125-.188	8T 10T	1T 1.5-3T	.6
Aluminum	6061-T6	42 Typical	37 Typical	0-.031 .031-.062	8T 10T	1T 2T	.8

NOTE:
1. **The formability factors listed in the chart are based on a factor of 1.0 for mild steel with a tensile strength of 60,000 p.s.i. and a yield strength of 40,000 p.s.i.**
2. **The ASTM specifications listed are those in effect January 1, 1987.**
3. **High strength low alloy steel and stainless steel should not be formed at temperatures below 50 F (10 c).**

Fig. 6-1. Bending factors chart (continued). (Courtesy of Cincinnati Incorporated) there are so many types of aluminums that can be made, only a few of the popular types can economically be inventoried by most supply houses. Some of the most commonly used types include: 3003-0, 3004-0, 2024-T3, 5052-H34, and 6061-T6. An aluminum called 1100-0 is dead soft. It will form like aluminum foil and it is very easily marked or damaged.

The following criteria will allow the user to determine what press brake or tooling may be needed to obtain good bending results. If the type of aluminum grade selected ends with "-0," the material will be very soft. Very little springback will occur after the part has been formed down into the vee die and the pressure is released. If the designation ends with "H34," the parts will spring back similar to a piece of mild steel of the same thickness. If the material designation ends with "T3" or "T6," abnormal springback should be expected.

Handbooks published by Alcoa Aluminum Company or Reynolds Aluminum Company specify the minimum inside radius that can be formed in each type of aluminum without causing the metal to fail. To obtain this information, call your supplier or the aluminum company that provided your material. Aluminum seems to form the same way as high tensile steel, since it will also try to stick reasonably close to the radius of the upper die as opposed to forming its own inside radius. In some tests with thicknesses of the "H" and "T" grades, it was seen that the inside radius is not true but slightly "loopy."

The tooling selected to form aluminum will depend on the type or grade of material specified. Material having a "0" after the four-digit type should have minimum springback. Most users select air bend tooling having an included angle of about 85° on both the upper and lower dies. The vee opening of 8 times the metal thickness is recommended. Smaller vee die openings cause the part to move upwards more rapidly, which can cause backbending in the softer metal. A larger radius machined at the top corners of the vee die opening will reduce die marking (see Chapter 10 for additional information). Other grades have unique considerations. If the heat treatment designation is "H," the part will bend the same as the forming of a similar thickness of mild steel. The same type of vee opening should be chosen. Check the aluminum chart to determine the correct inside radius that can be formed for the thickness to be bent, and make sure the upper die chosen has a nose radius close to the one recommended. The reason, as pointed out previously, is that the material sticks to the upper die radius. If the radius is too small, the material could crack. The normal springback will be about 5°.

If the material has a "T3" or "T6" heat treatment designation, the recommended inside radius will normally be larger than metal thickness. In most cases, the springback will be much greater than experienced with mild steel. It is not uncommon for the springback to vary from 5° to as much as 20°, depending on the inside radius and thickness of material being formed. It many cases, the press brake operator will select 30° dies in order to have ample room for the part to be overbent sufficiently, and to allow for the anticipated springback.

Note: The thicknesses of thinner sheets of steel are normally designated by "gauge" number. This designation was originally based on the weight of steel in lb/ft^2. The resultant thickness of any specific gauge thickness is a unique decimal which is not a common fraction or simple decimal. Many years ago, the aluminum industry used a separate "gauge" system for aluminum thickness using the same numbers as the steel industry. For example, 16 gauge steel is 0.0598" thick. The aluminum industry thickness for 16 gauge aluminum was 0.050" thick. The differences between the steel gauge and the aluminum gauge thicknesses resulted in a great deal of confusion and many mistakes. Since the 1970s, the industry has specified the thickness of aluminum only by the use of a decimal thickness. The "gauge" system was discontinued. Since this method has been installed, the use of incorrect material thickness in processing aluminum parts has been minimized.

For over 50 years, it has been possible to obtain aluminum sheets that have had the surfaces anodized. When the sheet is anodized, using a chemical or electrolytic process, the surface of the material that is affected will develop a very hard coating, approximately 0.001" thick. This surface will be as hard as a kitchen paring knife. Anodized aluminum can also provide unique color to the treated surface. The aluminum that has been anodized can be dead soft 1100-0 (almost pure aluminum) except on its surface. If this anodized aluminum were formed into a simple right angle bend, the bend angle of this type of material will react as if the material were "T"-type aluminum. It would spring back a great deal, and would result in abnormal angular variations, making it difficult to obtain good, consistent parts.

C) Stainless Steel

Stainless steel has a higher tensile strength than normal mild steel. Stainless is normally available commercially in the thinner gauges (10 gauge or thinner). It will form using the same basic forming rules as mild steel; however, some planning must be done.

Stainless will spring back several degrees more than a comparable thickness of mild steel. The same air bend dies required to do general purpose bending may not work well for stainless steel. Since an additional 2° to 4° of springback should be expected, either new dies may be required or the operator may use a set of 30° dies to form a 90° bend. Stainless requires 25% to 50% more forming tonnage than is listed on a standard "air bend" tonnage chart developed for mild steel. If, however, an available 85° air bend die is used to form a stainless part that requires 6° or 7° of springback, the part will be somewhat "bottomed" by the operator in an attempt to form an acceptable part, which will require additional machine tonnage. This "bottoming with springback" method may not cause any problems. Occasionally, when low tonnage press brakes are used for this type work, the press brake will not produce a good part without the operator "rehitting" the part several times. This practice is not acceptable if the part is considered "high production" or if the part will be made many times in the future.

For parts with more springback than is allowed in the die, the press brake operator may find it necessary to restrike an already formed piece of stainless to improve the resultant angle. Stainless steel work hardens; this means that the radiused area of the formed angle becomes very hard compared to the rest of the part. An increase in tonnage will be needed if the part is rehit. Additional time is then needed to inspect the part. Added tonnage will cause the ram setting to be changed, requiring skill on the part of the operator to obtain consistency from part to part.

Stainless steel can also present several tooling problems. The first to be noticed is that the vee die corners radii may wear at an alarming rate. The only answer will be to have the lower die vee area remachined. If more stainless is to be formed on that vee opening, the lower die corner radii must be "flame hardened." This will solve the wear problem. It may also help solve the following concerns. If stainless is formed using a conventional die steel vee die that was "hardened" to the normal 260 to 305 Brinell hardness prior to machining, the vee die corners will be subject to the pressure of the added forming load of the stainless steel. If the vee die had been used for a sufficient period of time forming mild steel parts, the corners of the vee opening may have become **work hardened**. This means that the corners will be harder than originally supplied at the time of purchase. As a result, any wear on the die from forming the stainless steel could be minimal. If the die is reasonably new, or if the die has just been reworked (which will remove any work hardened surfaces), the stainless, as it moves over the corners of the vee opening, will quickly create wear on the die.

The microscopic particles of the worn die steel often imbed themselves into the surface of the stainless; they are not visible. If the stainless piece is used outdoors, those small particles of die steel will soon rust. The lines of contact that the part had with the vee die will stand out. Obviously, this creates a major problem with the customer, and no easy corrective action will be available. In order to prevent this problem in the future, parts that have this particular problem should be formed only with dies that have been flame hardened or hardened in some other manner.

The other major concern that forming stainless steel parts entails is that many stainless parts must be formed without die marks or markings that can easily be polished out in a subsequent operation. Refer to Chapter 10 on "Die Marking" for assistance.

D) Exotic Materials

The aircraft industry has spent years developing methods to form materials like titanium, vanadium, and other exotic types. Some of the metals require large inside radii to keep the material from cracking. Some types (such as titanium) can require heated dies, or must be formed "hot" to eliminate fracture. It is best to consult with the supplier to determine the best method for forming their material using simple press brake tooling.

E) Painted Stock and Galvanized Steel

Many industries (and most "job shops") require the occasional forming of painted stock or galvanized steel. Most of the material is mild steel, although some aluminum painted stock is used. When painted stock was initially made available to the marketplace, there were severe limitations on the inside radius of any bends, and the painted surface was easily damaged during the forming stroke. In most cases, this is no longer a problem. The painted side will form to a reasonably small outside radius without cracking or the paint peeling. Die marking can be a problem, especially if the tooling has been used for a variety of other work prior to being used for painted material. See Chapter 10 for ideas on reducing die marks.

Galvanized steel, if 14 gauge or thinner, is normally produced using a "continuous dip" process. This type of material forms well with minimal problems. If thicker galvanized materials are to be formed, the plating is often done using a "hot-dip" process. This method can have limitations for some bends. If the inside radius is too sharp, the galvanization will crack and peel off of the base material. The galvanized coating is only several thousandths of an inch thick, but can flake or peel if stretched too far. Consult your supplier prior to purchase to ensure that a problem will not occur during the manufacturing cycle, and thus cause delivery problems plus the lost time and cost of attempting to salvage parts.

Galvanized coating will build up on the corners of the vee die. It is important that the operator be made aware of this, since the extra material on the lead-in radii can cause angular variation as well as gaging problems. The operator must have instructions and time to periodically clean the vee die opening.

Chapter 7
Gaging

A major manufacturing concern is the constant pressure to use small lot sizes and to obtain high-quality, close-tolerance parts. A good press brake and an experienced operator are essential—but not sufficient—to completely solve the problem. Other factors are: determining tolerances based on realistic angular variations, establishing the press brake tonnage required, selecting the best sequence of bends to form the part, and accurate gaging. To determine the proper gaging sequence, consider the following steps.

A) Determine the Inside Dimension of Each Bend Line Prior to the Part Being Bent

Most part drawings show "outside dimensions" that are subject to material thickness variations. If a flat blank gage position is to be determined, the first bend line is found by subtracting the actual material thickness and the inside radius specified from the outside flange dimension. To this figure add one-half of any calculated bend allowance to obtain the location where the upper die will contact the flat blank (Fig. 7-1). (This method does not apply if the inside radius is less than metal thickness.) The flange to be gaged is normally a short leg, which allows the majority of the blank to remain toward the front of the machine. If the inside radius is specified as less than metal thickness, it may be best to use a bend allowance formula developed by the company's fabricating area to arrive at a close gage position. Review "bend allowance" calculations in Chapter 8 to understand the problems of calculating a gage location with less than metal thickness inside radius.

Parts often have multiple bends on one side. The gage would be located against the outside leg of the first bend. Since this is the type of gage position that many companies leave to the press brake operator to determine, most operators simply index the part forward, after the first bend, the amount of the inside dimension shown on the part drawing plus one metal thickness. In most cases, the part will be very close to what is wanted, and only a minor gage correction

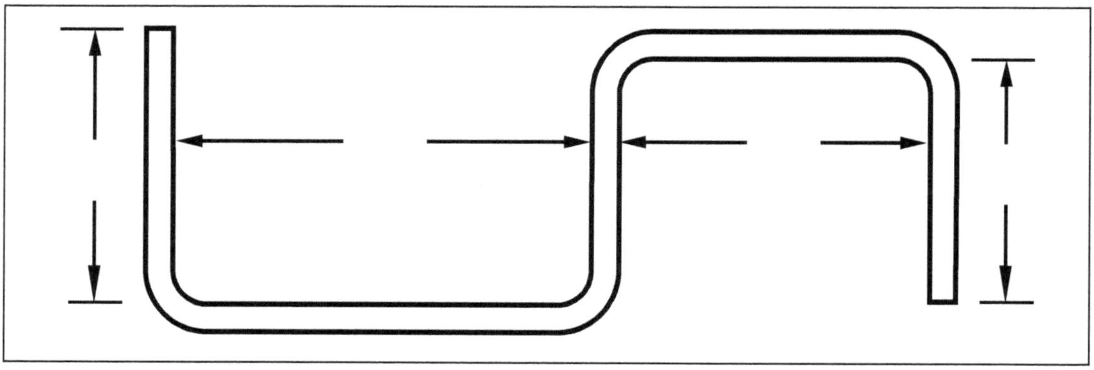

Fig. 7-1. Formed part illustrating inside dimensions used to calculate a flat blank size.

is made for the remaining parts in the run. If there is only one bend in the part, the gage location will be set to contact the edge of the material at a distance equal to the inside flange length shown, to the point of contact with the upper die. If the inside radius is close to metal thickness or even slightly smaller, it is common practice for the press brake operator to simply look at the outside dimension shown on the drawing, subtract the nominal metal thickness, and set the back gage at the operator's calculated leg distance from the centerline of the upper die. Normally, this method will be very close and only a small correction will be required to obtain the correct leg dimension.

B) Determine the Quality of the Actual Flat Blanks Prior to Forming

Blank errors or distortions can, and often do, occur due to a variety of factors. Any cut or sheared edge is not square. It will reflect the amount of clearance that was used between the top and bottom knives. All manufacturing processes have some acceptable "blank" tolerance. As a result, the part may not be cut square (have a shape like a parallelogram) or have the effects of bow, twist, or camber.

Bow (Fig. 7-2) can be easily seen. The blank does not lay flat, but appears to have a large radius formed in its length. Either the outer two edges of the part could be touching the floor with the center of the part slightly raised, or the center of the part is touching while both ends appear to be up in the air. This problem is often caused by "coil set" not being completely removed when the part was processed from coil to the required blank. Often, bowing is caused by the stresses introduced in the material by the rake angle of a squaring shear.

Twist (Fig. 7-3) is another distortion in the blank that can affect the formed part. To understand this concept better, take yourself back to the first grade. When you cut a piece of heavy construction paper near the edge, after the first cut was made, using your rounded nose scissors, you moved your hand forward to make the second cut. As your hand moved forward, the knuckles of your hand forced the wider part of the paper upwards, while the narrower strip being cut was forced downward. The narrower the strip being cut, the more likely it was that the strip would not lay flat but showed the effect of twist from our hand being forced near the scissor cutting line. The same thing happens when a piece of metal is slit or cut using a squaring shear. The knives of a slitter cause the metal to go up or down while the material is being cut. The table piece lays flat on a squaring shear, but the backpiece is forced downwards and slightly away from the lower knife. The narrower the backpiece, the more "twist" can be expected. Wider backpieces are stiff enough to overcome this processing distortion and will

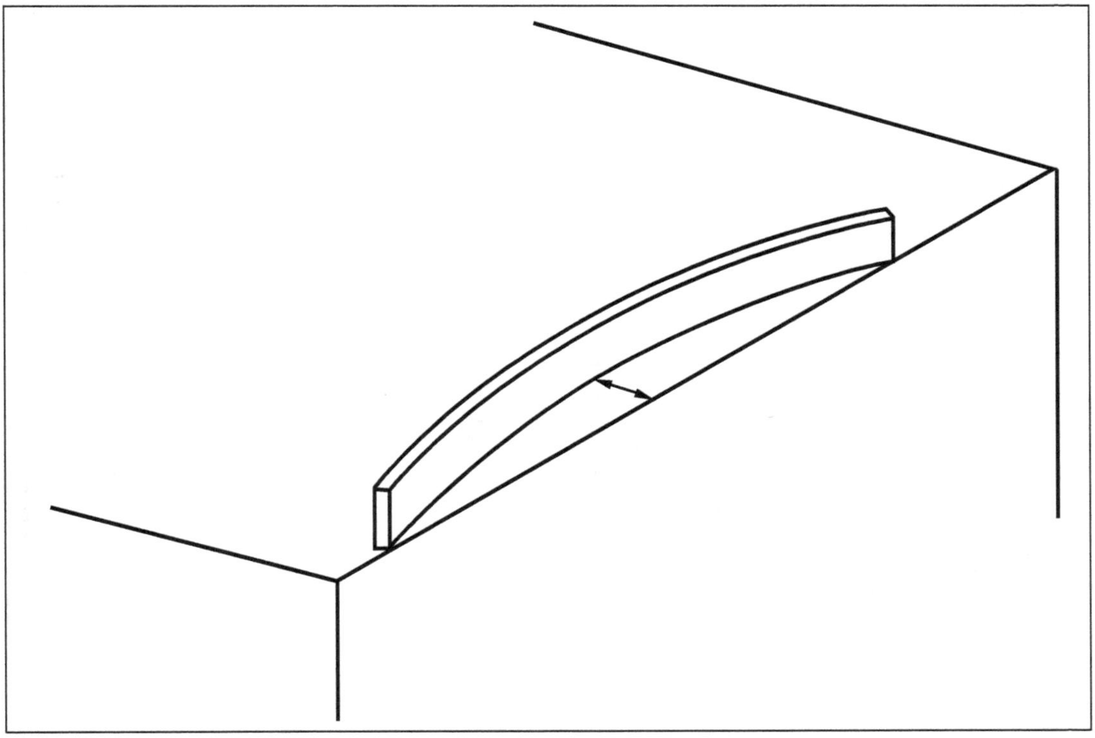

Fig. 7-2. Bow in a flat strip.

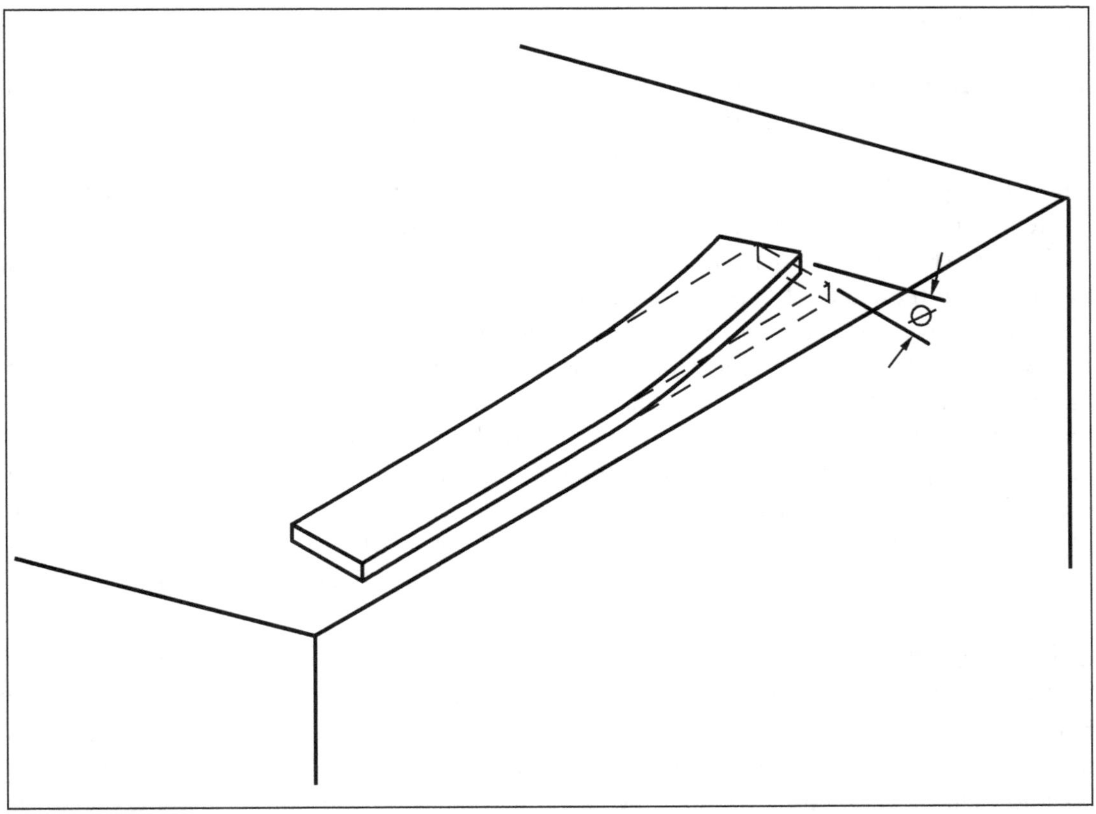

Fig. 7-3. Twist in a narrow strip.

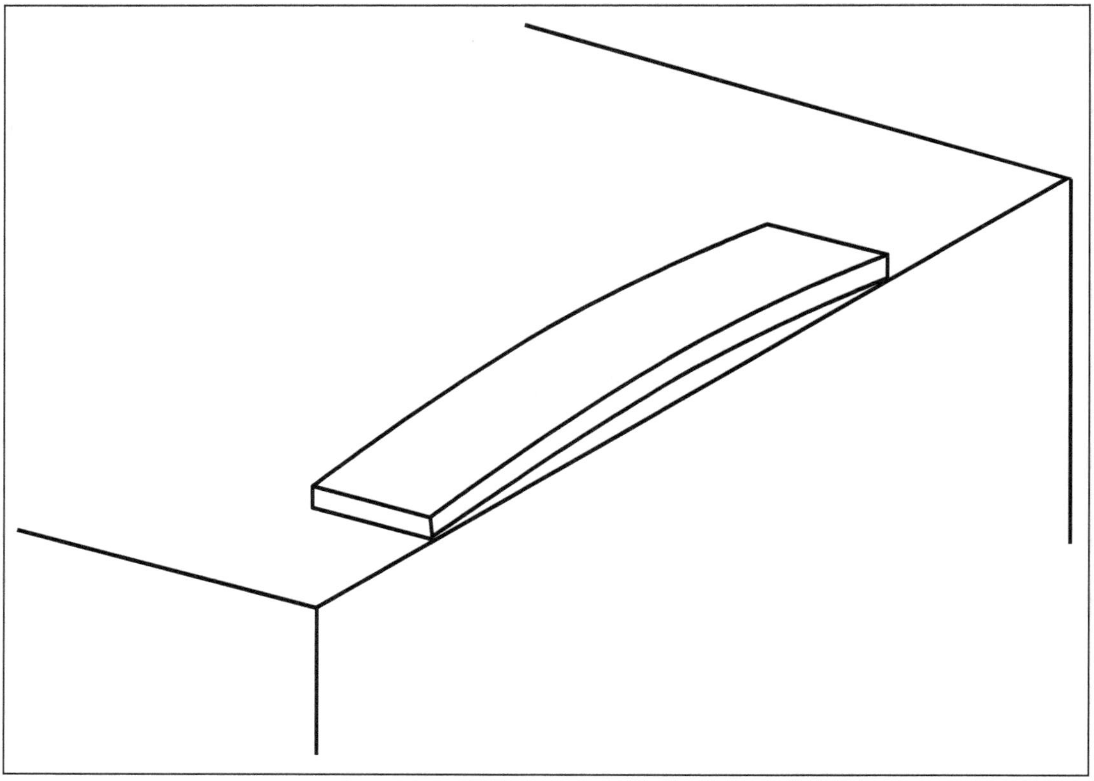

Fig. 7-4. Camber in a flat part.

lay flat after the cutting operation. The edges of rolled material are known to have abnormal stresses that often require a trim cut prior to the material being used for a formed part. It seems that, when the part is rolled, many of the stresses inherent to rolling are forced to the material edges. When the trim cut is made, the remaining body of material is much more stable. Abnormal twist is usually observed in the narrow trim cut.

Camber (Fig. 7-4) is a distortion along the edge of a flat blank that is caused by the release of stresses in the material where it has been rolled, slit, sheared, or cut. Normally, it is not visual to the eye, and can only be measured when the material is lying flat and does not show any effect of bow or twist. If a width of material is sheared, when the table piece and the backpiece are laid together at the cut line, some small gap can be expected. The longer the cut line of the blank, the greater the measurable gap. With a good quality of material, the resultant amount of gap that can be measured at the cut line may be only several thousandths of an inch. The steeper the cutting angle of the upper shear knife (called rake angle) is made, the greater the possibility that an unacceptable amount of camber will occur. Camber problems are normally a problem in parts made from sheet metal (i.e., 10 gauge or thinner).

C) Determine the Location of the Upper Die Centerline

When dies are changed, the tolerance range allowed for normal commercial dies or homemade dies will probably affect the location of the upper die centerline (Fig. 7-5). This means the location of the centerline may have moved toward the operator or away from the operator. This is very important since the location of the bend line is located where the upper die nose contacts the part. If the press brake is equipped with a numerically controlled (NC)

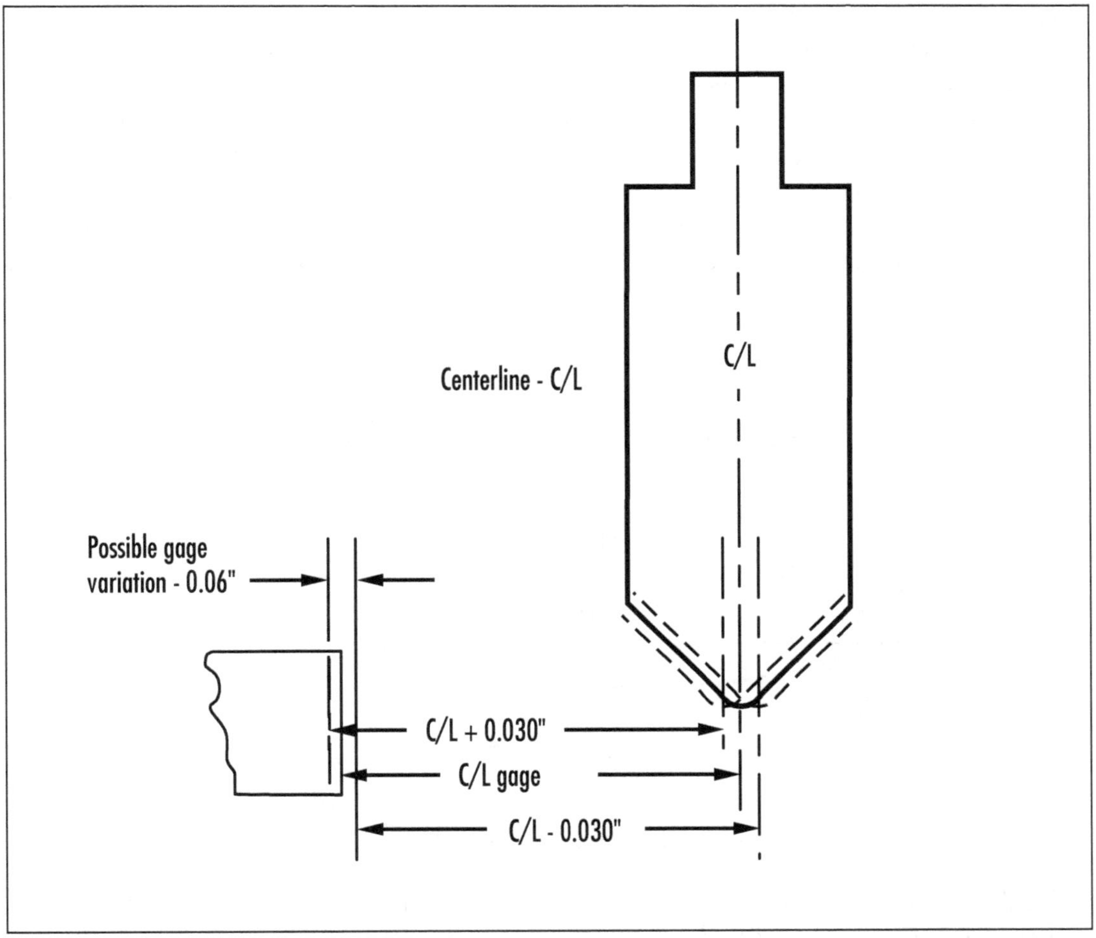

Fig. 7-5. Tolerances possible in the nose location of the upper die affecting the gage location or introducing dimensional error into the formed part if not corrected.

back gage, it will have to be recalibrated to obtain good gaging locations. Remember that the lower die must be realigned with the upper die.

D) Use Gage Points to Locate a Flange

A two point gage contact eliminates edge camber errors (Fig. 7-6). If a straight, full length bar is used for gaging, camber will allow two points of contact on one side of the blank (Fig. 7-7). When the part is turned over, the radiused edge side created by camber might allow the part to rock along the full length bar and cause variations.

Using a two-point gaging method, each finger is located approximately 25% in from each end along the part length. This allows about 50% of the part length between the gage points. By using this method, any amount of camber present along the gage edge, regardless of its direction, will be averaged. This often allows the formed flange and subsequent bends to maintain tolerance.

If a long narrow part requires a flange at the end, two-point gaging is impractical. In this case, it is best to set up one back gage location plus a side squaring arm (a side squaring arm bar for a shorter piece or two pin locations or longer parts) to locate the bend properly.

Parts having unusual shapes, where the bend line does not allow uniform support on both sides of the lower die, will often "kick" or move during the forming stroke. Unbalanced parts, where the bend line is off to one side of the part being formed, often will move toward the heavier section that is not being formed. These problems often require special gaging to allow for the known part movement in order to obtain acceptable consistency. Skilled operators have learned to allow for this movement. However, if the part is to be made numerous times, it is good management to have special gaging for known movement of the parts. Carefully analyze parts with unusual shapes to determine if special gaging with adjustable gage points is required.

Two-Point Gage Avoids Edge Camber Errors

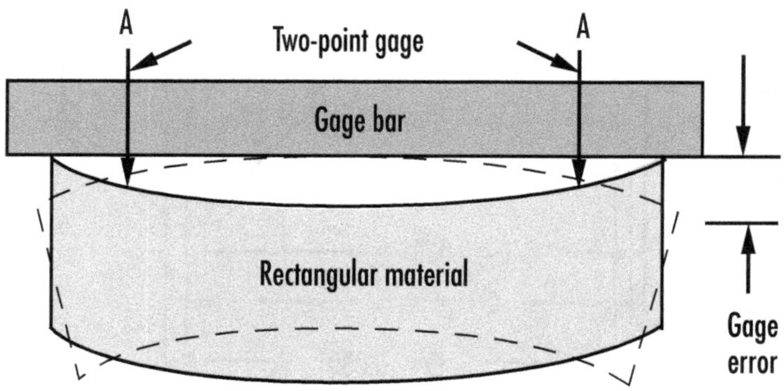

A straight, full-length bar for a press brake back gage can cause gage error. Camber in the workpiece results in good two-point contact at the outside edge of one side. Turn the part over and the radiused edge created by the camber allows the part to rock along the full-length bar, causing position variations and gage error.

Fig. 7-6. Straight bar gage shows the problem when camber is present in the blank.

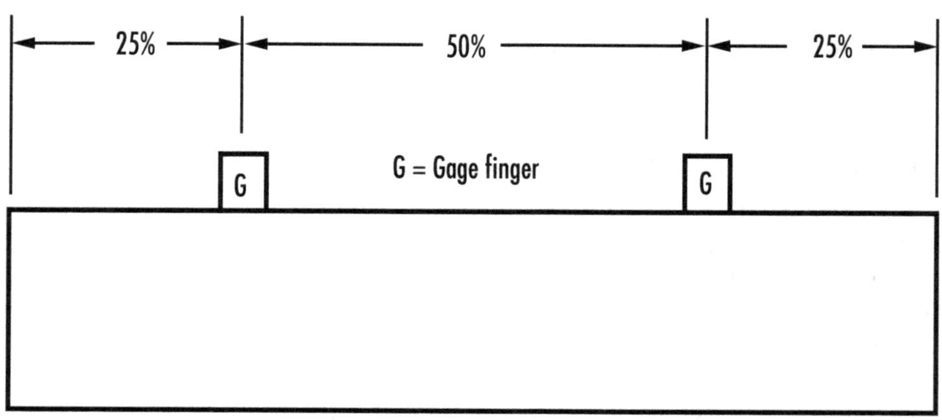

Fig. 7-7. Location of gage fingers to minimize the effect of camber.

E) Blanks that Have "Bow" or "Twist" May Not Form Within Tolerance

Making a bend in a part that has bow along the bend line may form an acceptable angular bend, but the material can return to a bowed condition after the angle has been formed. There is no guarantee that making the bend will produce enough part stiffness to eliminate the bow. If the formed part does have bow, the gage location of other bends to be made in the same part may be affected.

Parts that have twist in them prior to forming will produce legs that will often be out of tolerance. By looking down on an unformed blank that has twist, it can be seen that the part does not have a straight edge to be properly gaged. A bend line that is made, gaging to the twisted edge, will have different heights along the flange which, in most cases, would not be acceptable. A reasonable inspection of the part blank prior to the forming operation will minimize the scrap rate as well as reduce lost production time. It is also important to determine why the blanks are not acceptable, and work to correct the problem.

F) Other Factors that Influence Gage Locations

In addition to the concerns already listed, consideration must also be given to some common parts that may affect gage locations. The most common is a part that has a short flange. Short flanges imply that the majority of the material will be located out to the front of the press brake with the short flange and gages located to the rear of the die centerline. When the part is being formed, the majority of the material will move upwards at a speed dictated by the forming speed of the ram, the width of the vee die opening, and the ability of the operator to lift the material extending toward him that will move upwards during the forming stroke. It is common for the rear flange (which is not affected by any measurable amount of friction from the lower die) to simply slip downwards into the vee die faster than the front leg (Fig. 7-8), resulting in an incorrect leg length. The only way this can be corrected is to measure the short leg after a part has been formed, and adjust the backgauge position to compensate for the error.

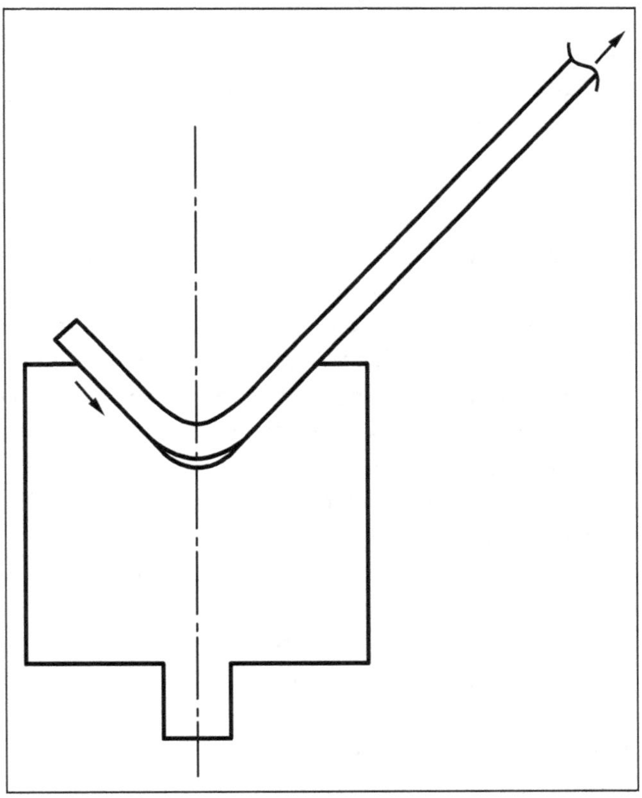

Fig. 7-8. Short flanges often slip into the vee opening during the forming stroke differently from a longer front leg of the part. As a result, the back gage position may have to be modified.

Notches and holes that are located anywhere within the vee die cavity after the part has been

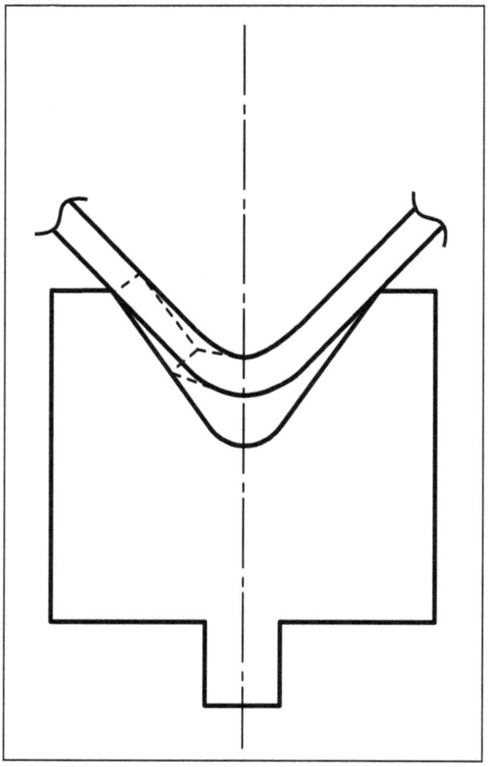

Fig. 7-9. Part distortion caused by a hole located in the vee die area.

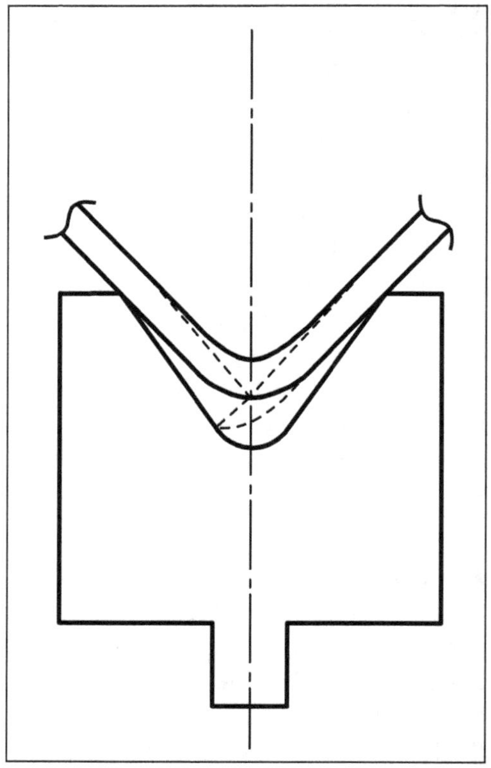

Fig. 7-10. Part distortion caused by a notch located in the vee die area.

formed will be subject to flaring (Fig. 7-9). Since the hole or notch is not supported during the forming cycle, the edges that are formed in space will stretch or bulge out in a variety of shapes (Fig. 7-10). When the formed part nears the bottom of the vee die, the flared area may cause movement of the entire part, as the sloping surface of press brake vee die contacts the protruded material. This will affect the gage location.

Tapered or oddly shaped parts, where the material weight is not balanced front to rear, may twist or turn during the forming cycle, resulting in a distorted flange length. In many cases, the backgauge will have to be tapered to correct for the material movement. Flanges that taper to a point, such as required when forming some cone shapes, can cause many gage location problems. Since it is possible that the forming lines are not uniform across the vee die opening (which will cause the bend to kick or move during the forming cycle), only testing can determine the most consistent, workable forming method. After developing the unique gage locations, the operator or setup person should record the setup method and gage locations so that the next run can be completed without repeating the initial time consuming process.

G) Back Gaging Basics

Most parts are back gaged. It is simple to set up, and normally requires minimum thought about how to form the part. If the blank is flat, the simple two point back gage will locate the part to provide an acceptable flange. Any error that is present in the blank—if the distance between the gage and the centerline is correct—causes all of that error to be in the front of the bend line toward the operator. The error can be attributed to shearing or cutting variations

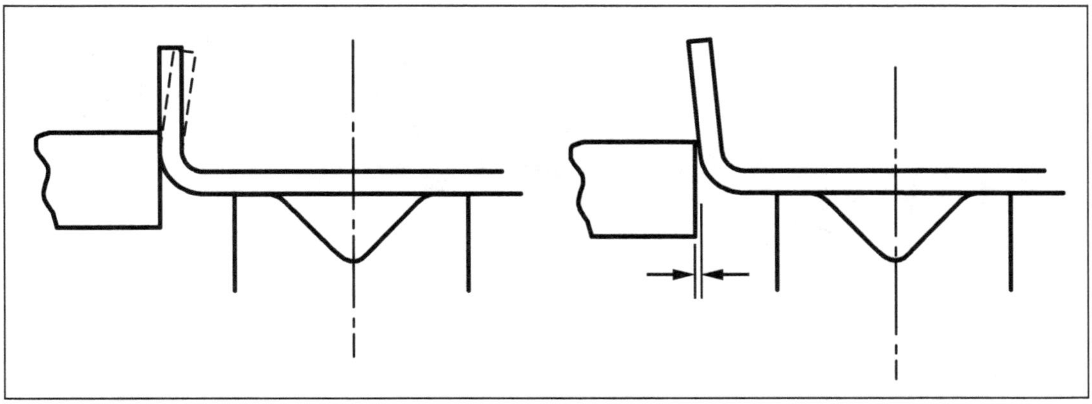

Fig. 7-11. Gaging to a previously formed flange requires 1 to 1 $^{1}/_{2}$ metal thickness of flat surface past the outside radius of the bend angle. If the formed angle is square or overbent, a good gage position can be held. If the angle is open, gaging error should be expected.

or incorrect blank size calculation. If a part is back gaged, and a second bend is required by turning the part 180° to complete a channel shape, the blank error will now be located between the two bend lines.

If the part is back gaged to form a 90° bend, then pushed inward to a second stop location where a channel shaped edge is formed, the gage finger height becomes very important (Fig. 7-11). The gage fingers must contact the flat section of the formed leg approximately 1 to 1.5 metal thicknesses past the outside radius of the first bend. If the first bend is square or overbent, the gage location will be consistent. This does not consider any minute errors caused by material thickness variations. If the first bend angle is open (greater than 90°), the formed flange will contact the gage finger in the wrong position and create an error. The higher up on the flange leg the gage touches the material, the greater the potential variation.

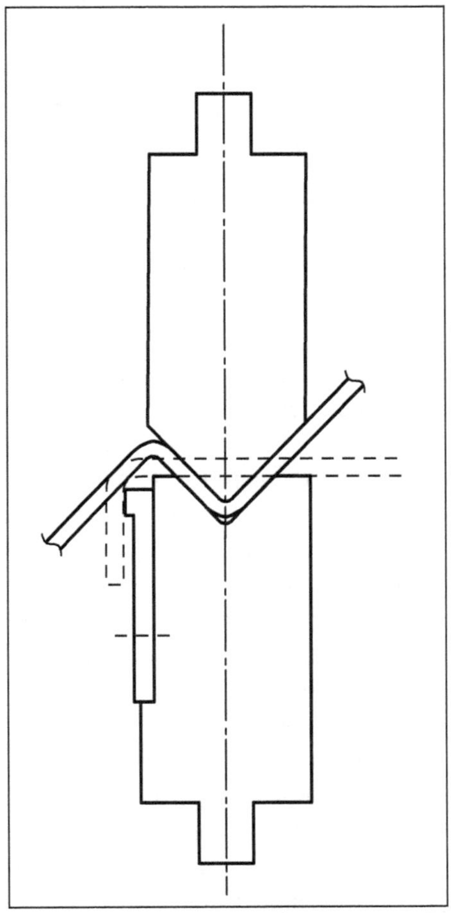

Fig. 7-12. Forming an offset in two strokes. The fixed gage for the second bend is attached to the lower die.

Parts that have been turned over to form an offset shape will have two gage setups. If necessary, a second gage can be located on the rear of the lower die (Fig. 7-12). The first bend is gaged by using either a front gage or a conventional back gage. A front gage will put any gaging error into the short leg located in the rear of the lower die. When the part is turned over to form the second bend, if a back gage was selected, it must be moved to provide clearance for the first bend leg as it swings upward toward the first gage's original location. The gage used for the second bend is located so that the leg of the first bend is placed over both the vee die and gage position. Then the part is pulled back toward the operator until

Fig. 7-13. NC back gage. (Photo courtesy of Cincinnati Incorporated)

the inside flange of the first bend contacts the gage block. Often the gage blocks are mounted directly to the vee die body, since limited space may restrict the use of conventional gage fingers.

Some manufacturers offer various types of "flip finger" back gages that are supposed to allow the previously formed first leg of an offset shaped part to move upwards when the second bend is being formed. This allows the moving leg to hit the gage finger and lift it up without damage to the part or to the gage finger. These fingers are often available with NC back gages (Fig. 7-13). This method may have some appeal to management, but for the operator it will create many problems (refer to Chapter 11-B). The flip finger design normally requires the edge of a flat part to contact the gage finger below the pivot pin centerline in order for the finger to stay in its proper location. If a part is pushed against the top of the gage finger, which is above the finger pivotal centerline, the finger will often move upwards. This causes the part to accidentally move too far forward which results in scrap parts, plus possible operator safety concerns (Fig. 7-14).

When the offset part, as described above, is located for the second bend, the back gage must be set low enough to allow the gage finger to contact the previously formed leg on a flat surface in order to ensure proper gage location. As a result, the first gage position forces the part to contact the gage finger above the gage pin centerline. It is recommended to use the flip finger back gage design only after careful review.

The best method for forming a second bend when gaging to the outside leg of a previously formed flange is to use an NC back gage that has a "clamp/retract" feature. The first bend is made by pushing the part against the back gage finger and forming the part. The part can then be turned over, and the gage finger will automatically move to the correct gage location. After the piece is located properly against the back gage finger, the top die is cycled downward. When the top die touches the part, locking it in position, the ram motion immediately stops. At that point, the back gage moves back to provide clearance for the first bend leg when the second bend is made. The operator must activate the ram to make the second bend.

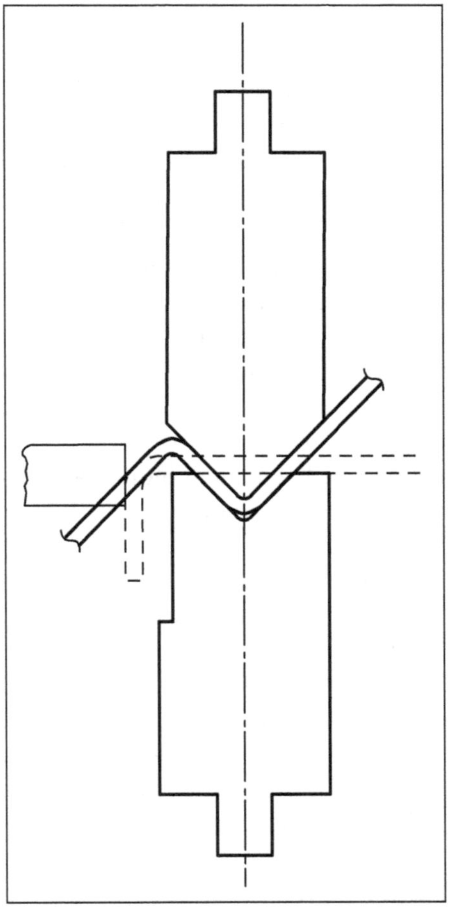

Fig. 7-14. Interference of the material with the back gage while forming an offset shape in two strokes.

H) Types of Gages

There are four basic types of gages.

1) Fixed Gages

The best gaging is obtained by mounting permanent gage blocks or pins on the lower die. The setup requires proper upper and lower die alignment to hold the required gage dimensions.

2) Numerically Controlled (NC) Gages

Most NC gages are designed as back gages. The gage moves back and forth to a programmable gage position. It can be programmed to move to the proper position for each bend in the part. Some companies offer an NC front gage. Unless there is a good reason to have one for specific jobs, they often get in the operator's way and interfere with the material handling needed for all of the other jobs normally run at that press brake.

The accuracy and speed of modern NC gages permit high production plus good quality metal forming. Gage positions depend on proper gage calibration with the upper die nose and the gage fingers. NC gages with memory allow repeat setups by recalling a previous program, thereby minimizing costly setups. Sheet supports should be considered to help ensure proper material location.

Today's sophisticated press brakes offer programmable four-axis, five-axis, or six-axis back gaging and other combinations designed to minimize costly setup time. The gage fingers can automatically move in and out, up and down, as well as left to right to provide possible gaging for parts to be made in one handling, as opposed to the multiple setups and handlings previously required. Heavy duty NC back gages are also available for plate gaging.

3) Manual Gages

Many press brake beds provide some method of securing support brackets in the front and rear of the bed. The most universal method is having dovetail slots machined in the front and rear of the bed to allow the support arms to be moved anywhere along the bed length. If the press brake that is currently being used at your facility does not have a universal mounting method available, a bar with a dovetail slot machined full length can be mounted in the front and rear of the bed to allow commercially available gage support brackets to be used.

The simplest brackets have a 2.5" wide × 1.25" thick top bar with a dovetail slot machined full length (normally 18" to 30" long), welded at 90° to a similar size vertical bar (18" to 24" long). All support bars are provided in pairs since all press brake gaging requires two points of contact at the edge of the metal to properly locate the bend line. A 2" to 3" high × 12" long welded 90° angle made from thinner 2.5" wide stock is be used as a gage stop by mounting one of the angles on the top of each support bracket. In order to keep the surface of each gage angle parallel to the forming dies, the bottom of each of the 12" mounting legs of the gages can be machined with a $^1/_8$" deep key. This fits into the top of the dovetail slot of the support arm. The 12" horizontal gage leg normally has two or three holes drilled in it to allow commercial dovetail bolts to be used to clamp the 12" leg to the horizontal support arm. Each of the support arms requires a mounting block that allows the vertical part of the arm to be clamped to the front of the bed, but still allow easy vertical adjustment. The block must provide enough guiding, and have sufficient length, to keep the arm from twisting. The other side of the mounting block has a $^1/_8$" key that locates into the dovetail slot in the bed or the dovetail bar that has been mounted on the bed. This key keeps the support leg vertical and also prevents twisting.

Other commercial designs provide a dovetail top bar that can move back and forth and can be clamped in any required location. On the top of the dovetail bar is mounted a 0.5" round bar that has been bent 90° to provide a 2.5" high gage surface. The surface of the round bar is machined to a 0.25" thickness at the long end to allow the gage bar to be located on top of the vee die. This allows gaging of short flange legs and minimizes the problem of the angular surface of a standard upper die touching the gage finger at the bottom of the stroke.

Most manual gaging is mounted in back of the lower die to provide a good solid locator for the operator. Additional arms are often mounted in the front of the bed to help support the flat blank prior to the part being formed. The support arms should be adjusted so that the top surfaces of the arms are the same height as the top surface of the lower die; this assists the operator by ensuring that the gage finger has proper contact, and that the part is not accidentally tilted and pushed under the gage fingers (Fig. 7-15). Special blocks or squaring arms can be added as different parts are designed, and require unique gaging to produce acceptable shapes. Some parts may need both front and back gaging to hold tolerances. Manual type gaging in the front plus NC back gaging may also be considered.

The main disadvantages to manual gaging include longer setup time, difficulty in setting both gages to the identical gage location, and the difficulty of having the exact same gage points for future setups. Without good setup sheets, it is unrealistic to expect different setup personnel to use the same method to form the same part.

The problem with setting manual gages, especially the common commercial gages that are made with a round 0.5" bar, is trying to set the gages to the correct distance from the bend centerline to the actual gage position. Most operators or setup personnel use the machined surface normally found on at least one side of the lower die as a starting point. A measurement from the machined surface to the centerline of the vee opening provides a close fractional distance if a standard combination square is used. When the die is placed into the filler block, the machined surface that was used for the starting measurement should be placed facing the gages.

To locate the gages, the operator mounts the two back gages at the desired location on the bed and loosely clamps them in place. The gages are then raised to a height where the gage fingers will locate the edge of the material that will be placed on the top of the vee die. The

Heavy duty gage and support arm for material up to $^1/_2$".

Simple manual gage and support arm.

A double dovetail bar with disappearing fingers mounted on a support arm.

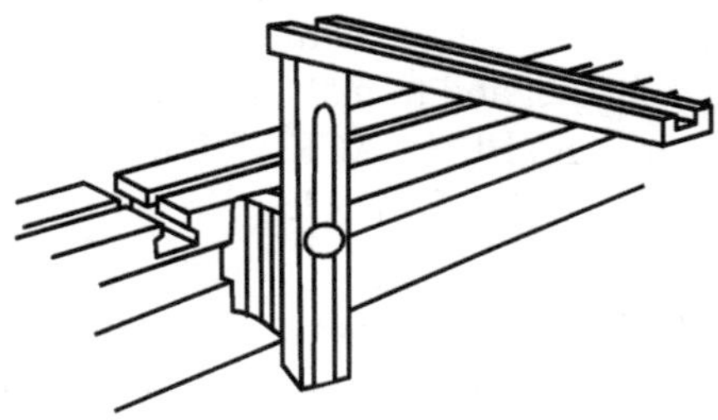

Support arm to help locate the informed material at the same height as the lower die.

Fig. 7-15. Manually set gages and support arms. *(Photos courtesy of Cincinnati Incorporated)*

gage height must allow the flat blank to safely contact the gage surfaces without risk of the material being accidentally pushed under the gage if the operator inserts the part tilted instead of flat. The location of the gage bars can now be clamped to the bed. Each bar is fastened in place by some type of simple clamping bolt.

By using the combination square with the end of the 12" scale set to the location that establishes the centerline of the vee opening, the square is held against the machined surface of the vee die. The scale is placed at a location next to one of the gage bar fingers. While holding the square with one hand, the other hand can adjust the moveable gage in or out to locate the gaging surface of the bar to the desired dimension. The combination square is then put down and, trying not to move the gage bar, the clamping bolt is tightened. The same method is used to locate and clamp the second gage bar. When the combination square is put down to tighten the clamp, DO NOT put the square on the top of the lower die. Safety concerns will be discussed in a later chapter.

After the first bend has been made to check the gage dimension, one or both gage locations may have to be changed slightly. Many times, the gage that has to be changed only a small amount is simply tapped with a hammer in an attempt to move the bar only a fraction of an inch. In this instance, the gage bar is not unclamped but the rod will move slightly when hit. Often this method corrects the problem. Trying to determine which gage to adjust, and in which direction to move it, is the major problem. The part is formed with the operator standing in the front of the machine. The gages must be adjusted in the back of the press brake. When the setup person moves to the back to adjust the gage fingers, it is very easy to move the gage location the wrong way, or to adjust the wrong gage. It has been found that a great deal of practice is required to make gage adjustments with minimal errors.

Some manufacturers sell what is called a micrometer-type adjustment for the simple universal-type gage discussed above. This is not a micrometer like the standard micrometer found in a machine shop; it is a graduated wheel which allows the round 0.5" gage bar to be moved backward or forward several thousandths of an inch to allow minor corrections to be easily made without loosening the bar. Since loosening the bar and reclamping it can introduce new errors, the micrometer gage option is a very appealing way to minimize gage setup time.

Manual-type gaging has its place in all press brake facilities, but the availability of NC back gaging has greatly reduced the problem and time required to properly set manual gages. Whether manual or NC gaging is used, all adjustments that must be made to gage fingers located in the rear of the press brake require the operator to walk around to the rear of the press brake, and to make the needed corrections, or to use a special tool to move an NC gage finger. NEVER PLACE YOUR HAND BETWEEN THE DIES TO MAKE ANY TYPE OF GAGE ADJUSTMENT.

4) Scribed Lines

One of the oldest gaging methods is to scribe a line on the top of the part that locates the bend line. Normally a "soapstick" or a piece of chalk that will give a visible narrow line is used. The operator jogs (activates the ram so that it moves very slowly) the upper die down so that it barely touches the scribed line. Movement of the ram should be stopped just above the part to allow the operator to align the scribed line with the nose of the upper die. If the part is properly aligned, the forming stroke can be completed. Since a scribed line has some thickness (even if made using a sharp pointed nail or scribing tool), the operator will have difficulty visually obtaining accuracy closer than ±0.015", and the results will not compare to the solid gage method.

The experienced operator will often move the scribed lines toward the front of the machine by one-half of the vee die. Aligning the scribed lines with a solid fixed lower die surface is faster because the stroke can be made without stopping and checking the location of the scribed line to the nose of the upper die. The location quality is also improved. Using scribe lines often requires a helper in addition to the operator to help properly locate the correct scribed line at each end of the part.

5) Front Gaging

There are NC front gages available from some manufacturers. They have never gained much popularity since back gaging is usually easier to set up, and in most cases is simpler to use. The NC front gaging systems, by necessary design, are large and bulky. When the gages are not being used, the operator pushes each arm out to the end of the bed. Because of their size, they occupy the same space where many operators locate their unformed blanks on one side, and where they stack their formed parts on the other.

The most common type of front gage uses the manual gages described previously. The gage arm is mounted on the front of the bed, and the gage is set the same as it would be when used in the rear of the machine. Some suppliers offer a disappearing style front gage. A bar having a dovetail slot machined on each side is bolted to the top of each standard heavy duty gage bracket. When multiple front gage positions are required, flip finger gage blocks are mounted to either side of the double dovetail gage bar. The weight of the material resting on the top of the gage fingers pushes them down, out of the way, and allows the dovetail bar to provide a reliable sheet support. As the part is moved forward from position to position, two gage fingers, one on each gage arm, pop up at the proper location to provide the next gage location.

Pin gaging is also a possible option when front gaging is used. The pins are used to locate the proper bend line by placing two pins in prepunched holes in the unformed blank. The pins must be especially designed so that the part does not "hang up" on the hole as it moves up away from the pin during the forming cycle. Suppliers that make this type of gage often mount them on the same double dovetail bar that can be used for the flip finger gage blocks described above.

I) Gaging Reminders

1. Gaging accuracy depends on initial blank quality, gaging methods, and bend sequence. Thicker material will cause greater variation. Some thin materials are being fabricated to a ±0.005" tolerance using good gage methods. Holding this type tolerance is rare. Normal tolerance of ±0.020" should be expected when bending 16 gauge and thinner material. **Note**: The tolerances given are variations expected when locating the part using gages. Variations caused by gaging the material are added to the angular tolerances. As the material becomes thicker than 16 gauge, tolerances of ±0.030" may have to be considered for most gage locations. Forming plate (greater than $^{3}/_{16}$" thick) can increase the tolerance from ±0.060" to "We can make this part work."

2. Attempting to gage to a leg of a previously formed angle that has been bent to less than 90° (an open angle) is a waste of time and material. For example, if a part has a 3" leg formed to a 45° open angle (135° obtuse angle) that may vary ±1°, the resultant gage position can vary ±0.060". Attempting to gage to any sloping surface will produce poor results (Fig. 7-16).

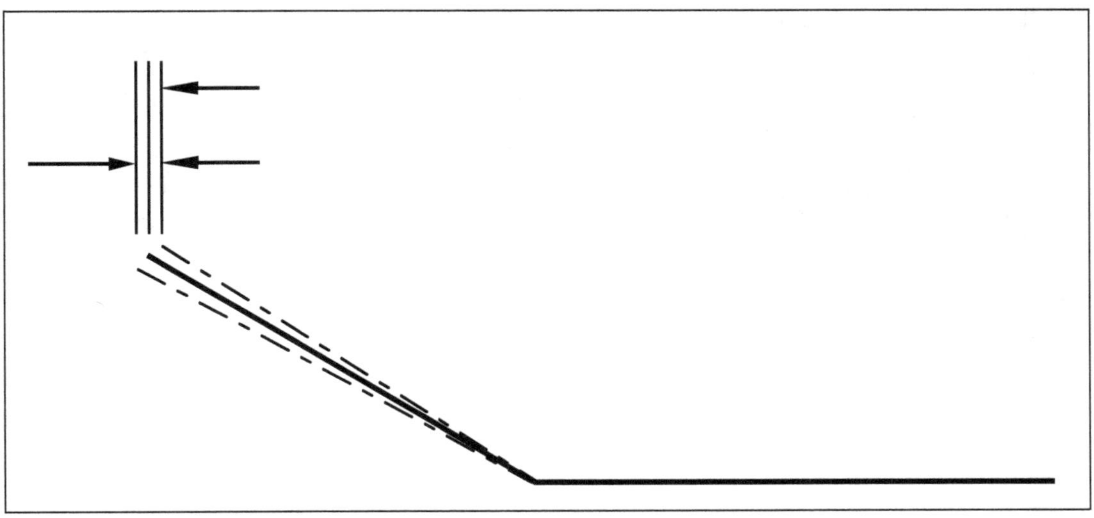

Fig. 7-16. Gaging a part to the edge of a flange greater than 90° can introduce the variation encountered when standard angular variations occur. The position of the edge of the material may cause an unacceptable bend location.

3. Review the die selection, bend sequence, blank size, gage methods, and operator safety procedures that were used to produce acceptable parts, and ensure that they are available for the next production run. Make sure that any minor corrections are included. Make part development a one-time event. If necessary, update drawings and routing sheets (Fig. 7-17).

4. The basic rule for operator control and safety is to form the part by keeping the weight to the front of the press brake. This allows the operator to control the part position and to ensure consistent gaging positions.

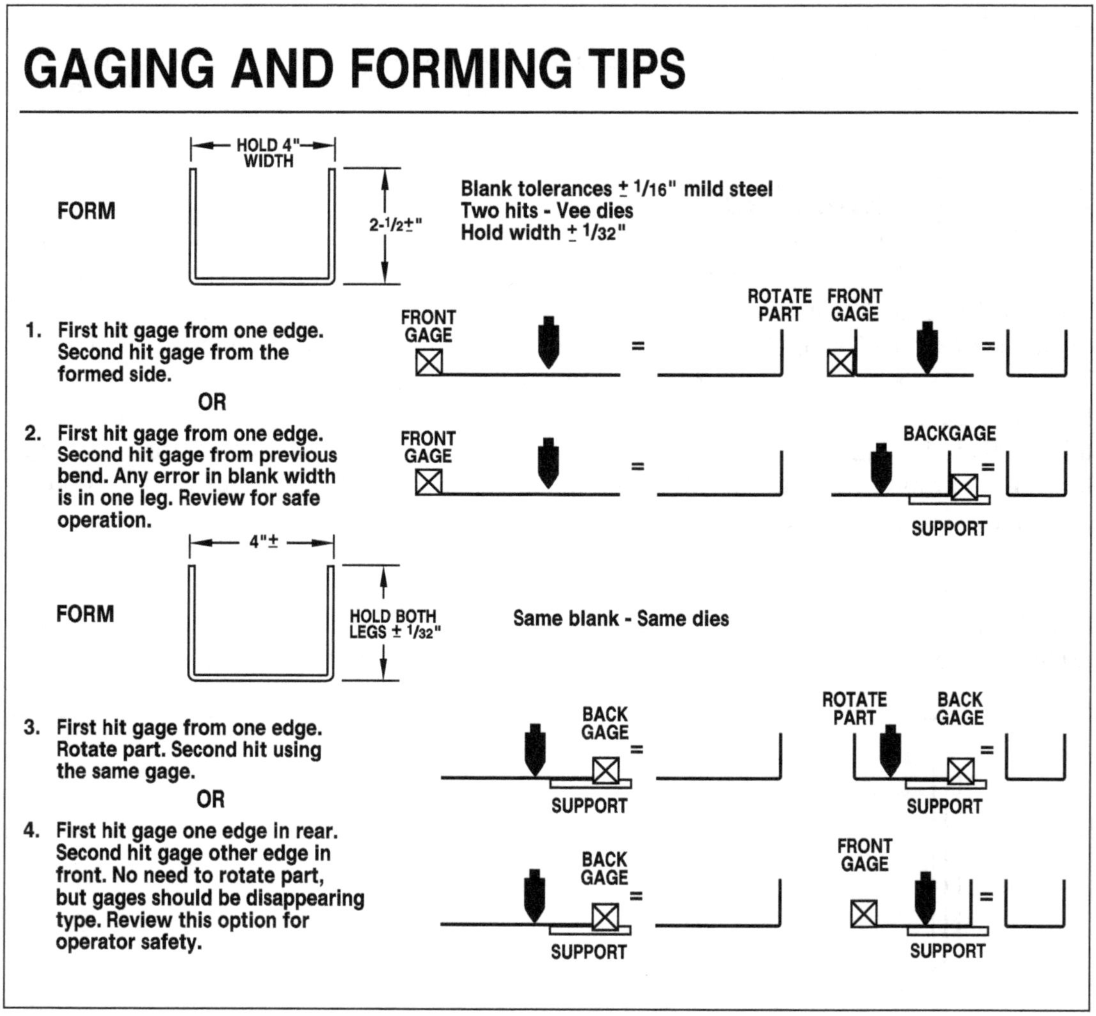

Fig. 7-17. Gaging reminders to insure any "blank error" is located in the best place. (*Courtesy of Cincinnati Incorporated*)

5. The press brake manufacturer can be a prime source for commercial type gaging that can be used on the press brake. If gaging is purchased at the same time the press brake is purchased, responsibility for the machine and gage performance will be consolidated with one source. In some cases, this will minimize potential manufacturing problems, warranty concerns, and lost time (Fig. 7-18).

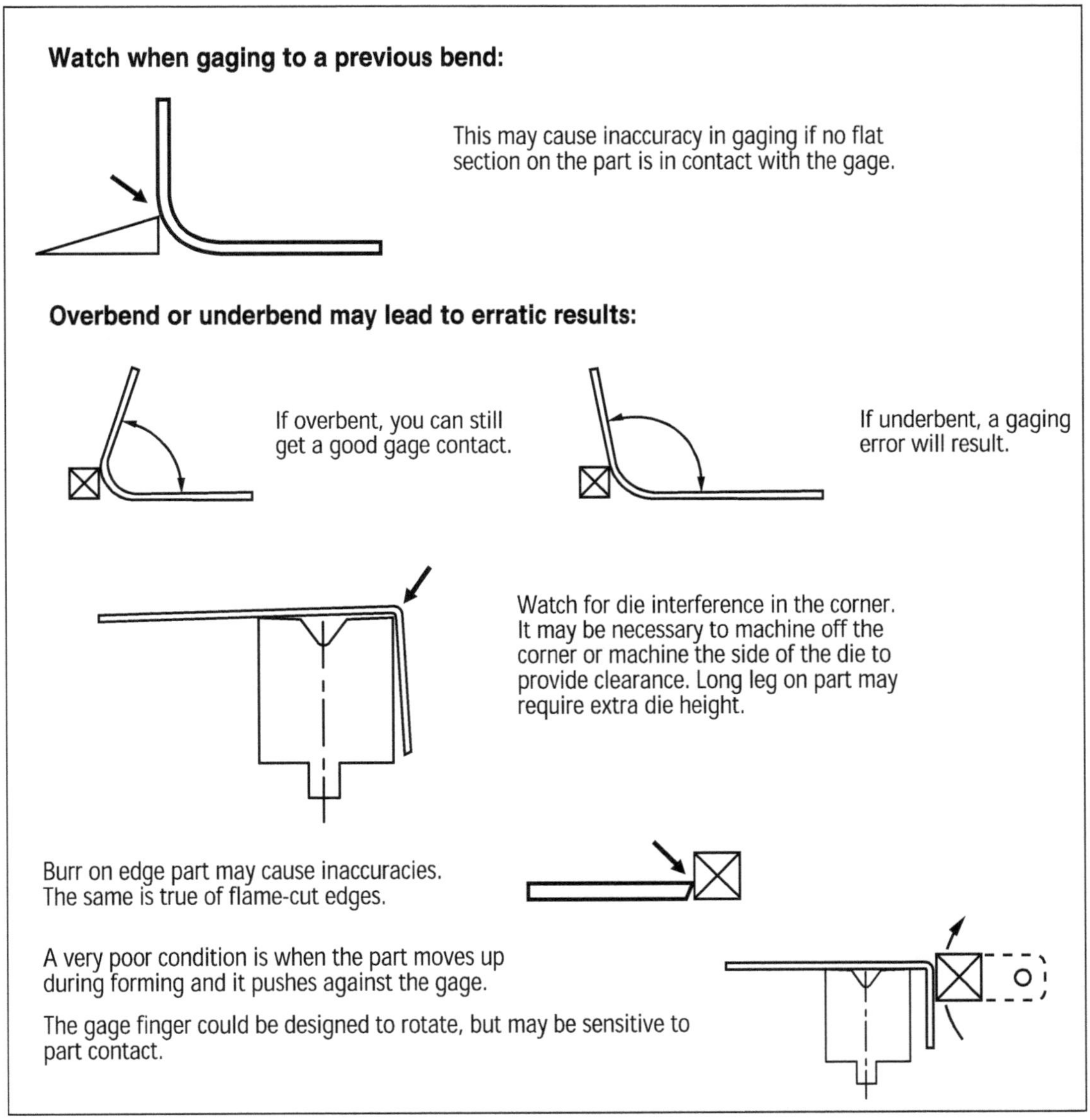

Fig. 7-18. Review the gage setup to insure the gage method does not cause problems. (Courtesy of Cincinnati Incorporated)

Chapter 8
Bend Allowance (What Size Should I Cut My Blank?)

Prior to forming a new part, management must provide the operators with a blank size that takes into consideration the material thickness, type of material, and the length and width of the part. Information concerning holes and notches is sometimes also required.

To calculate using the basic method, add the length of each of the straight legs of the part, and then add the distance required to form each radius (Fig. 8-1). The radius distance is found by calculating the length of the neutral axis in the radius area. This is the theoretical length of line found between the area at the inside of the bent shape that is in compression and the outside area that is in tension (Fig. 8-2).

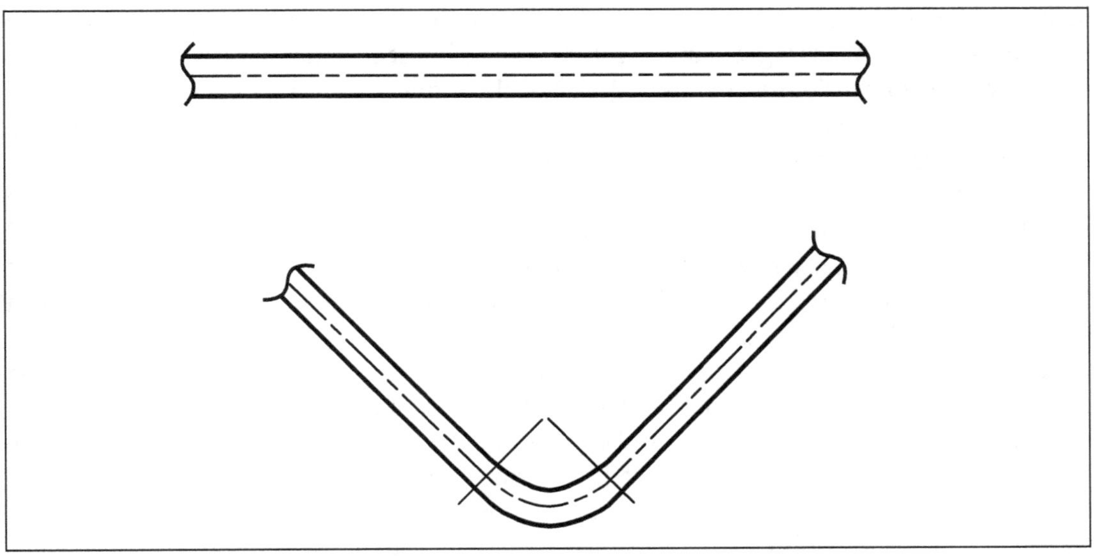

Fig. 8-1. The theoretical "neutral axis" is in the center of an uniformed blank. After forming, the neutral axis becomes an elliptical shape in the area of the bend radius.

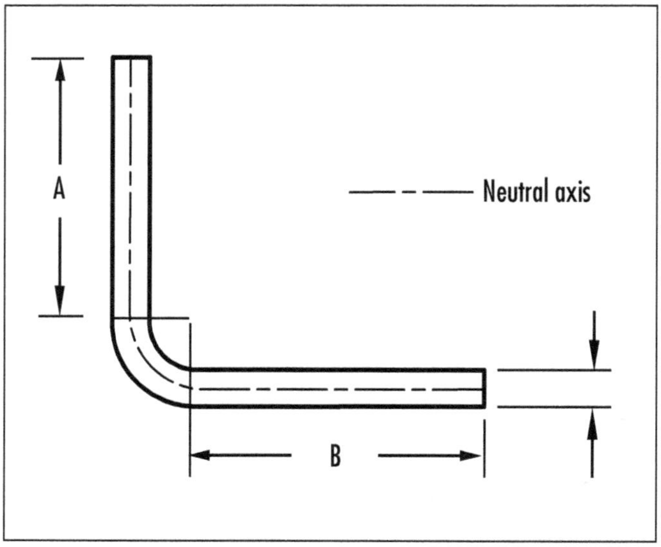

Fig. 8-2. To determine the size of the flat blank, the straight lengths of each leg must be added to the length of the elliptically shaped neutral axis in the radiused area of each bend.

The action of the metal can easily be pictured—again using a beveled rubber eraser as discussed in Chapter 5. By holding the eraser at each end, form it into a 90° angle. The inside radius of the eraser will be in compression, which can be visually observed by the rubber being forced outward throughout the inside arc of the bend. The outside radius of the eraser is in tension, as observed by the flattening or stretching of the rubber on the outside of the arc, plus the "peaking" of the rubber along the front and rear of the top surface. When the eraser is viewed by looking at the formed angle along the width of the eraser, the inside of the eraser bulges outward from the compression on the inside of the arc. A unique line from the inside surface to the top of the peak on each side of the outside surface can be seen. A similar shape occurs when a piece of metal is formed to a 90° angle. If the formed part was sectioned to give an inspection of a 90° bend, and if the sides of the sample were polished to allow the grain structure of the material thickness to be examined, an elliptical line would be found between the compressed and tensioned areas of the formed part. This line is called the neutral axis.

To calculate the blank size, the length of the neutral axis of each bend must be determined or estimated. Attempting to calculate the length of an elliptical line when the exact dimensions are not available prohibits the determination of an exact length. A research paper done by this author in 1964 found at least 12 published authoritative formulas, all providing an engineering "answer" to the problem (Fig. 8-3 and Fig. 8-4). Each of these formulas gave a different answer when the same dimensions were inserted. To test the formulas, several hundred premeasured blanks having different material thicknesses and blank widths were then formed. To obtain the dimensions of each formed part, optical comparators were used to measure actual leg sizes. Conventional radius gages were used to estimate the closest measurable radius of each part. This data was then inserted into each of the 12 different formulas to obtain a calculated blank size. An exact answer was calculated only twice, but from two different formulas. At last count, as many as 15 blank size formulas may be available; but it is safe to assume that an exact answer will seldom be calculated using one formula.

A second major problem is found on the engineering drawing. The engineering part drawing released to the shop often does not have an inside radius on the bend angles that conform to good press forming practice (see the air bend forming section in Chapter 3). For example, the engineering drawing will call for a 0.125 I.R. (inside radius) in 11 gauge mild steel. To obtain the best air bend accuracy, the recommended vee opening to form the metal should be 8 times the metal thickness. In this case, 8 times the 0.125 thick materials would suggest the operator use a 1" recommended vee die opening. The problem is that the resultant inside radius will be 0.156 × the vee die opening per the "rule of thumb" formula discussed in Chapter 3. This

1) **L = E + F + .4T** (for sharp corners only)

2) **L = A + B + 1.57 (.4 + R)** If R is greater than or equal to T

3) **L = C + D - F** If R=T; F=1.7T; If R=2T; F=2T; If R=3T; F=2.5T

4) **L = A + B + Number of Degrees (.01745R + .0087T)**

5) **L = C + D - .8T**

6) **L = A + B + Angle/360 (.4 T + R)**

7) **L = A + B + 1.57 (1/3T+ R)**

8) **L = A + B + Number of Degrees x .01745 (R + .5T)**

9) **L = A + B + (1.57R + .064T)**

10) **L = E + F**

11) **L = A + B + 1.57 (R + .2T)**

12) **L = A + B + Number of Degrees (.01743R + .0078T)**

Fig. 8-3. Reference list of bend allowance formulas published by a variety of engineering books.

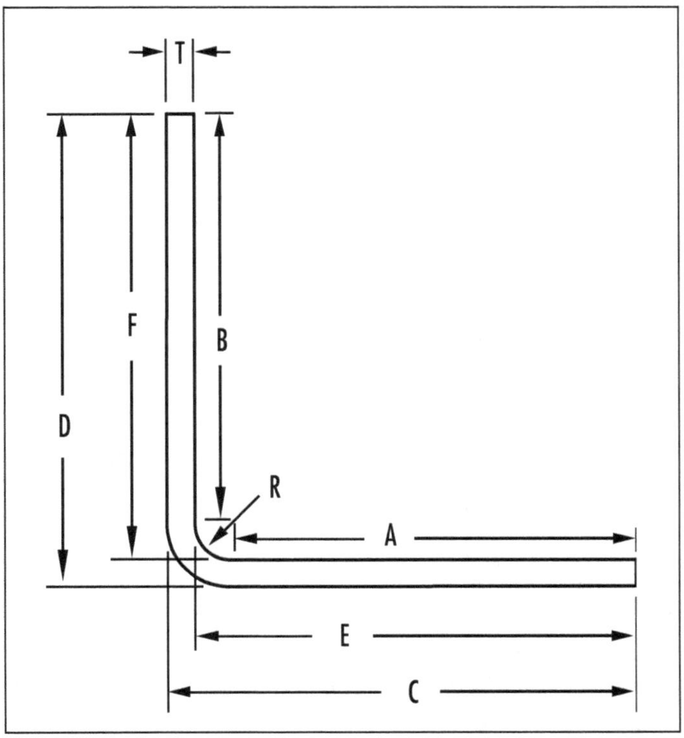

Fig. 8-4. Simple right angle showing letters used in the Fig. 8-3 formulas.

means that the closest measurable inside radius (although the actual shape is elliptical) will be 0.156", which is visually larger than the 0.125" inside radius called for on the drawing. If the 0.125" figure is used to calculate the blank size, and the 1" vee opening is used to form the part, the blank will not be correct or, in some cases, not even close (Fig. 8-5).

Is this part with a larger inside radius usable? Probably, yes. The problem can be solved by encouraging engineers to use an inside radii that conforms to recommended shop forming practices. This concept is important since, in most cases, only a limited number of vee die openings are available for the operator to use without purchasing additional dies. The majority of engineering schools do not provide their students an opportunity to become familiar with the basics of metal forming. As a result, poor practices develop. Little helpful information is available in handbooks to correct the mistakes before they are made.

If the drawing is calling for one inside radius, and the part is actually made with a different inside radius, even though it appears to be very close to what is shown, the blank size will be wrong. As a result, the proper part dimensions or tolerances may not be held. Naturally, the shop will question the blank size called for on the drawing or on the routing sheet. If the engineering department rechecks their figures, using the same formula to calculate the blank width using the same incorrect inside radius dimensions, they will still have the wrong blank size. To correct the problem, the engineering drawing needs to be changed to show the actual inside radii that are being formed in the shop, and the correct blank size listed on the drawing or routing sheet. This will minimize forming costs and the operator skill needed to produce future parts.

Most of the bend allowance formulas do not attempt to calculate an unknown elliptical shape, but try to estimate an arc of a circle that may be close to the length of the neutral axis in the radius area. The method is as follows.

1) The perimeter (or circumference) of a circle is calculated as ($2 \times \pi \times r$), which is usually stated as $2\pi r$

where $\pi = 3.14159$, and r is the radius of the bend.

2) To form a 90° bend, the arc length would be ($2\pi r \div 4$).

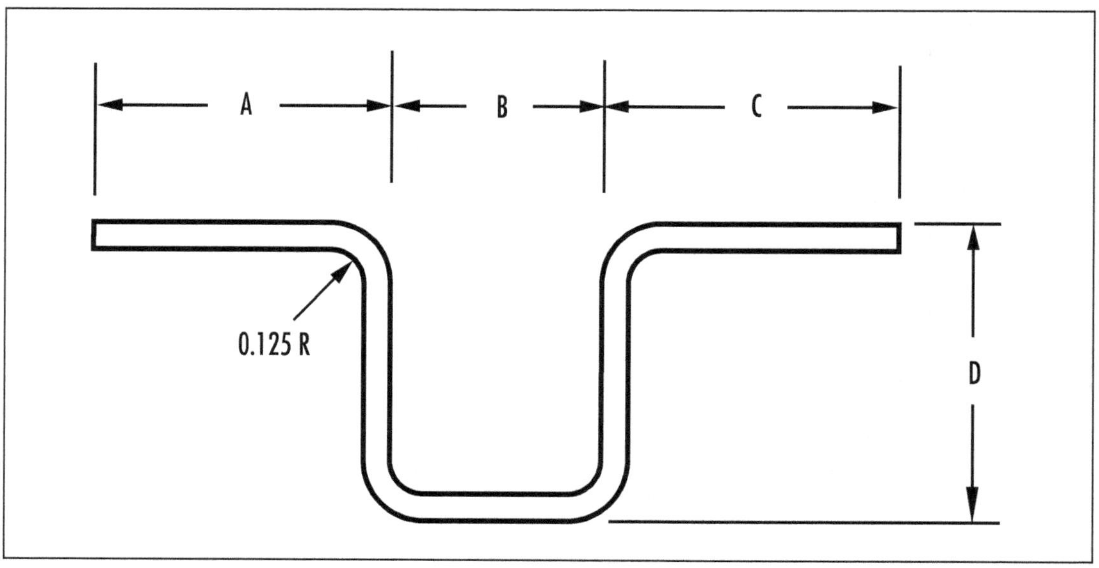

Fig. 8-5. Typical 11 gauge part calling for inside radii that do not conform to the radii produced if the recommended vee opening is used for forming the part.

3) If an angle other than 90° is to be calculated, use

$$(2\pi r \times \alpha) \div 360$$

where α is the angle, and 360 is the total degrees in a circle.

4) The above formula can be expanded by defining r as the theoretical inside radius (ir), and by identifying the distance from the surface of the inside of the metal to a point which is believed to approximate the location of the neutral axis. First, the 90° formula is modified by dividing 2π by 4 to obtain a constant of 1.57. Handbooks will often show the resulting generic formula for determining bend allowance (BA) as

$$BA = 1.57(ir + kt)$$

where t is the thickness of the material, and k is a value normally estimated to be in a range from 0.2 to 0.4 of the thickness of the material for a simple 90° bend, again using the recommended vee opening. As the inside radius is required to be made larger for a unique design, or if the inside radius approaches 4 times metal thickness, the k factor will increase to up to 0.5. However, the result of the tests done in 1964 determined that the published formulas that seemed to produce the closest calculated answers are those found on the chart shown on Fig. 8-6. Note that one formula seemed to work best for materials 15 gauge and thinner, and a second formula is recommended for 14 gauge and thicker.

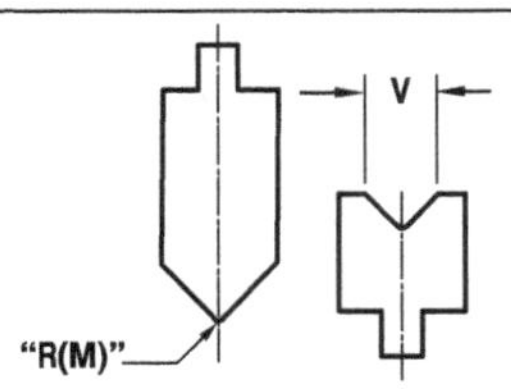

Bend Allowance for Sheet Metal Air Bends
(Mild Steel Only)

Blank Width - L1 + L2 + Bend Allowance (BA)

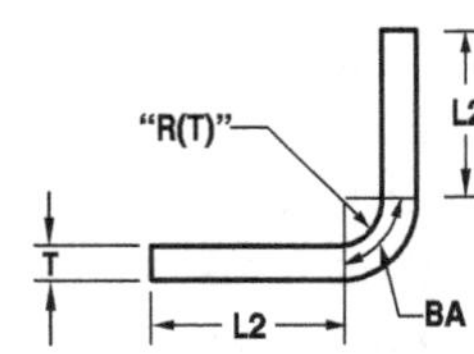

Material Thickness "T"	Recommended Die Width "V"	Recommended Die Radius "R(M)"	Theoretical Bend Radius R(T)=5/32 x "V"	Allowance for 90 Bend "BA"	Formula for Basis of Calculations
5/8 (.6250)	6	15/16	.9375	1.8644	BA=1.57(R+.4T)
1/2 (.5000)	5	3/4	.7813	1.5406	
1/2 (.5000)	4	5/8	.6250	1.2952	
3/8 (.3750)	3	7/16	.4687	.9713	
5/16 (.3125)	2 1/2	3/8	.3906	.8094	
1/4 (.2500)	2	5/16	.3125	.6476	
3/16 (.1875)	1 1/2	3/16	.2344	.4857	
10 GA. (.1345)	1 1/8	1/8	.1757	.3605	
11 GA. (.1196)	1	1/8	.1562	.3203	
12 GA. (.1046)	7/8	1/8	.1367	.2803	
13 GA. (.0897)	3/4	1/16	.1172	.2403	
14 GA. (.0747)	5/8	1/16	.0976	.2001	
15 GA. (.0673)	1/2	1/16	.0781	.1649	BA=1.57(R+1/3T)
16 GA. (.0598)	1/2	1/16	.0781	.1539	
18 GA. (.0500)	3/8	1/32	.0586	.1181	
20 GA. (.0359)	1/4	1/32	.0390	.0802	
22 GA. (.0299)	1/4	1/32	.0390	.0770	
23 GA. (.0269)	1/4	1/32	.0390	.0754	

Note: 1. Calculated bend allowances are approximate, not exact. Actual dimensions can only be found by making the bend and measuring the results.

2. The above formulas are for right angles only. For gaging, remember to use only one leg and one-half of the bend allowance to approximate the distance.

Fig. 8-6. Chart showing the formulas that seem to provide the closest actual blank size when the recommended vee die is being used. (Courtesy of Cincinnati Incorporated)

One of the easiest methods for calculating blank size (only if the recommended vee openings are being used) is to add all of the inside dimensions, ignoring the inside radius. This works very well in light plate (7 gauge up to 0.5" thick). For example, if the outside legs of a 90° angle made from $^{1}/_{4}$-inch plate are each 3" long, the inside legs would each be 3" minus 0.25" of material thickness, or 2.75" long. The estimated unformed or flat blank size would be 2.75" + 2.75" or 5.5" wide. If a specific formula has been used to calculate the blank size of a unique part, assuming about one metal thickness inside radii, adding up all of the inside dimensions of the part will provide a good check to ensure that the formula calculations were correct (refer to Fig. 8-5).

It cannot be emphasized enough that all of the published bend allowance formulas were developed for parts having an inside radius of at least one metal thickness. If a part is designed with an inside radius of less than metal thickness, there is no guarantee that the formula will work or even be close to accurate. The bend allowance formula attempts to calculate the length of the neutral axis. If the inside radius is less than metal thickness (Fig. 8-7), the neutral axis is actually pushed down toward the outside radius of the part. The radius is no longer a radius but a distorted shape; as a result, the actual blank size becomes larger. The key is to listen to the operator's comments about a specific blank.

If an operator refers to "growth" in the bend radius, he is saying that the original blank is too small. What is actually happening is that the dies being used are coining dies. This is causing the neutral axis to be forced down into the material into a unique shape that is longer than the engineering calculation. If the operator feels that the blank size is too large, that means that the inside radius is slightly larger than the dimension used for the engineering calculation. This is normally the result of the engineer designing a part with an inside radius that is not in line with good press brake forming practice.

As previously pointed out, the engineer normally knows very little about what inside radius will be formed. As a result, the normal response to the shop fabricating area is: "The blank size is correct, just make it like the drawing." Without a change in the drawing, all future parts made using that blank size will not be made like the drawing, and the operator will always have to adjust the part gaging to accommodate the blank error.

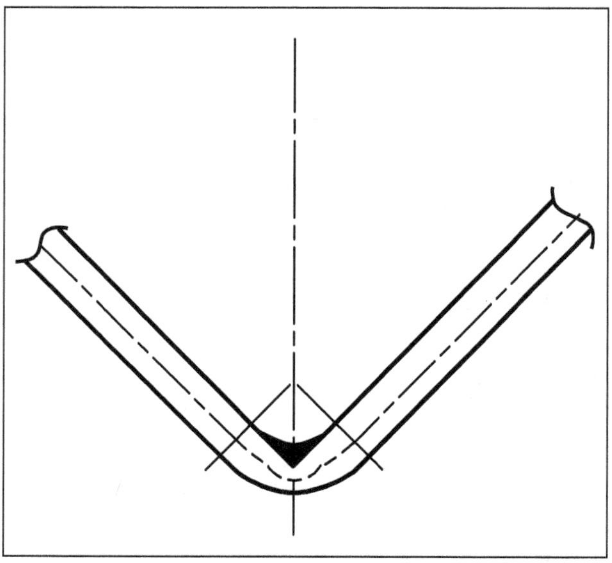

Fig. 8-7. The normal elliptical shape of the neutral axis is lengthened if a smaller than metal thickness inside radius is formed.

Management can often reduce part costs by insisting that initial blank sizes be tested with the tooling that will be used to form the part. That should reveal if the drawing needs to be changed to reflect what will actually be formed in production. Rigid adherences to standards that do not work reflect a lack of common sense. The engineer should not only have the responsibility to calculate the initial blank size, but also should be willing to change the drawing and blank size dimension to correspond with what will actually be formed.

This means that the tooling that is to be used in production should be coordinated with the engineering drawing to ensure future part consistency.

When parts are being reviewed by companies that do subcontract work that requires forming parts on their press brake, the person reviewing the drawings must know what tooling their company has available; what inside radius they can supply; and they must have a good knowledge of probable blank size requirements. In many cases, the subcontractor will take exception to the drawing tolerances and inside radius dimensions. If those written exceptions are not acceptable to the user prior to the subcontractor accepting an order, a great deal of time and money will be in dispute since any formed parts sent to the customer would be rejected because they don't meet drawing dimensions.

Remember: Good management will insist that drawings be changed to reflect the acceptable part dimensions that are actually being produced.

Chapter 9
Safe Use of a Press Brake

A) Basic Safety Considerations for Press Brake Users

A famous saying that has been repeated many times over the years asks this question: "If common sense is so common, why don't more people have it?" This statement certainly applies to the owners of press brakes, setup people, and press brake operators.

Before tooling is installed between the press brake ram and bed, even if the ram is cycled, a minimum distance of 4" to 12" will remain between the bed top and the ram nose. Even without tooling installed, some individuals have managed to injure themselves by climbing up onto the ram while the equipment was running, or reaching into areas where parts of the machine were turning (often having removed the guards). Some have even attempted to ride a moving ram up to the top of the stroke and then hurt themselves by falling off. Our discussion of press brake safety will be limited to the normal use of a press brake by setup personnel and press brake operators.

When the dies are installed into the press brake, in most cases the machine must be turned on in order to activate the ram (up or down.). The filler block will be on top of the bed unless it has been removed for a special forming operation. When the ram has been adjusted to allow sufficient space for the installation of the upper and lower dies, the operator's foot controls (foot pedal on a treadle or any electrical foot switches) must be made inoperative by the use of locking devices or on/off switches. As described previously, the lower die slides onto the filler block from one side of the machine. The operator or setup person will use a crane, or be aided by a forklift if the die is too heavy. The tongue of the die is located in the slot of the filler block. The die is pushed from the end into the die area until the end of the die is located at the edge of the filler block. The upper die is then loaded into the die space (Fig. 9-1).

Common sense safety precautions include: wearing protective gloves, not pushing a set of dies that requires more force to push into place than is available from the person doing the work, wearing safety shoes, and ensuring that the floor is not slippery. Care must be taken to keep fingers from getting caught between the upper and lower dies as they are being pushed

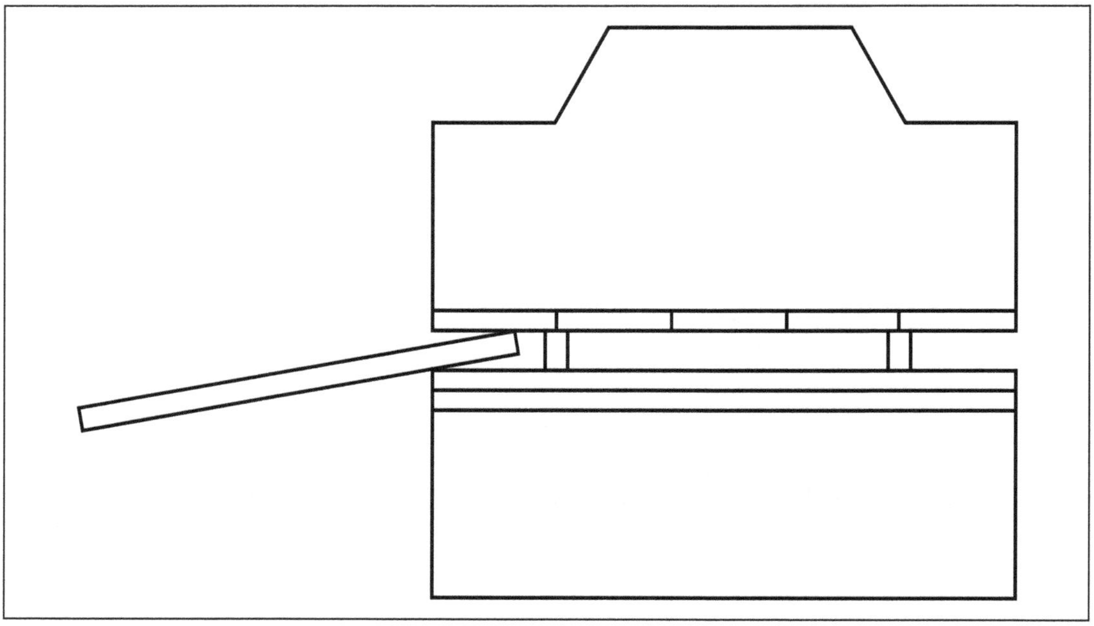

Fig. 9-1. An upper die shown being installed into the press brake from the end of the machine.

into the die area. After the upper die is located with the tongue in the die slot of the ram nose, and the ram does not have to be adjusted to obtain the proper clearance, any operating switch allowing the press brake to be cycled must be turned off. It is also important to follow similar safety steps when the tooling is removed from the machine. After the dies are securely clamped in the ram and filler block, the dies are aligned, and a test bend is made to obtain a correct angle, the gages may have to be set. If a gage has to be adjusted in the rear of the machine, make sure that the safety on/off switch is turned off so that the ram cannot be activated by a helper or another person. It cannot be emphasized too much: hands must not be placed in the die area. As was discussed in Chapter 7, the adjustment of all manual back gaging is done by the operator from the rear of the press brake. **Never reach through the die area to make an adjustment**.

When a part is being formed using a press brake, the part being bent will normally move upwards during the forming stroke; rarely will the part in the front move downward. If the blank is lying in a horizontal plane, unless the flange on the operator's side is very small, the operator will have to assist the material by hand to keep the unformed leg from distorting due to its own weight. Additional support provided by a helper during the forming stroke will minimize backbend and bellying in the part. (Backbend is an undesirable bend line located at the front corner of the vee die opening caused by the unsupported weight of the material. Bellying is a distortion in the middle of a long wide sheet that is supported at each end during the forming stroke. The air pressure of the unsupported middle results in a conical shaped part).

In order to prevent an injury to the operator's thumb, the hands supporting the material must be kept flat under the front edge of the part. Thumbs should never be hooked over the top of the part. As the material moves upward, hands and a thumb on top will not keep up with the motion of the material, and the operator's thumbs may be cut as they are stripped away from the part. Most operators learn this after the first bend. If the front leg of the part is more than 6" long, operator assistance may be required. For some parts that are very flimsy because of cutouts, notches, or many holes, a front leg as short as 4" may require assistance during the forming cycle.

Since operator training is no longer the lengthy training period or apprenticeship that was demanded for many years, many safety accessories have been designed and tried, with mixed results, in an effort to provide the operator with what is considered a safe operating environment. All devices that have been used have some limitations. One device will not work for all press brake bending operations.

One of the biggest problems is actually the operator. Some people possess a certain mentality that they feel that they know how to do things safely and do not need any type of safety device. Common examples include the individuals who resist wearing seat belts, motorcyclists who refuse to wear a helmet, and drivers who disobey the speed limits. In the factory, many operators do not use the safety devices provided. The only answer is for management to provide strict supervision with authority to ensure that safe operating procedures are used. Even under the best of circumstances, some operators will not follow the rules, and thereby expose themselves to serious injury.

B) Machine Safety

One of the first safety devices used on mechanical press brakes was a moveable pin that, if left out of its locking hole, would allow the treadle along the front of the bed to be pushed down and activate the ram motion. When the treadle was released, the ram motion stopped. Just as in an automobile, the clutch and brake had to be maintained to provide the expected machine reliability. When parts were not being run, the operator could push the locking pin into a hole, preventing the ram from being activated even if someone stepped on the treadle (Fig. 9-2). When electric foot switches were incorporated to activate the ram, simple push button "on/off" switches were used to deactivate the circuit. The ram elevation buttons were

Fig. 9-2. A lockout pin used on a mechanical press brake to keep the treadle from being activated is shown. (Photo courtesy of Cincinnati Incorporated)

still active to allow the tooling to be removed and new tooling to be installed, aligned, and clamped into place. In the 1920s, The Cincinnati Shaper Company painted a bright red stripe across the nose of the ram to visually alert the operator when the ram was in motion. Operator safety in those years was taught by requiring that all new operators serve an apprenticeship. During the apprenticeship, they were not allowed to run the press brake without a well-experienced operator being present.

C) Palm Buttons

One of the first exterior safety devices used on press brakes was the development of the palm button (Fig. 9-3). This option could only be used on press brakes that had an air-electric clutch (a press brake operated by foot switch, not a treadle). The first palm buttons were installed on a press brake in the 1940s. The main reason that palm buttons were used was for hole punching and some small press operations that were carried out on some press brakes. Since the parts that were punched were often narrow and did not have to be held by the operator, palm button operation required the operator to activate the ram by pressing the two palm buttons down, one with each hand. Palm buttons cannot be used if the parts being formed have to be assisted by the operator during the forming cycle.

Original palm button designs were soon modified to keep operators from trying to increase production at the expense of their safety. The first palm button stations were made as two separate boxes to allow the operator to locate each one at the most advantageous position since each job was different. It was soon found that some operators put both boxes next to each other, and cycled the ram by using one hand to activate both palm buttons at the same time while leaving the other hand free to hold on to the part. This was corrected by placing both buttons in one box with a reasonable distance between each palm button. Some operators with long arms tried to span the distance between the palm buttons in order to activate the ram while still holding on to the part. Others devised boards to activate both buttons. This created a need to put individual guards around each button to minimize this abuse.

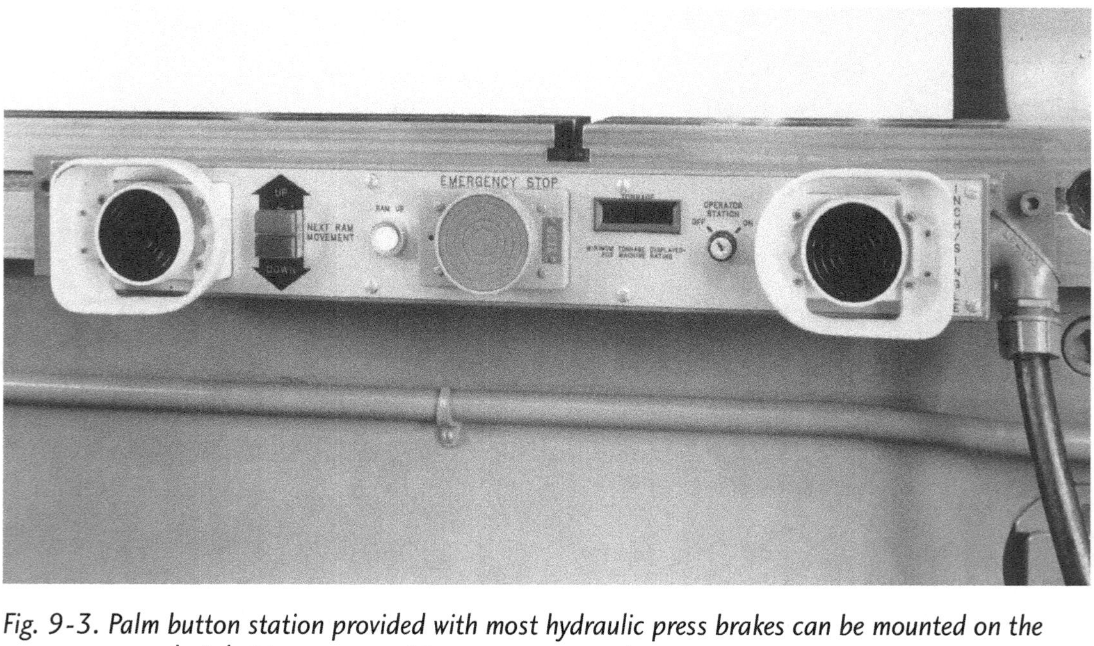

Fig. 9-3. Palm button station provided with most hydraulic press brakes can be mounted on the ram or on a pedestal. (Photo courtesy of Cincinnati Incorporated)

Some longer parts, or parts that required two operations in dies that run side by side, needed two operator stations. The press brake was then equipped with two palm button stations that required all four buttons to be activated by both operators. When the press brake was to be used for other work that needed only one operator, either one of the palm button stations or both stations could be turned off by a simple on/off switch built into the palm button electrical box. To provide additional "supervisor" control, the on/off switch had a key lock control added. Due to occasional operator negligence, lights were added to the palm button box to visually show if the palm buttons were active or shut off.

By the late 1950s, palm buttons were included as a standard feature on hydraulic press brakes to assist in machine setup. The reason is that a hydraulic press brake does not have a ram elevating adjustment like those found on mechanical press brakes (which requires the operator to move to one side of the ram and reach up to adjust the ram setting). A palm button station, mounted on the ram or separate stand, serves the same purpose of requiring the operator to use their hands to activate the ram during setup. Within several years, the emergency stop button that was included on all hydraulic press brakes was moved to a location in the palm button box and located between the two palm buttons. An "emergency stop" button will immediately stop ram movement and power on a hydraulic press brake. Mechanical press brakes may or may not have an "emergency" stop, depending on when it was built, and whether it has a linkage type control or an "air-electric" clutch control. Shutting off the power on a mechanical press brake cannot stop the flywheel motion. Management has to ensure that the operator has sufficient training to know what safety locks are on their press brake and how to use them.

The development problems of palm buttons over a 40 year period emphasize the need for proper operator training, and the continual need for management to demand that operators practice good safety procedures and not try to "beat the system."

D) Other Exterior Safety Devices

Over the years, a variety of safety devices have been used on press brakes, with each method having unique limitations.

1) Moveable Drop Gate Guards

Moveable drop gate guards, which moved up and down during the forming stroke, were attempted for several years. Since the designs did not allow for the longer lengths needed for many parts, the guards (or gates, as some called them) would sag and not move fast enough—they were often destroyed by the part that was being formed. The guards did not allow sufficient clearance for many acute angle parts or for some parts that had previous bends. This design comes back every few years, but has never had good success as a general purpose press brake guard due to its many limitations.

2) Pullbacks/Restraints

Pullbacks are ropes or cables that attach to the operator's hands using "wristlets," with the other end of the rope fastened to a mechanical "caming" mechanism that is attached to the ram (Fig. 9-4). The ropes are adjusted by a third party in such a manner that the operator can hold on to the edge of the part and assist in the forming cycle, but is restricted from reaching any part of his hand into the forming area as the dies are closing. A variation of this method is a **restraint**. A restraint will not allow the operator's hands to enter the die area at any time while they are being used (Fig. 9-5). For some users, these methods have worked very well.

Fig. 9-4. Pullbacks showing cables attached to the wristlets the operator uses. (Photo courtesy of *Cincinnati Incorporated*)

Fig. 9-5. Restraints are being used to keep the operator and helper a specific distance away from the point of operation. (Photo courtesy of Cincinnati Incorporated)

For other users, it has created a sense of false security. Every time an operator changes, dies are changed, or the forming method is modified, the setting of the ropes (cables) must be rechecked or changed. Often, the operators are careless or assume that the ropes are adjusted correctly, which may or may not be true. This has resulted in serious injury to the unsuspecting operator.

These methods can work, but they do require a great deal of proper supervision. Some operators do not like pullbacks or restraints because they feel they are "tied" to the location, which may limit them in handling the blank or the formed part. Pullbacks and restraint safeguarding methods are normally used for a specific job or family of jobs, and are not considered useful for general-purpose press brake work.

3) Fixed Guards

Fixed guards can be used for some jobs to keep the operator from inadvertently entering the unused section of the die (Fig. 9-6). For example, if an operator and assistant are needed to move a long piece incrementally through the die area in order to bump form a radius, the location of each bend is often aligned to scribed lines marked on the part being formed. The part is located by the lines that are visible, and located so that the top die will hit the proper line on each end of the part. To locate the part, both individuals will stand at either end of the part with their hands and heads fairly close to the die area. A simple, clear, plastic barrier, bolted to the front of the bed, will allow the lines to be seen and positioned, but will keep all hands and other parts of the body from accidentally moving into the die area. Some dies lend themselves to fixed guards, but each setup will have some limitation. General-purpose fixed guards are not possible without constant setup, and must be reviewed to ensure that they are really helping operator safety.

Fig. 9-6. End guards mounted on the bed. Flat sections of similar guards can be mounted along the front of the bed to help protect the unused sections of full length dies. (Photo courtesy of Cincinnati Incorporated)

4) Warning Signs and Safety Manuals

Warning signs and safety manuals are normally supplied with new equipment (Fig. 9-7). If the press brake(s) that are currently being used do not have any safety signs or other pertinent information available, management has an obligation to contact the original press brake manufacturer to obtain the latest current information. Simple warnings such as **Do Not Put Hands in the Die Area** are very important to have posted on the front of the machine (Fig. 9-8).

The Occupational Safety and Health Administrative (OSHA) directive concerning press brake guarding is found in Sub-Part "O," Section 1910.212. The ANSI B11.3 Press Brake Safety Standard (American National Standards Institute, Inc.) is available in booklet form. The write-up goes into great detail explaining terminology, giving simple examples, and providing sketches of what the different guards look like. If your facility is ever visited by an OSHA inspector, having the proper warning signs, a copy of the B-11.3 booklet available, safety guards in place, and the required safety devices being used will minimize any inspection concerns. The address of the American National Standards Institute, Inc. is 1819 L Street NW, Washington, DC 20036. Some additional information is also available from the National Safety Council, 1121 Spring Lake Drive, Itasca, IL 60143-3201.

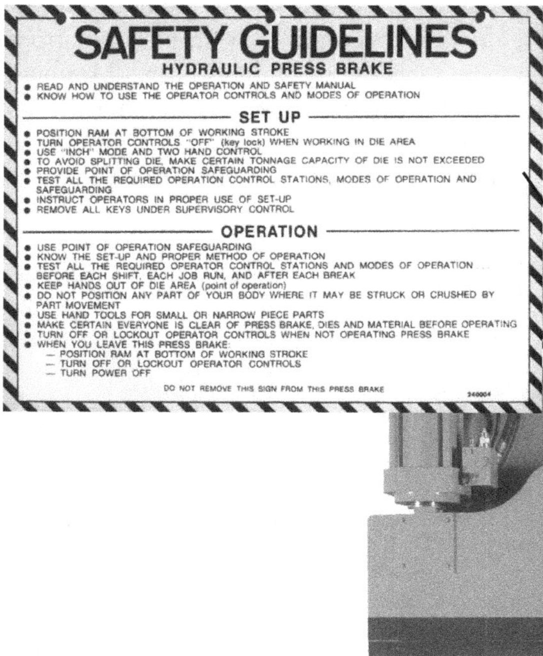

Fig. 9-7. Example of a warning sign that may be placed on the front of the press brake.

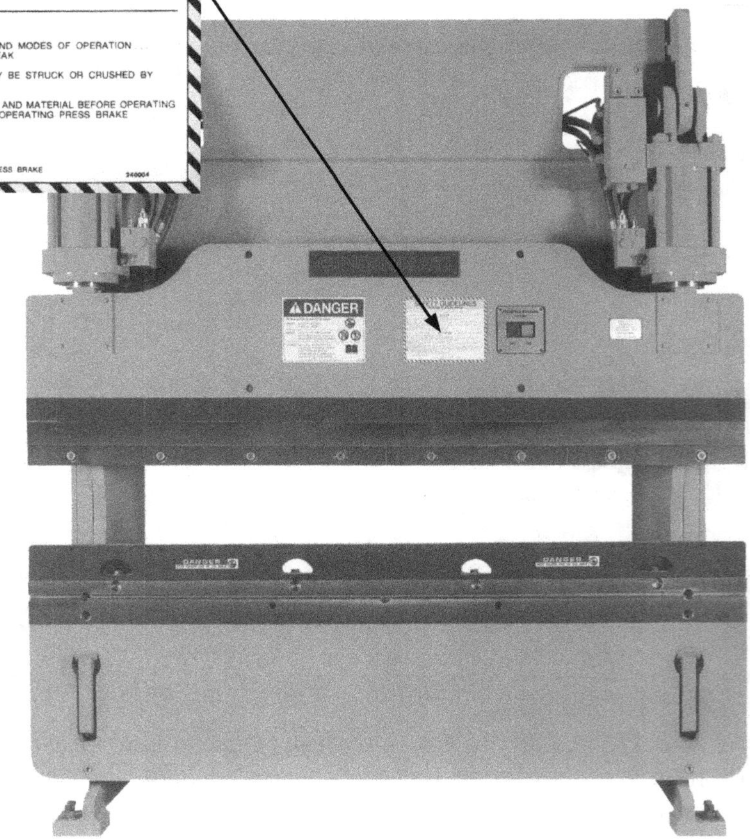

Fig. 9-8. Illustration of how warning signs might be located on a press brake. (Photo courtesy of Cincinnati Incorporated)

5) Operator Training

As previously stated, operator training is mandatory. If your facility does not have the capability to properly train an operator to know how to run the equipment and to operate it safely, contact the press brake manufacturer for their recommendations. In some cases, they can provide proper training either at their facility or at your plant. An operator who knows how to properly operate a press brake, set up the tooling, and what is required for his or her own safety is a very valuable employee.

6) Hand Tools

Hand tools have some merit for removing certain parts after they have been punched or formed. In some cases, hand tools can be used to load the part into the die area. Normally, only small or narrow parts fall into this category. A review of each part must be made to determine if hand tools are practical, and what design should be selected (Fig 9-9).

7) Work Zone

The front of the press brake should be considered a **work zone**. It is not a place to stand around with friends and discuss the latest sports event. If the floor in the front of the press brake is painted a noticeable color or has warning stripes, the supervisor can easily see if that area is being used for machine setup or for making parts. Any other person should not be in that area. Accidents have occurred when the press brake is under power and unauthorized people walk up to the machine to talk to the operator while it is being run. While the press brake is running, nobody but the operator should enter the work zone unless they have a good reason and the operator is aware of their presence.

Fig. 9-9. Hand tools being used to load a narrow part onto the top of a special channel die. (Photo courtesy of Cincinnati Incorporated)

E) Presence Sensing Device (PSD)

During the 1960s, several companies experimented with **light beam sensing devices** (light curtains). These were adapted as a method of providing operator protection when forming many varieties of parts on a press brake. Reliability of the systems was suspect since the refined electronics available today had not yet been invented. By 1971, a presence sensing device (PSD) became available in the United States which performed well and was safe to use on press brakes (Fig. 9-10). That was the about the same time that OSHA was becoming law. Many companies started to investigate what press brake guards were available, and what might have to be done to comply with the new law. New designs and manufacturers of PSD units became available in the 1980s, along with lawsuits from operators who had accidents using press brakes that often had no safeguarding. It soon became apparent that the PSD (normally a light beam safety device) provided operator safety for many of the various parts that can be formed using a press brake. If a PSD device is being used, it should be maintained and properly set up.

One of the major limitations of a PSD is that it cannot be used for all parts. Some complicated parts that have flanges on all four sides, some parts that are bump formed into a radius, special double deck type dies, and press brakes running in tandem (arranged to run side by side in order to obtain a long forming surface) are typical of tooling that do not work well with a standard PSD unit. As a result, some PSD units are purchased with an on/off switch (Fig. 9-11). If this option is supplied, it must only be used with strict supervisory control, and the PSD must be turned back on after the special job has been completed. A PSD unit must be installed using electrical circuits supplied by the press brake manufacturer as well as the

Fig. 9-10. Part being located on the lower die when presence sensing device (PSD) is being used.
(Photo courtesy of Cincinnati Incorporated)

Fig. 9-11. Part in the formed position with the PSD turned off. The operator is supporting the workpiece during the forming operation, and the "guard-off" warning light directly in front of the operator alerts him to this condition. (*Photo courtesy of Cincinnati Incorporated*)

PSD builder. There are so many machine PSD unit combinations that instructions cannot be supplied for each of them in this book.

Important Reminder: Providing for operator safety is the responsibility of the employer or company that owns and is using the press brake. Only the owner of the press brake can determine what safety options must be used to provide for their operator's safety. See the information concerning OSHA and ANSI B11.3 in Section 4D of this chapter. It is also the responsibility of the personnel using the equipment to follow all of the procedures established, and to use the equipment provided for their safety.

Chapter 10

Minimizing Die Marks on the Material

What can be done to eliminate the marking caused by vee dies when forming sheet metal in a press brake? This question is heard daily in shops that process parts that require a quality finish, especially "show" parts that are visible after final assembly. Polishing to remove die marks is expensive; it adds processing steps, time, and labor to the part cost. The following is a brief review of various press brake tooling modifications that will minimize and sometimes eliminate die markings on sheet metal parts.

A) What Causes Marking?

Simple vee dies are normally used to bend light sheet metal parts (10 gauge or thinner). The recommended lower die will serve as a general-purpose die, which means that the lead-in radius at the top of the vee die will be 0.030", 0.060", or 0.125". To make a 90° bend, the flat, unformed material must span the vee opening, including the corner lead-in radii, and be wide enough that the edges of the material do not move into the die opening during the forming operation. During the forming cycle, the material is pushed into the vee die cavity. The only contact that the material should have with the lower die is at the corner radius. The forming pressure (measured in tons/foot)

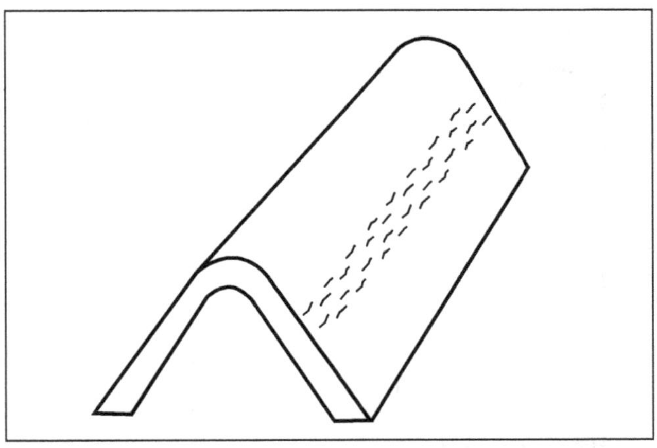

Fig. 10-1. Typical burnish marks often found on the outside surface of a simple 90° bend. The marks are caused by the pressure of the material moving across the corners of the vee die opening as it is being formed.

causes a burnish mark to be made where the bottom of the material (or outside surface of the part) touches the corner radii as it moves into the vee die (Fig. 10-1). The thicker the material and the smaller the corner radii, the more severe the die marks. The greater the distance the part is formed into the lower die to obtain the required angle, the longer the burnish mark. Air bending, which does not allow the part to touch the bottom of the lower die, limits the die marking to the contact of the part with the corner radii. Bottoming the part will not only produce marking from the corner radii, but also from any contact points where the outside of the part actually contacts the sloping surfaces of the vee die cavity.

B) Basic Steps to Minimize Marking

Completely eliminating marking is difficult, but a variety of steps can be taken to minimize die marking and produce acceptable parts without further finishing operations.

1) Provide a Good Finish on the Lower Die

To reduce the marking caused by normal machined surface finishes, the vee die opening and lead-in radii may have to be polished. The standard machined finish that is supplied on most commercial press brake dies is a 90 microinch or higher finish. A microinch finish is a measurement of the surface roughness using comparison templates or special "surface finish" inspection machines (Fig. 10-2). A 16-microinch finish, which is very smooth, can be achieved by buffing or hand polishing. Grinding is an alternative in some cases.

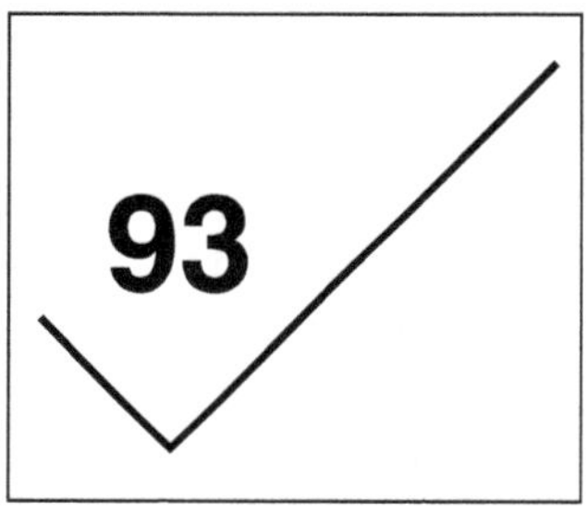

Fig. 10-2. Symbol showing the microinch finish required on the machined surface of a press brake die. A 93 microinch finish is a very good machined finish obtained by using standard "high speed" tooling. A ground finish may have a 16 or lower microinch finish.

2) Lead in Radii

The first step is to increase the corner lead in radii to 1.4 to 2 times the metal thickness (Fig. 10-3). This will increase the material contact with the die radii, and decrease the unit pressure by spreading the load over a larger area. The reduction in load reduces the depth of a possible die mark and can minimize or even eliminate a polishing operation. Painting the final part often covers the remaining burnish so that it is not visible. When forming stainless steel, the time required for a final "polish for appearance" is greatly reduced.

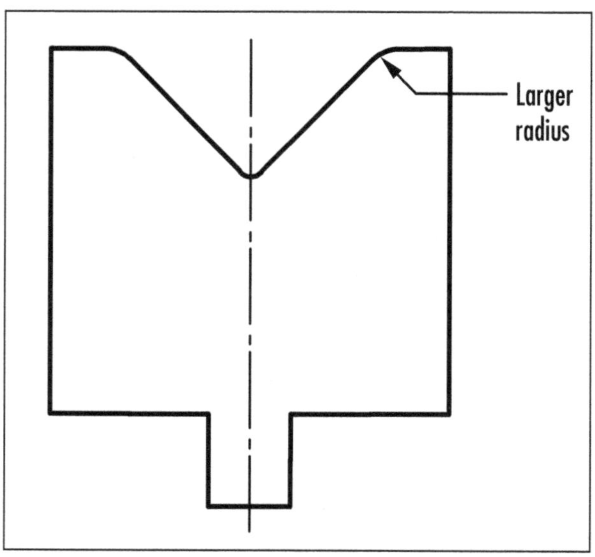

Fig. 10-3. The lead-in radii of the vee opening can be increased to minimize any friction from the material moving over the corners during the forming process.

C) Harden Radii

The corner radii can have its normal hardness increased by flame hardening (Fig. 10-4). For some materials, such as stainless steel, this step can be taken to minimize marks only after the corner lead-in radii have been increased to an acceptable size. Hardening the radii reduces the friction as the material moves downward into the die cavity. Since most press brake dies, hardened to 250 to 300 Brinell prior to initial machining, can be easily scratched by the burrs on the corners of the parts being formed, flame hardening the lead-in radii eliminates this concern.

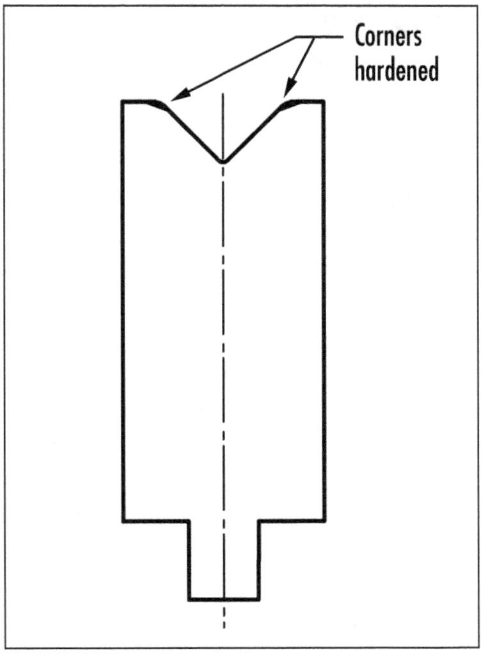

Fig. 10-4. The corner radii can be flame hardened to about 55 Rockwell C. Hardening helps to minimize friction.

Not having harder than standard die steel lead-in radii magnifies the problem when forming stainless steel. Commercial grade stainless steels have a higher tensile strength than mild steel, but both have comparable yield strengths prior to forming. The initial bend made in a stainless steel part often results in an open bend. The problem occurs if the part is rehit. Because stainless steel springs back more than mild steel, and the angle machined on the press brake dies may not allow for the additional springback, the operator may rehit the part. To accomplish this, the part is placed back on top of the lower die and is centered on the lead-in radii of the vee opening. The operator will reset the ram elevation and rehit the part, attempting to correct the initial angle. In some cases, the press brake being used has limited tonnage to make the desired bend. The operator will start making multiple hits on the part attempting to obtain the correct angle. In some cases, this will work. The problem is that stainless work hardens. Work hardening means that when the stainless is bent, the area that makes up the inside radius will have its tensile (and often yield) strength change. The tensile strength of the material will change somewhat, but the yield strength will increase from about 35,000 PSI to somewhere in the 70,000 PSI range. Since the forming load reflects a relationship of tensile and yield strengths, if the part requires a restrike, additional forming tonnage is required (refer to Chapter 6-C).

When the flanges of a piece of stainless steel that are in contact with the corner lead-in radii of the vee die move down into the cavity, those burnish marks (or die marks) are work hardened. The surfaces of the stainless steel, in those areas of contact, actually become harder than the surface of the die steel. If the part is rehit in order to obtain the correct angle, the work hardened stripes of stainless (the burnished area) will cut into the radii and quickly cause an unacceptable wear pattern to develop in the die. The only solution is to harden the die radii to a surface hardness of over 50 Rockwell C. The only disadvantage to flame hardening is that the die can no longer be reconditioned to compensate for wear. Flame hardening will normally increase the life of a vee die by a factor of four compared to standard commercial heat treated die steel (refer to Chapter 2-E).

D) Use a One Piece Die

Since the lower die is normally the immediate source of marking, this die should be made in one piece whenever possible. Sectionalized tooling has been recommended by some suppliers to minimize setup time and reduce part cost. Cutting up tooling can also cause or increase marking on the outside surface of the formed part. When a part is formed on the press brake, the forming load will cause the bed and ram to deflect. The lower die will stay with the contour of the deflected bed top if the die is in one piece. If the die is sectioned, after a number of parts are formed, the lower sections may actually separate ever so slightly, which can cause a very small gap to occur between the sections (Fig. 10-5). There will be no mark where there is a gap between the die sections. This may not be objectionable, but the sharp corners at the end of each die section often create a deeper burnish mark that is more difficult to remove. A sectionalized upper die can cause similar marking; however, since the mark on the inside of the formed angle caused by the pressure of the nose of the upper die is seldom seen in a finished part, it is rarely a problem. Occasionally, if very thin material is formed using a bottoming or coining operation, the possible gap between the upper die sections will actually show through to the outside of the formed part in the form of a visual distortion. This distortion has also shown up when the part is coated with a high gloss paint prior to assembly or shipment.

E) Secondary Options

In addition to making changes in the design of the lower die, several other options might be used to minimize or eliminate marking. Some of these options have merit, while others have been tried but don't seem to work too well.

1) Chrome Plate the Lower Die

This option is effective when soft aluminum is being formed, but is rarely used otherwise; that is because the chrome is only 0.0015" thick and can easily be damaged. The edges of steel, or even burrs, often at the outside edges of punched holes, will damage or ruin the chrome plating as normal forming occurs. Hard chrome can only be applied as a thin coating

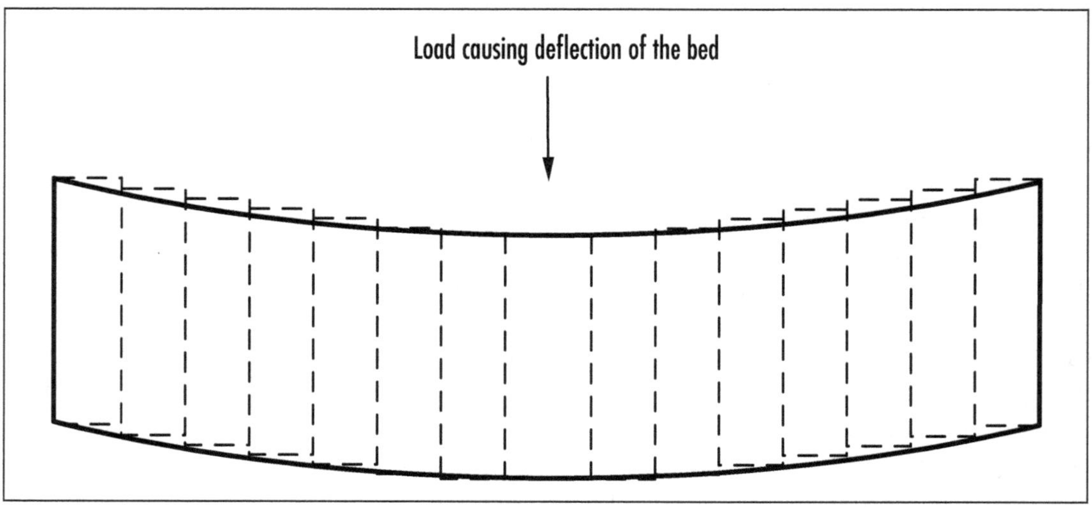

Fig. 10-5. A sectional lower die can cause marking of the formed part either by creating steps during any bed delection, or by the sections moving apart slightly, (also caused by bed deflection).

without being prone to peeling away from the base material. As a result, it can easily be scratched or damaged, even by handling the die in and out of die racks. Prior to chrome plating, the surface must be polished since every imperfection will be visible after the plating process.

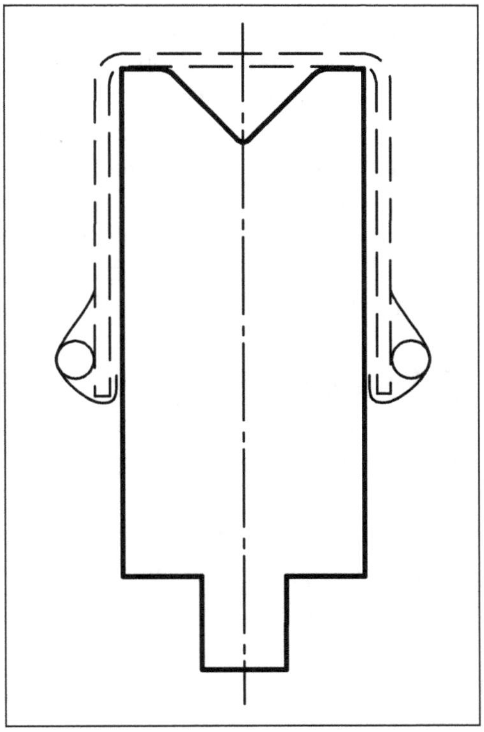

Fig. 10-6. Polyurethane strip covering the top of the vee die. The strip moves down into the vee opening during the forming stroke and is pulled back to the position shown by the front and back round weights when the formed part is removed.

2) Use Rubber or Polyurethane Strips

Thin strips of rubber or polyurethane are available that can be laid across the top of the vee die to protect the outside surface of the material being formed (Fig. 10-6). The strips can do a good job but have a limited life. In order to keep the pad in place, a $^1/_8$" diameter rod can be taped to each edge of the strip. When laid on top of the lower die, the weight of the rods will pull the pad down both at the front and in the rear of the die. This provides consistent location for the pad and also allows gaging of the part without interference.

3) Tape the Corners of the Vee Opening

Several companies market die tape, which adheres to the lead in radii of the lower die (Fig. 10-7). Many operators have used this tape with good success. Limited tape life is a shortcoming, but the process can solve a marking problem. After the tape has worn, it must be removed (often with some effort) and the die cleaned with a solvent prior to replacing with new tape. The cleaning process can expose the die surfaces to rust. Carefully follow the instructions supplied by the tape manufacturer.

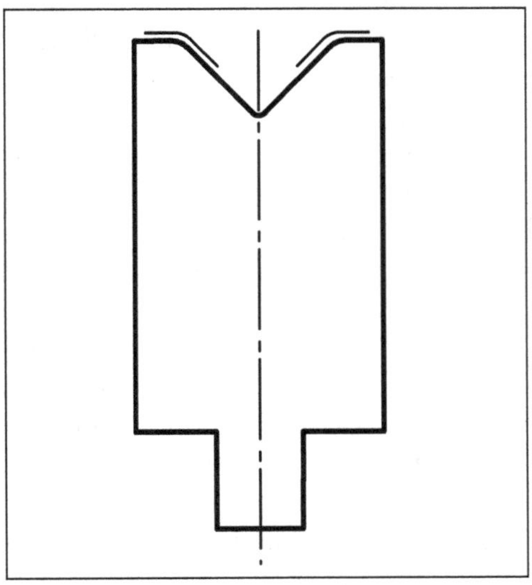

Fig. 10-7. Special die tape fastened to the vee die corners.

4) Cover the Lower Die with Plastic

Heavy duty plastic (about 0.007" to 0.010" thick) can be used with good success to eliminate die marks. The problem is that the plastic sheet must be indexed after each forming stroke. The pressure of the material being bent on the plastic during the forming stroke will cause the plastic to be damaged. Normally, the points of contact with the part will create a serrated line in the plastic (similar to those that separate sheets on a roll of paper towels) which appear after the forming stroke. To maintain a usable plastic surface, the sheet must be moved forward or backward 0.125" to 0.25". For low volume work this can be done, but generally it is impractical for normal production.

F) Special Situations

There are several specialized methods that offer varying degrees of success. They are typically used in low volume applications to solve specific problems.

1) Square or Rectangular Pads

Using the upper die to push the material down into a pad of rubber originated with some aircraft manufacturers in the 1930s. The idea was to reduce tooling costs and to eliminate die marking. The concept never became popular since the rubber could not generate sufficient pressure to form the complex shapes often required in the aircraft industry. In the 1960s, a new plastic called **polyurethane** was introduced into the United States from Germany. Although it has all of the characteristics of rubber, it has a much higher strength, and offers the ability to generate a much higher forming pressure than what is called “bumper grade” rubber.

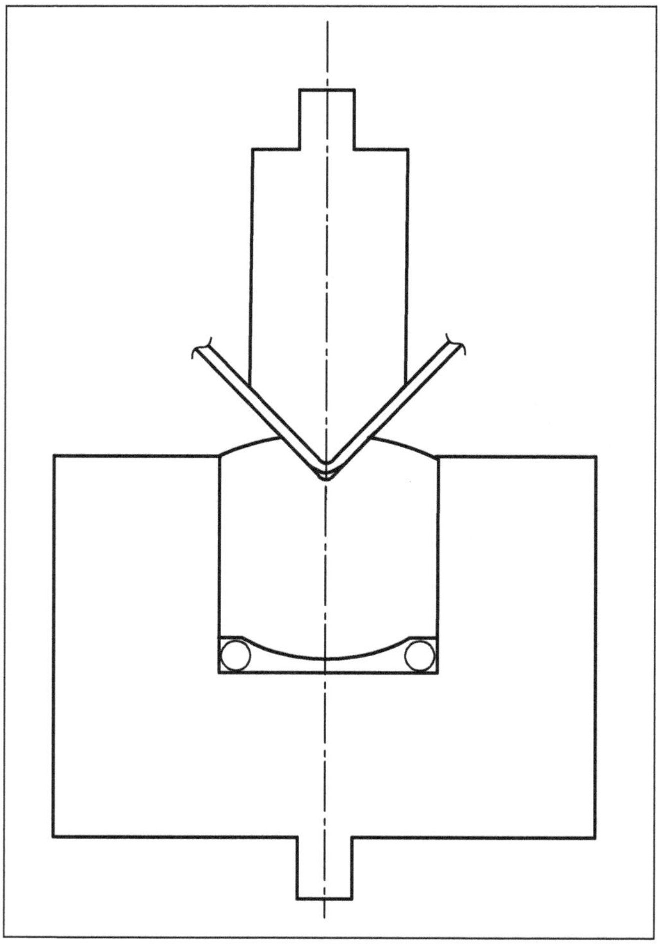

Fig. 10-8. Polyurethane pad used as a lower die in order to eliminate die marks.

The hardness of rubber is measured with a hardness scale called a durometer. Most rubber products used to form metal are limited to a practical hardness of 60 durometer on what is called the “A” scale. Polyurethane made into square or rectangular die pads starts with a hardness of 80 durometer on the “A” scale, and can be as hard as 90 durometer on the “D” scale. The actual shape of the pad can be any specified shape since the material is usually poured into a mold while in a liquid state.

When a pad is used to form parts, the pad must be placed into some type of holder in order to keep the material from mushrooming sideways under the forming load (Fig. 10-8). End plates are also needed to keep each end of the pad from elongating. This restriction of the pad ensures that an adequate amount of uniform forming pressure can be generated, allowing parts to be formed from thin materials. The best results, if polyurethane is used, are attained when the pad hardness is 90 durometer on the "A" scale when forming material to a 90° angle in material up to 0.125" thick. Refer to Fig. 10-9 for specific information about using polyurethane as a lower forming die, and to Chapter 24-G for a more compete discussion.

2) Triangular Pads

This polyurethane pad (Fig. 10-10) is made in a triangular shape and is designed to lay in a standard vee die opening. The vee opening is supposed to act as a retainer when the part is forced into the pad in order to make a 90° angle. For any one bend, the idea works well. The problem is that the pad is difficult to keep properly located in the vee die opening when more than one bend is to be made. Attempting to hold the pad in place with double-backed tape does not work too well, since the pad moves in all directions during the forming stroke and tends to pull the tape away from the die surface. As in the use of square pads, the forming tonnage will increase and must be considered in selecting a proper size press brake.

Polyurethane pads have been available for press brake use since the 1960s. Most of the pads are made using a hardness of 90 to 95 Durometer. Some successful applications have been attained despite many limitations.

The cost of the polyurethane is very high per stroke compared to conventional steel vee dies. While steel dies normally will provide hundreds of thousands of hits if used within their capacity, polyurethane pads have a long life at 40,000 strokes. Besides the pad cost, a holder, including end plates, must be provided. Small flanges do not form well but tend to dig into the pad, often damaging its top surface.

Gaging over the top of the pad is not too practical since the pad will move up during the forming stroke causing possible damage to fixed gages. Either special flip finger gages must be used, or the gage can be polyurethane strips fastened to the top of the holder that extends over the top of the pad.

Polyurethane tends to allow dirt and oil to embed into it's top surface which will cause marking if it becomes too abrasive. Some users lay a $^1/_8$" replaceable polyurethane strip on the top of the larger base in order to overcome this problem.

Polyurethane pads can produce a very sharp inside radius in the part but the forming tonnage will be at least bottoming type tonnage (approximately 3 to 5 times air bend tonnage) regardless of the radius on the upper die nose. If the upper die is full length, the tonnage will reflect the penetration that results from a full length part even if a much shorter length is being formed.

Simple vee bends or simple radius bends work best with pad type dies. Occasionally, a small channel shape can be made, but multiple bends such as offset shapes cannot be made. The pad cannot push the metal into the configuration of the upper die in order obtain a correct dimension since concentrated loading is not practical with polyurethane. In addition, a pad will not allow flanges to move into the area occupied by the pad or holder.

Fig. 10-9. Polyurethane pad considerations.

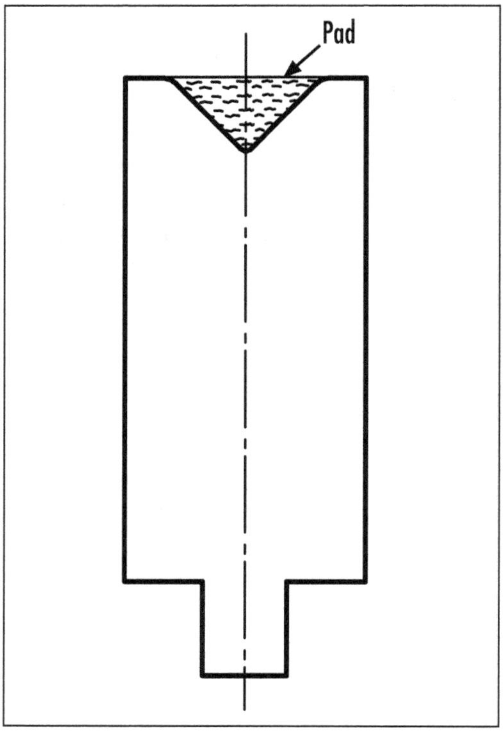

Fig. 10-10. Polyurethane pad in a vee die. A larger vee opening is needed to hold the trangular pad. The pad is often difficult to keep seated in the vee opening between strokes.

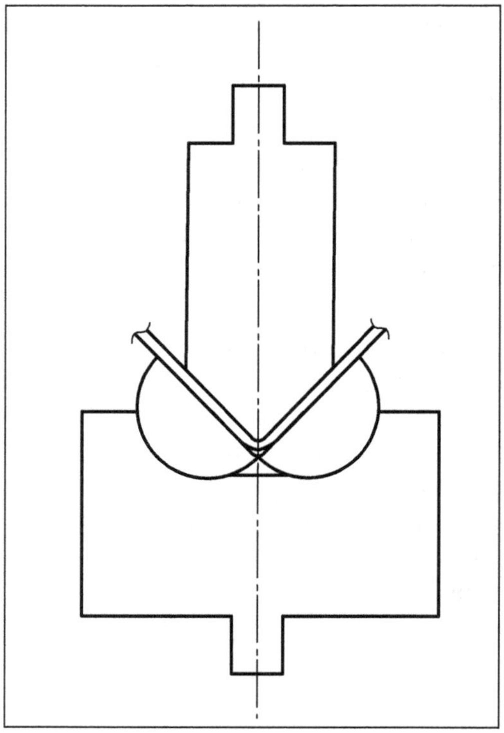

Fig. 10-11. Roller type die. The flat surface of the bars rotate from a horizontal position to the angle shown during the forming stroke. They return to the horizontal position while the part is being removed.

3) Polyurethane Lower Dies

Molded vee dies made from polyurethane are available in many shapes and sizes. For some jobs, they may offer the ideal answer. Limited life expectancy can increase part costs, thus preventing polyurethane lower dies from being an acceptable solution.

4) Roller-type Dies

A lower die consisting of a holder containing a semicircular bar in the front and a similar bar in the rear has been available since the 1940s. This type of die was originally designed to slow up the forming speed when the part was being formed using a mechanical press brake. It was found that die marking was minimal using this tool. Occasionally, some company will use this design for a specific part (Fig.10-11). The major problems limiting its use are the larger inside radius that must be expected, angular inconsistency due to the larger inside radius, minimum flange length possible since the rear leg must reach past the centerline of the rear roller, and the problem with back gaging over the rollers. Refer to Chapter 24-J for additional information.

5) Spraying Plastic Coating on the Material

Years ago, most highly polished stainless steel was covered with a piece of heavy duty paper glued to the polished side. This allowed the material to be handled throughout the plant operations—from shearing, punching or blanking, press brake forming, and final assembly—

before the paper was removed. The major problems that were encountered were the cost of applying the paper, plus the fact that the glue that was used "age hardened" and became very difficult to remove if the material had been in storage too long. As a result, in the 1980s, a change in methodology was made. The highly polished material was sprayed with a plastic coating on the polished side. The plastic accomplished the same results as the paper, except in the area of press brake forming. The paper was not compressible. The paper thickness was normally 0.007" to 0.010" thick. The plastic, ranging in thickness from 0.0015" to 0.005" thick, is compressible. When the parts are formed, the force needed to bend the metal often creates die marks through the plastic coating. If this is a problem, refer to one of the previous ideas obtain a solution.

Chapter 11

Offsets, Joggles, and "Z" Bends

A) Definitions

It is often necessary to offset a piece of material to provide clearances, proper shape, or structural strength. An offset consists of two reverse 90° bends. This provides two parallel legs held together with a web. If the two parallel legs are farther apart than five metal thicknesses, the web can be formed to a good 90° angle. This measurement is made by taking a combination square, placing the base plate on one of the flat surfaces of the part, and measuring the distance from that surface to the top surface of the parallel leg that has been offset (Fig. 11-1).

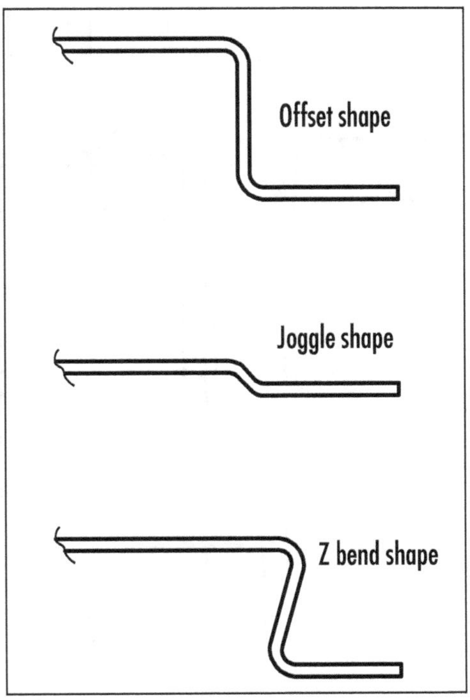

Fig. 11-1. Offset, joggle, and Z bend shapes.

When the parallel legs are less than five metal thicknesses apart, the web will be difficult to form to a 90° angle, and will often remain open several degrees. If the offset legs are closer together, the outside radius of one flange will blend into the inside radius of the reverse bend, with the resultant web angle approaching a 45° appearance (or a 135° formed angle). This type of shape is called a joggle. Normally a joggle is a type of offset with the parallel legs close together. In some cases, the open angle design requires the two parallel legs to have a greater distance to accommodate a specific design requirement. The shape is still called a joggle to indicate that the web angle is greater than 90°. Some parts are designed so that the web is actually overbent to an angle less than 90° (e.g., 60°). This might be done to provide clearance so that a mating part does not have interference with the inside radius at the web and flange. Because of its visual appearance, this shape is called a "Z" bend.

B) Forming the Offset in Two Separate Strokes

If the depth of an offset type part is sufficient to allow the shape to be made using two forming strokes, simple vee dies can often be used for the bending operation. If sufficient press brake tonnage is available, most operators will select bottoming dies to hold good parallel legs on the finished part. Note that after the first bend has been made, the part must be turned over to allow the second bend to be made (reference Fig. 7-12).

Care must be taken to ensure that the first leg, now located in a downward vertical position, does not hit the lower die, the filler block, or the top of the bed of the press brake (Fig. 11-2). Gaging the second bend also creates several concerns. If the gage fingers are to be used to locate the outside flange of the first bend, the fingers must be low enough to contact the material underneath the outside radius of the first bend. The gage fingers must then move out of the way during the forming stroke to allow the first flange to move up in an "arc" motion to complete the offset shape. (Review the initial discussion in Chapter 7-G concerning offset gaging.)

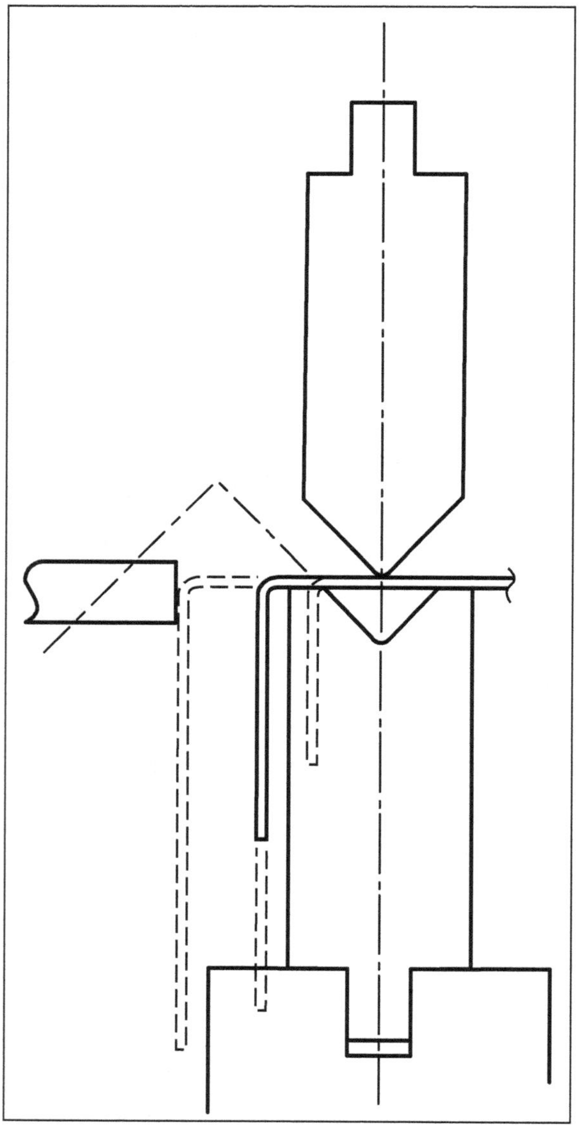

Fig. 11-2. Interference possibilities when making an offset in two strokes.

The information supplied in the following paragraphs is provided because the problem of gaging offset shapes made in two strokes has been very difficult for many operators to understand. The press brake options listed are not available for many users, but may be considered if new equipment is added in the future. All production personnel involved with this type of forming should be aware of the limitations and advantages of the different methods of forming the offset, joggle, or Z bend shapes since they are very common for most shops.

New NC-type gages may have a "Clamp/Retract" option that allows the upper die to move down until the nose radius touches the top surface of the material to be formed. At that point, the ram is programmed to stop. The gage finger that is used to gage the part then moves back out of the way, providing clearance for the vertical leg of the first

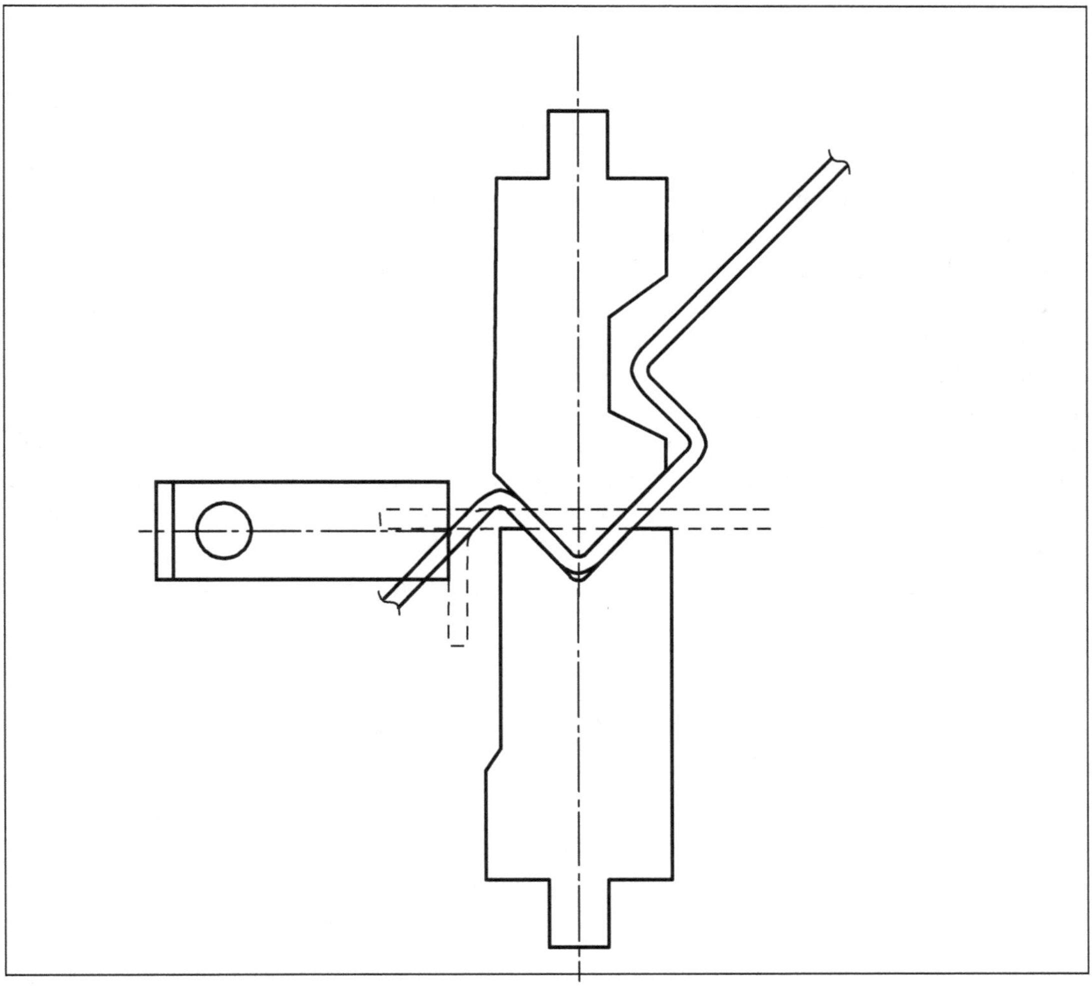

Fig. 11-3. Risks in gaging with a flip finger gage.

flange when it moves upward as the second bend is made. This option is not needed for simple vee bends where the gage finger is used to contact the edge of the material. It is a problem solver if the required gaging surface for a bend is located in the rear of the lower die at a point lower than the top surface of the vee opening.

When forming an offset, the first bend is turned over and the initially formed flange is facing down in the rear of the lower die. The NC gage fingers can be programmed to move down to obtain a gage surface on the vertical leg of the part if that feature is available on the gage. If the NC back gage only has a back and forth gage motion, the gage fingers used on the back gage must be about 1" high to allow the fingers to properly contact the part for both bends. If the gage fingers were not moved back out of the way when the second bend is made, the vertical leg of the part will move upwards in an arc like motion and will strike anything in its way. The "clamp/retract" feature allows the gage fingers to move back, thereby eliminating the problem.

Some users have tried "flip finger" gage fingers on the NC back gage with questionable results. The gage fingers often have to be located low enough that the part does not slip underneath the gage when the flat blank is being located on the top of the lower die. The working height of most commercial "flip finger" gage fingers is about one-half inch. When the part is turned over to form the second bend, unless the material is very thin, the outside radius requires that the NC gage be set lower in order to have a proper gage location to complete the part. If the back gage does not have any vertical programmable ability, the flip finger gage will require the first bend gage position to locate the material edge near the top of the one-half inch "flip finger" gage (Fig. 11-3). There is a high risk that the gage finger could inadvertently swing upwards during the gaging action, causing the part to move inwards, and creating a potential safety problem (reference Chapter 7-G).

If manual back gaging is to be used, the gage finger location must be between the lower die and the vertical leg of the first bend. This allows the leg to move up and away from the gage finger when the second bend of the offset is made. When sufficient space is available, commercial universal type gages can be selected. Some parts have the legs of the offset too close together for a gage finger to fit into the space between the die and vertical flange of the part. Special thin gage blocks can be mounted to the back of the lower die, or the die itself may be modified to provide an acceptable gage surface. The operator can then position the part with the previously formed flange located against the gage surface on the lower die. It should be pointed out that any gage location should only be about 0.25" high, located close to the radius of the first bend, and allow adequate clearance under the gage surface to compensate for any angular variation that occurred in the first bend.

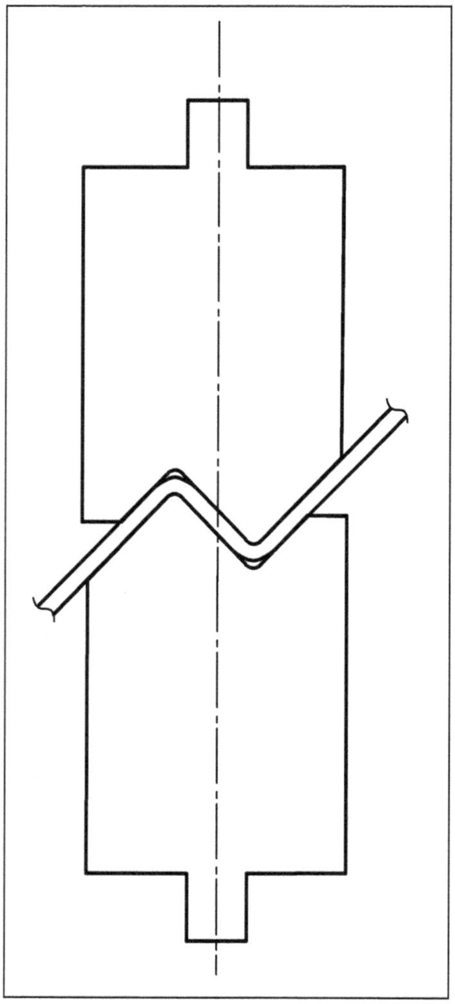

Fig. 11-4. Standard offset die.

C) Forming the Offset in One Stroke

Many offsets are designed with the web too small to allow the part to be formed in two strokes. The special die required to make the part is called an "offset" die. This includes both the upper and lower die needed to make the part. Many problems can be expected when offset dies are required. They include: a major increase in forming tonnage compared to making a simple vee bend, difficulty in holding both legs parallel and the web vertical, and obtaining the inside radius shown on the drawing (Fig. 11-4).

If possible, offset dies are initially made by machining one bar long enough to cut the bar in half, providing both the upper die and lower die. The bar is machined with the web of the part made on a 45° angle. The centerline of the die is located at the center point of the web dimension. After the die has been machined, it is cut in half. One-half of the die bar die is turned 180° to make the upper die. The other section will be used for the lower die. The die chosen to be the lower die is located in the filler block with the vee die opening closest to the operator. The rear of the die will have a surface that slopes down away from the operator at a 45° angle. The upper die is installed having a sloping side toward the operator. Its vee area will be at the rear. The purpose is to allow the part flange extending out toward the operator to move upward during the forming stroke. Like any type of standard commercial die, the operator may have to align the lower die with the upper die to ensure metal thickness clearance between the upper and lower dies.

An attempt must be made to allow the longer flange of the blank to extend toward the operator so that the part does not have a tendency to fall down toward the rear of the press brake. The common sense rule is to keep most of the part weight on the operator side of the tooling. If the weight were in the rear, the operator would have to "hold on" to the part to maintain gaging and location, instead of simply keeping his hands under the flange in order to lift it upwards during the forming cycle. That may present a safety concern unless special operator safety options are provided.

Some parts require the dies to be turned around to allow the flange toward the operator to move downward. If the flange does move downward with the operator trying to initially hold the unsupported flange, the forming stroke should be set for a very slow speed. Only a skilled operator should be expected to do this type of work.

If you examine a set of offset dies from the end of the press brake after they have been installed, you will observe that the vee opening on the lower die is considerably larger than a vee die opening normally used for the material thickness being formed. The same vee opening is present in the rear of the upper die. The same rule applies here as when mild steel is formed into a conventional vee (refer to Chapter 3-A.). The inside radius is a function of the vee die opening being used. For offset dies, the vee opening that affects the initial inside radius of the part can be measured by using the distance from the corner of the lower die opening, and measuring to the center of the sloping leg on the other side of the vee shaped opening. This should be the centerline of the die bar (Fig. 11-5). The opening for the upper die will be

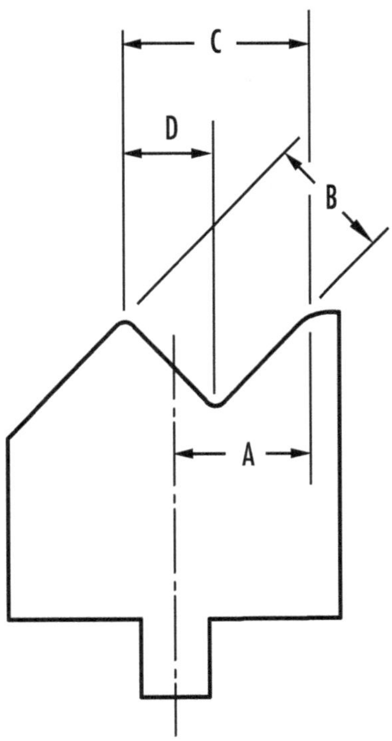

A - Vee die opening determines the air bend inside radius.

B - Dimension of formed offset.

C - Common measurement for the vee die opening. Note the top surface of the die must be on the same plane.

D - Forming noses of the upper and lower dies closer than the depth of the offset will produce a "wire draw" effect if the offset is greater than $^1/_2$".

Fig. 11-5. Important dimensions on a one-hit offset die.

identical. Estimate the approximate inside radius that will form on each bend of the offset by multiplying the measured distance times 0.156. The answer will always be much larger than one metal thickness.

When the dies are being observed from the end, one can see that the two nose radii used for forming the offset shape are closer together than the length of the middle sloping center length of the upper and lower dies. This means that, when the part is formed, the two forming radii will cause the material to move over their upper and lower corners in order to have sufficient material to obtain the offset dimensions. If this **drawing** of the material is more than two or three material thicknesses, a forming problem can occur that will affect the shape of the part. This problem is called **wire draw.**

Many commercial offset dies are sold without testing for shape or required forming tonnage. The dies are made by machining the offset shape with the depth of offset as shown on the drawing. Some suppliers call this die a general purpose tool. This means that: if the supplier cuts the die to a shape, but takes no responsibility as to the part shape, the depth of the formed offset, or what forming tonnage will be required to make a part; the cost of the die can be greatly reduced. One of the decisions made by the supplier will be to make the forming corners of each die with a sharper radius (i.e., $^{1}/_{32}$"). Small radii on the die magnify the problems of the larger inside part radii needing additional forming tonnage to hold parallel legs and the web having a "Z" shape. In addition, maintaining the depth of offset and parallel legs requires some compromise and skill by the press brake operator.

Many purchasers assume the sharp radii on the die will be the actual radius on the part. As pointed out previously, the engineers who design parts in sheet metal normally have little or no training in what is needed to form it. They magnify the forming problems by calling for inside radii of metal thickness or even sharper on their drawings, even when special tooling is required to make the part. The sharper the die corners, the higher the risk of **wire drawing** affecting the part.

Here is an easy way to demonstrate how wire draw works. Take a piece of 8.5" × 11" paper and hold it securely at each 8.5" end. Find a file cabinet or similar object that has a relatively sharp straight edge. If you take the paper and start pulling it over the edge of the sharp radius along the 11" length, when the paper is released, it will curl up into a tube shape (Fig. 11-6). If you had enough strength and did the same test with a piece of metal, the metal would also curl

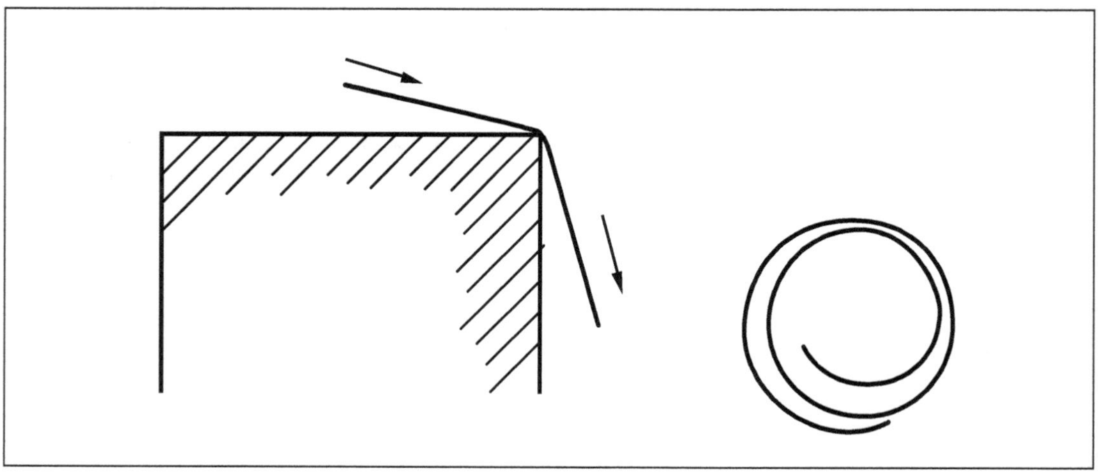

Fig. 11-6. Pulling paper over a sharp corner will produce wire draw. The paper will roll up into a tube shape when released.

into some type of radiused shape. This distortion is called **wire draw**. The part that is formed in an offset die with a too-sharp forming radius—and where the material is required to move over each radius similar to the paper over the sharp edge—will produce a part having some "wire draw" in the web area.

A larger-than-metal-thickness inside radius, caused by each vee opening in the upper and lower dies, will occur in each of the bends. If the die radius is sharper than the radius formed in the part, the following distortions can be expected: the offset will have a depth deeper than what is machined on the die (the result of the larger radii), the web angle will be less than 90° (or appear to be "Z" shaped), and the inside radii will be elliptical. The operator will inspect the part and see that it is "not acceptable," and then do what all operators do in that case—they lower the ram elevation of the press brake and rehit the part. The additional tonnage will reform the inside radii into a unique shape, change the depth of the offset slightly, and perhaps make a part that will meet the outside range of the part tolerance. The web will still be overbent slightly. In some cases, where the operator is using a hydraulic press brake, the maximum tonnage of the machine can limit the tons/ft needed to form a good part. Management then needs to decide how to modify the dies, and possibly the part dimensions, to allow the parts to be made within their press brake capacity.

The simplest modification method is to increase the die radii to at least one metal thickness. This will reduce the "wire draw" effect and change how the formed radius fits into the die, which will also reduce the depth of the formed offset. This change may also reduce the necessary forming tonnage to the point that the part can be made to be acceptable, i.e., within the press brake tonnage available. The best solution is to solve the problem before it starts. If the part is properly designed, having radii that will not lead to wire draw concerns, or require forming tonnages that exceed the press brake capacity available, it is probably advisable to purchase tooling that is made, tested, and guaranteed by the supplier. Although the delivery time for this type of tooling is not as fast as the dies cut to an arbitrary shape, there is the confidence that when the tooling is received, production of the parts will have minimal problems with scheduling, part quality, or assembly.

In order to control the consistency of the part depth and parallelism of the legs of the offset, the contact points of the lower front radius and the top rear radius on the straight legs of the part must be at least 2.5 times the metal thickness past the tangent point of each outside radius. This provides control of the angle the same way a simple vee die requires the same type of flat to control the formed angle (reference Chapter 3-B). To obtain a square offset, this means that its "depth" will exceed 5 metal thicknesses. In order to obtain the 2.5 metal thickness flat required to control the bend, in cases where the offset is small (often a joggle shape), the forming surfaces are rotated so that instead of the flat surface of the dies being at a 45° angle, the front forming surface is changed to a 15° to 30° angle. This requires more setup time by the operator to ensure that sufficient clearance is present in the web area, to allow the part to form with minimal "kicking" or movement of the dies front to back during the forming stroke.

D) Estimating the Forming Tonnage if Offset Dies are Used

The forming tonnage required to form a good, consistent offset is a major concern. Many companies seek a magic formula to solve this problem. Back in the 1950s, the chief tooling engineer for a large press brake manufacturer prepared a paper showing that if the material thickness of the mild steel to be formed was multiplied times a factor, the resultant answer would give the required forming tonnage in tons/ft. In some cases, the method worked; in other cases, the answer was not even close. The factors were numbers with nice round figures like 300, 400, or 600. Regardless of the fact that many of the answers gave tonnages that were not useable, it was a "magic" answer that sales people were looking for. As a result, copies of the factor list quickly spread all over the country, and are even published by some tooling manufacturers today. The major reason why this type of tonnage calculation seemed to work was that all of the smaller size press brakes made at that time were mechanical. Those press brakes had a reputation that they could be overloaded without damaging the machine. There were often disastrous results: broken housings, damaged rams, and twisted drives were common problems. The overloads continued, however; and, as a result, the tonnage factor formula survived.

During the 1970s, hydraulic press brakes became popular because they cannot be overloaded. If the forming load requires too much tonnage, the machine simply reverses direction and refuses to complete the forming. Obtaining a correct tonnage estimate now became an important consideration, especially if the user was contemplating buying a new press brake. The machine buyers needed assurance that the press brake they purchased would make their parts. The best way to have that assurance was to have the machine manufacturer provide the tonnage estimate of a critical part to insure that the correct size machine was being purchased.

For users, the following thoughts may help. The sharper the inside radius shown on the drawing, the higher the tonnage required to form the part (see the discussion between air bending and bottoming in Chapter 3). The question that has to be answered is "Why do we need such a sharp inside radius?" The reply "Because we always have done it that way" is not very good. In most cases, it is not needed. It may be a good idea to measure a current part to see if the part dimensions match the drawing.

The purchaser's problems are magnified when different builders give very different estimates. Who do you believe? Who will guarantee an acceptable part at the tonnage they have estimated? Can the questionable part be subcontracted in case sufficient tonnage is not available? If only a tooling quote is required because you already have a press brake, have the tooling supplier provide a tonnage estimate for making the required part. Review any exceptions the supplier may take to ensure that the parts made will be usable.

The above discussion is invaluable. If the tonnage estimated for forming an offset is required, remember that the "magic" factor formula does not account for the offset depth or the inside radius that may be needed. Both size and radius will affect tonnage. Normally, a specific thickness will have a range of tonnages that must be anticipated. For example, 16 gauge will require anywhere from 20 to 40 tons/ft to form a good square offset shape with an inside radius of one metal thickness or larger. Standard tonnage estimates assume bottoming-type dies. After forming, no springback should occur in the part. The higher the tonnage used, the better the part consistency.

About 5% of the offset shapes formed are made using an air bend tool. To push both bends down into an air bend offset die usually requires only two and one-half times the tonnage needed to form the same thickness material using the recommended vee die opening. The resultant offset shape and part depth depends on how consistent the springback will be from part to part. Some parts require bottoming offset dies that use a certain amount of tonnage in order to produce an acceptable part. A good example is the $^{5}/_{8}$" deep offset required to make door bucks (door frames) from 16 gauge mild steel. To make a good part, the tonnage must compensate for a part that has holes or notches inside the forming area of the die, which are needed for striker plates and hinge locations. Some "door buck" designs have tapered notches at one end of the long narrow part to produce a mitered fit when the frame is assembled. If less than 35 tons/ft is used to form these offsets, the flaring that occurs in the cutout areas will cause the part to be scrapped (Fig. 11-7).

What is interesting about the door frame offset is that, when the parts were being formed using a mechanical press brake, the majority of manufacturers used offset dies with forming radii of $^1/_{64}$" or $^1/_{32}$". This required forming tonnages far in excess of 35 tons/ft. As these companies began replacing their old press brakes with new hydraulic machines, they found that the hydraulic press brake required to make the same part needed much more tonnage than their current mechanical press brake rating. They also found that the larger hydraulic machines run much slower, increasing the time needed to make each part. As a result, parts were redesigned to increase the inside radius and reduce the forming tonnage to the current 35 tons/ft figure. It is also noted that the tonnage estimate for the "magic" factor formula is to multiply the material thickness times 300. Using the 16 gauge normally used for the frames, the estimated tonnage would be $0.060 \times 300 = 18$ tons/ft. This estimate would provide insufficient tonnage to make an acceptable $^5/_8$" offset in door frames having the notches just described.

Fig. 11-7. A special offset die where the lower die moves back and forth to produce a "hat shaped" channel in two strokes. This shape is common in most door frames. *(Photo courtesy of Cincinnati Incorporated)*

Be careful when obtaining special tooling, either from an outside source or tooling that is built to your specifications. Confirm that you can make the parts that are needed at the forming tonnage that is available. Specific tonnage information can only be found by testing the dies that will make an acceptable part, and by reading the tonnage required from a press brake that has a calibrated tonnage gage. Most offsets from 20 gauge to 12 gauge will tend to fall into the ranges below, provided that the following parameters are met. The resultant offset is between five material thicknesses to 0.75" deep, the inside radius of each bend is one metal thickness (or slightly larger), and the web angle can be slightly overbent or appear to look like a drawn-out "S." The tonnage range for the above offsets in mild steel are as follows:

20 gauge – 12 to 25 tons/ft
18 gauge – 15 to 30 tons/ft
16 gauge – 20 to 40 tons/ft
14 gauge – 25 to 50 tn /ft
12 gauge – 30 to 60 tons/ft

Remember that forming tonnages will always be higher if a sharper inside radius is made.

Chapter 12
Simple Channels

A part with a channel shape has a base of a specific width and a 90° flange bent upwards at each edge. The 90° flange height must be high enough to allow a good 90° angle to be formed in either a simple vee die using two forming strokes, or a special one-hit channel die that will form both 90° angles in one stroke of the press brake. The main forming considerations are:

- Can the part be made without interference with the upper die?
- Is the upper die strong enough?

This can present a problem if the channel shape is to be formed in two strokes using an upper die, cutout on one side, which will provide clearance for the shorter of the two 90° flanges during the second forming stroke. The die used for this type of forming is called a **gooseneck** die in the United States; if you are in the United Kingdom, it may be called a **swan neck** die (Fig. 12-1). If the bottom flat of the channel is too small, or the flanges on either side are too high, it may be impractical to use a simple gooseneck die.

What are the tonnage limitations? If the part is to be formed using a gooseneck die, the die may not be strong enough to use a bottoming type forming method. Each gooseneck die will have a much lower maximum tonnage rating than an ordinary shouldered, bar-style upper die. An air bend die might not offer sufficient clearance to allow the second bend of the channel to be formed. The first flange that is already formed may strike the inside relieved portion of the gooseneck die and cause the bottom flat section of the channel to distort. If a special one-hit channel die is being used, the forming tonnage will increase, similar to the increase in tonnage that is expected when an offset die is used. If the part is to be formed using two forming strokes, some thought should be given to the forming process to be used prior to the part showing up in the press brake area.

The tooling required is the first consideration. Is the proper size gooseneck available in the current tooling inventory—one that has sufficient clearance to allow the second bend to be made after the first 90° angle is formed? What angle is machined on the nose of the gooseneck? Often, a gooseneck that has sufficient clearance was purchased for bottoming

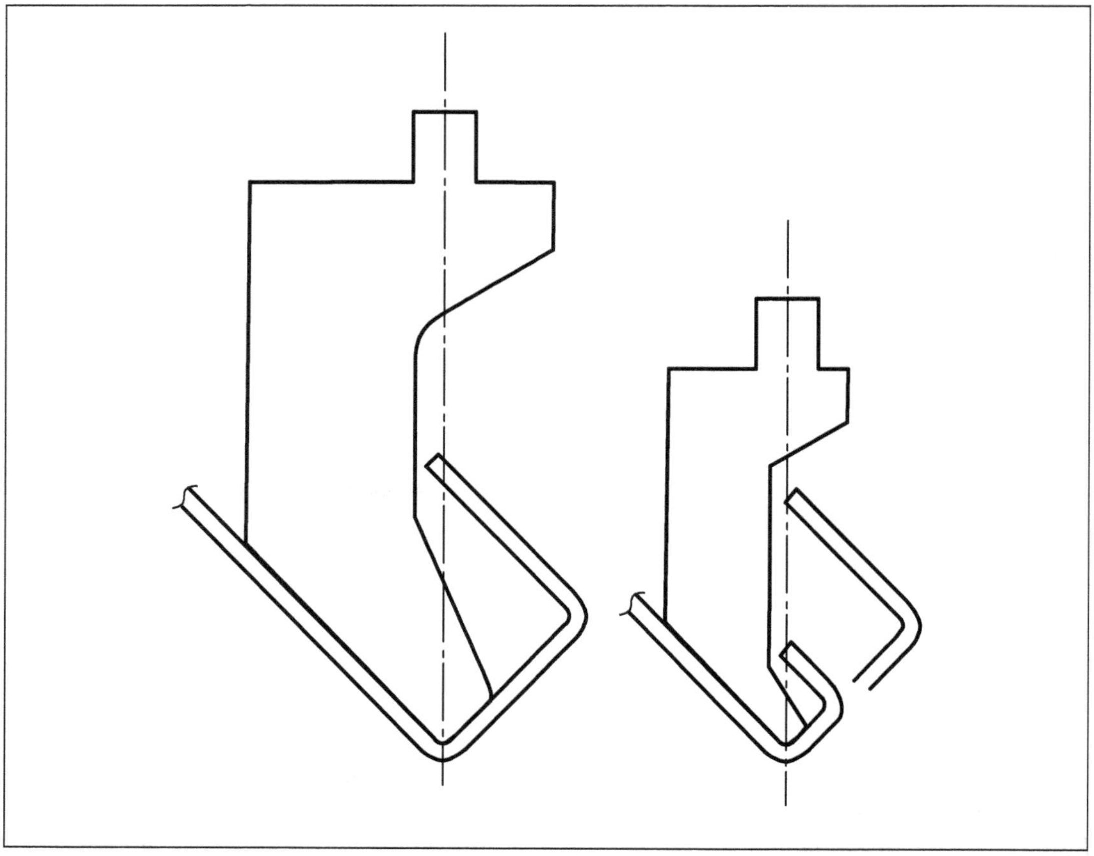

Fig. 12-1. Gooseneck die used to form a channel shape in two strokes.

20 gauge mild steel. However, the new part being reviewed may be 16 gauge. To use the bottoming method, 16 gauge steel will require significantly more tonnage than a 20 gauge part. Also, the gooseneck in inventory may not be strong enough to produce that tonnage without a high risk that the nose of the die may distort. Instead of buying a new die, it may be decided to modify the angle on the nose of the present die so the forming load can be reduced and the die still be used for the 20 gauge part. If the upper die is modified, make sure the lower die that will be used has the same angle that is on the gooseneck.

After the upper die selection is made, the next question is "How will the operator actually handle the part while making the bends?" The operator can take the blank, push the edge that is in the rear of the lower die against a properly set gage, and make the first bend. If the gage is an NC-type gage, the finger will move to the correct location so that the operator can simply push the part up against the new gage location and complete the second bend.

Tight dimensional tolerances can be held by placing any blank error in the flange leg that was made by the first bend. The part may have to be front gaged for this bend. After the first bend has been made, the part might be moved forward to another front gage, which locates the second bend. In this manner, not only are the dimensional tolerances held, but any possible gaging error that may have been caused by angular variations in the formed leg of the first bend are eliminated. (Review Chapter 7; Section H-5, to see if disappearing gages should be considered for this type of application.) If the part has the same outside flange leg on both edges, the part can be made by using a simple back gage, forming the first bend, turning the part 180°, gaging the other side of the part against the back gage, and making the second forming stroke. Any error in the blank will be left in the flat base of the formed channel.

Gaging of the part is important, since many parts may have holes that have to match other parts. If the holes are prepunched into the blank prior to the channel being formed, any variation in the original blank size can create an incorrect hole location after the forming process. If this is a consideration, it may also be necessary to determine the following:1) how the hole locations were gaged when the blank was being processed, 2) which edge of the blank is the base side, and 3) if the same locations can be used to gage the part when it is being formed.

When production warrants—or in order to minimize manufacturing time—a simple channel can be made in one forming stroke using a **channel die**. The design of channel dies has evolved over the years. The first dies had an upper that was cut to the inside dimensions of the part. The lower die was made from a block of die steel large enough to allow a rectangular groove to be machined into the top surface of the die that was the width of the outside dimension of the part. The depth of the first grooves used to form the part was estimated. Several problems surfaced immediately. If the upper and lower dies allowed only the exact clearance for one metal thickness, or if the material was thicker than the nominal gauge dimension, the part would stick in the groove of the lower die. If the material was thinner than the nominal material thickness, the part might stick to the upper die. In either case, the operator had to reach into the working area to try to pull the part out of the lower die or off of the upper die.

The initial die designs were quickly modified. The lower die block was made wider, and another insert called a "wedge" was added either to the front or rear of the die block. The wedge was located so that it became one of the vertical sides of the groove. When a part was pushed down into the groove and the channel shape formed, if the part would stick in the lower die, the operator could reach into the die area and lift out the formed part. The wedge would also be lifted up to provide clearance for easy part removal (Fig. 12-2). At the same time, an upside down "L" shaped assembly was fastened to the rear of the lower die. One leg of the "L" was high enough vertically to allow the legs of the formed channel to fit easily under the top leg of the "L," which was spanning the rear of the die body and barely clearing the rear of the upper die. As the part was formed, the leg of the channel in the back would form against the rear of the upper die. If the formed part would stick to the upper die when the stroke of the ram returned to the top, the formed part would move upward until the rear flange would contact the top hook of the L assembly. As the upper die continued upwards, the channel section would be pulled down off of the upper die. The L shaped piece became known as a "stripper."

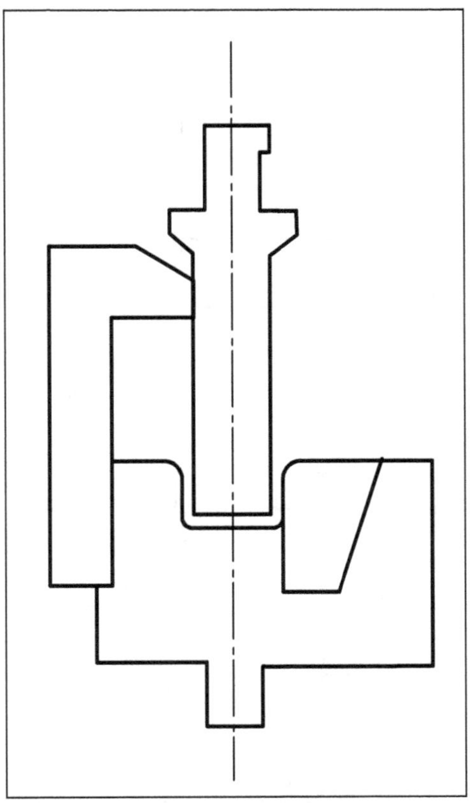

Fig. 12-2. Channel die using a wedge release and a hook-type stripper.

Only thinner materials were used in channel dies since the mechanical press brakes being used at that time had limited strokes for part removal—only the last $^3/_8$" of the working stroke had the full forming tonnage. One of the problems quickly addressed was the fact that channels that had a base width of $^3/_4$" or greater (especially if the material was 16 gauge or thinner) would have the part break away from the square nose of the upper die. This caused additional material to be drawn down into the die cavity. When the bottom of the stroke was reached, there was no place for the excess material to go. As a result, the bottom of some of the channels were not flat but were either concave or convex. This problem was solved by adding a pressure pad to the bottom of the rectangular forming slot in the lower die.

The pad, normally made from commercial quality cold rolled steel, was fastened to the lower die by special bolts called stripper bolts. Springs were spaced between the bolts to lift the pad up to the top surface of the lower die so that when the part was located for forming, the top of the pressure pad was slightly under the material. When the forming stroke was made, the material would immediately contact the top of the pressure pad. As the stroke continued downward, the springs under the pad built up resisting pressure which held the part being formed relatively close to the surface of the top die, minimizing the effect of material breakaway, and also reducing the amount of extra material that would be drawn into the bottom of the die. The secondary advantage of the pad was that it lifted the formed part up out of the lower die cavity, eliminating the problem of the operator having to reach into the die area to "lift" the formed part out of the die. Not only did this improve production, it also eliminated a potential safety problem (Fig. 12-3).

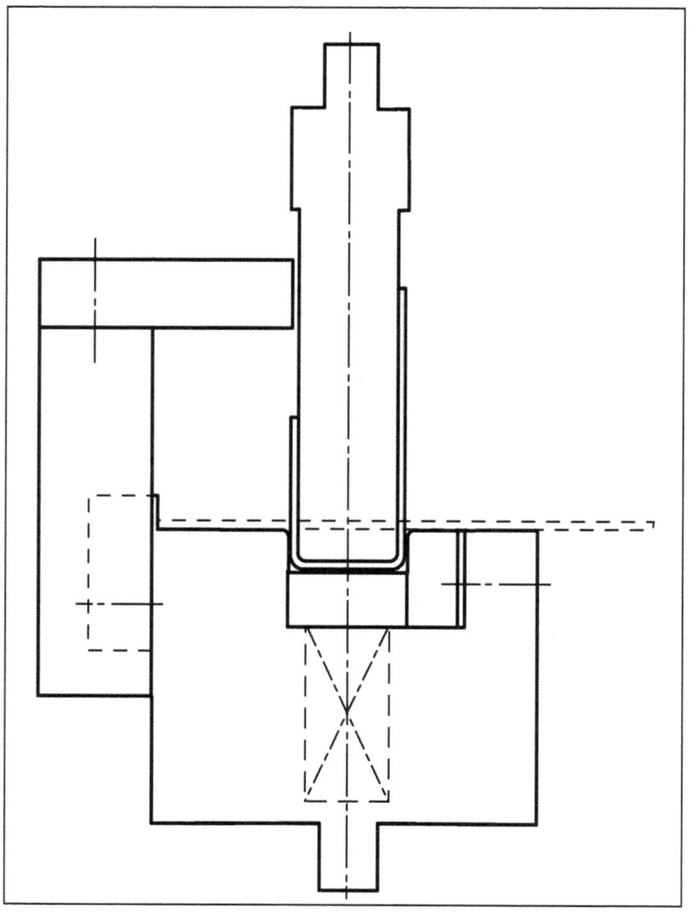

Fig. 12-3. Channel die using a shimmable insert on the lower die body allowing adjustment for material variation. Also included is a pressure pad to help hold material flatness during the forming stroke plus assisting part removal. A hook stripper plus a fixed gage are mounted on the lower die body.

As hydraulic press brakes became popular, heavier materials were able to be formed using channel dies. The greater depth of penetration required for the thicker metals was possible because a hydraulic press brake has uniform tonnage throughout its forming stroke. Heavier materials also meant more thickness variations had to be accommodated in the lower die cavity. The wedge insert design could not provide the die clearance allowance required. It then became standard practice to provide channel dies with a rectangular insert in place of the wedge insert. The new insert could be shimmed back and forth to provide the proper die clearance as material variations occurred.

The depth of the lower die cavity must be deep enough to accommodate the following: outside radius of the formed part, about three metal thicknesses of flat to control the angle of each vertical leg of the channel, plus the distance of the top "lead-in" radius of the lower die. The radius on each side of the upper die is normally made about one metal thickness. This radius is often modified during die testing to ensure that the legs will form "square" to the bottom of the channel and not be open or overbent out of tolerance. The overbending occurs as described in Chapter 3, Section C-1.

Since there is often a tonnage limitation due to machine size, material thickness, and possible inside radius that can be formed, the shape or flatness of the bottom of the channel may have to be evaluated. If the legs of the channel will not form to a 90° angle due to lack of tonnage and normal springback, the corners of the upper die are remachined to a smaller radius. When the part reaches the bottom of the stroke, and any additional tonnage used to attempt to set the corners is applied, the legs of the part are overbent (refer to Chapter 3, B-1). Not only are the vertical legs affected, but the flat bottom of the channel is also overbent on each side. When the stroke returns to the top, the formed channel often has its vertical legs held snugly to the upper die, requiring the part to be stripped off the upper die. The bottom of the channel, however, may have its corners slightly overbent, which results in the bottom having a slight "hollow." This hollow can vary somewhat as the material thickness changes from setup to setup, or as the operator changes the tonnage by adjusting the ram setting of the press brake.

The same considerations that apply to forming other parts also will be found when forming channels in one stroke. The only way that the quality of the channel shape can be improved is to increase the forming tonnage at the bottom of the stroke. The sharper inside part radius that is needed, and the flatter the bottom of the part, the higher the required tonnage. Simple channel shapes, like offsets, fall into general tonnage classifications. If the inside radii is about one metal thickness, if the bottom of the channel is not perfectly flat, and if the legs can vary ±1°, the forming tonnage will probably be as follows:

20 gauge – 10 to 30 tons/ft
18 gauge – 10 to 30 tons/ft
16 gauge – 15 to 35 tons/ft
14 gauge – 20 to 40 tons/ft
12 gauge – 25 to 50 tons/ft
10 gauge – 30 to 60 tons/ft
3/16" – 35 tons/ft to form
– 65 tons/ft to set
1/4" – 45 tons/ft to form
– 80 tons/ft to set

Chapter 13
Hems and Seams

When a piece of metal is required to have the edge folded back 180°, many tooling considerations arise. If the part is initially formed to an acute angle of about 35°, it can then be flattened so that the short leg touches the longer leg of the part in a secondary operation. This shape is called a **hem.** If the part is formed to some acute angle in an initial operation, and is then formed down in a secondary operation to complete the 180° bend, but the short leg is spaced a specific distance from the longer width of the part, the shape is called a **seam** (Fig.13-1).

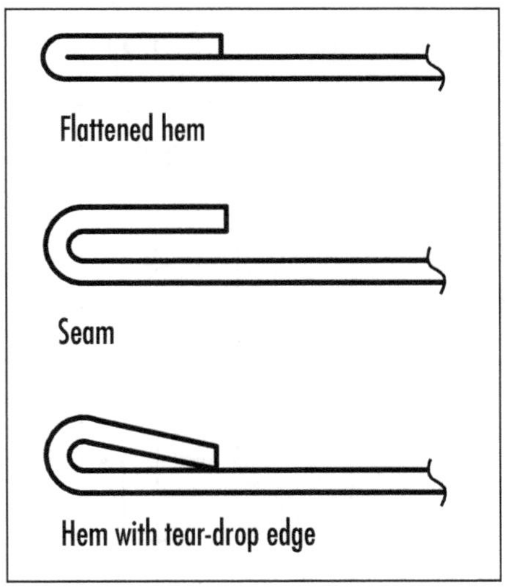

Fig. 13-1. Hem and seam shapes.

A) Hems

Most hemming operations are designed to be used on thinner materials to create an edge on the part that will minimize the possibility of the material handler being cut on a sharp edge. When the part is formed into a hemmed shape, no appreciable additional strength or stiffness should be expected in the design. Parts that are considered thin, and which thus may require a hemming operation, are 18 gauge steel or thinner. Materials 16 gauge and heavier are often deburred and have their edges rounded using a special tool designed for that purpose. Occasionally, parts may be designed with a hem even when thicker materials are being used.

1) Hems Formed Using Simple Press Brake Dies

In most cases, hems are made using simple press brake tools (Fig. 13-2). The first bend is made using an acute angle upper and lower vee die. The normal tools of choice are dies

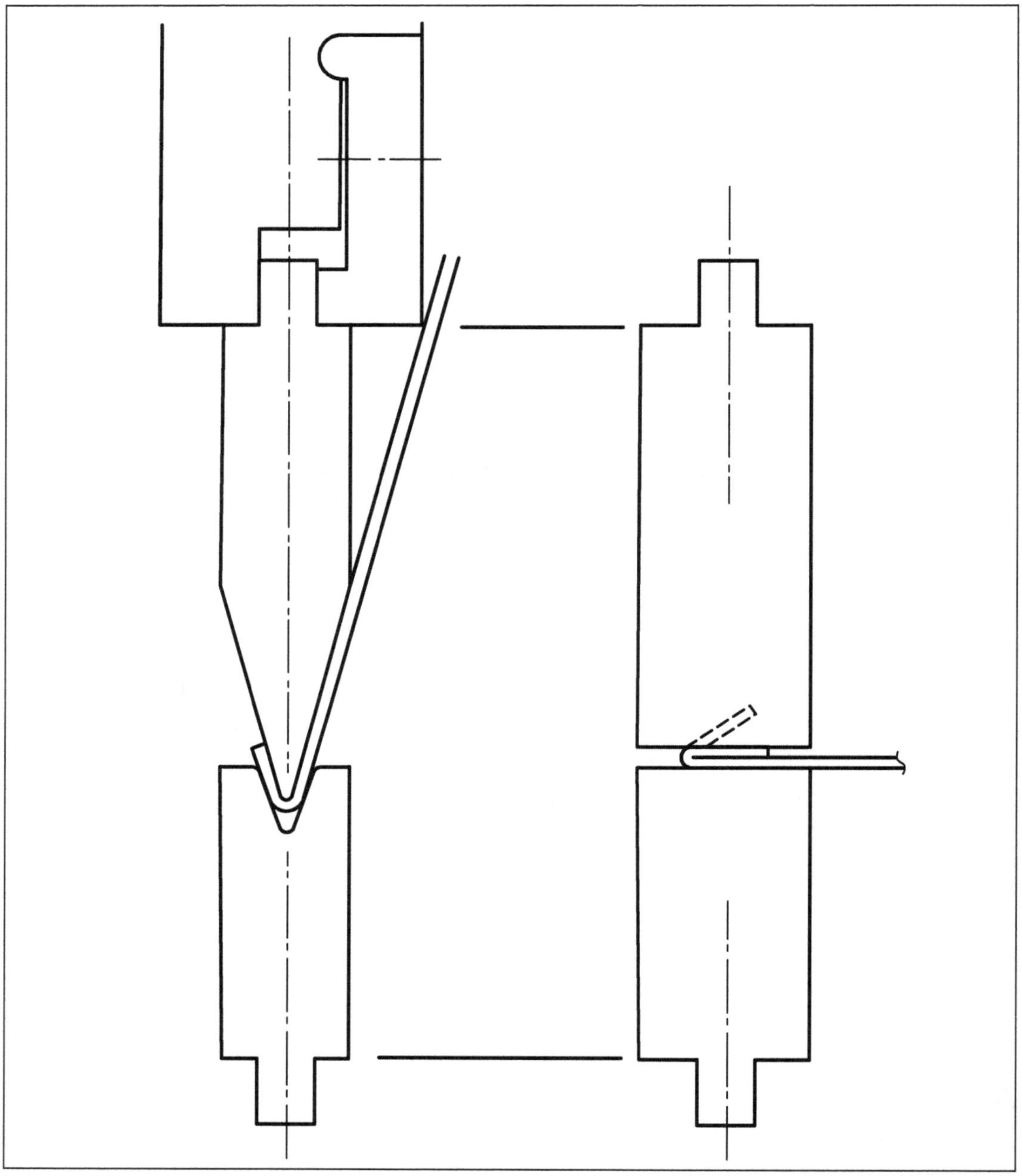

Fig. 13-2. Simple 30° dies and flattening dies running side by side.

cut to a 30° included angle (an included angle is the angle of the upper die). When the part is formed, the legs of the part are moving down into the vee die cavity. If the short leg that will serve as the hemmed leg is too short, it can be drawn down into the vee die opening and be very difficult to remove after the part has been formed. Many operators do not use the recommended 8 times the metal thickness vee die opening, but instead choose a vee die that is only 6 times the material thickness in order to minimize the possibility of the short leg being formed into the die cavity. Acute angle vee dies are not very strong and are subject to splitting if their safe tonnage rating is exceeded. It is not uncommon for the maximum safe tonnage rating of a standard 30° vee die to be limited to 12 tons/ft. This means that acute angle dies should not be bottomed. The wedge action at the bottom of the stroke builds the tonnage up very quickly, and is very hard to limit if the operator lowers the ram setting too much. Commercial acute angle dies are made from narrow bar sizes. They are considered air bend type tools.

When an acute angle is formed, both legs of the part move from a flat position upward, to form some angle. When the pressure is released, the part will spring back. At the bottom of the forming stroke, the longer flange must not strike anything or it will be distorted. Prior to forming the first part, the setup person should check to ensure that the upper die is high enough so that the material will not touch the ram nose when the part is in the acute angle position at the bottom of the stroke. If either leg is not too long, there is no problem. However, many parts are designed with the longer leg in excess of 5" long. The probability will be that the longer leg will hit the ram nose if the upper die is not high enough. The minimum die height that should be considered when purchasing 30° upper dies is 4.25". Most of the ram widths will be approximately 2.5" wide. The 4.25" die height allows a leg that has been formed up to a 15° angle (30° upper die) to barely miss the tip of the ram nose. Even if it does touch slightly, no distortion in the part should occur. Higher dies are available, but the available die space in many press brakes will not allow them to be used without changing the height of the lower die.

After the acute angle has been formed, the resultant angle will be approximately 33° to 37°. If sharper, the part may hit the ram, be extremely difficult to remove from the lower die, or may bottom out in the lower die and possibly damage the tool. If the part did stick in the vee die, the operator may have to hit the part with a gloved hand on the flange sticking up in front of him. Sometimes, this will cause the leg to vibrate and allow the part to dislodge from the tight fit caused by the part springing back in an acute angle cavity after it has been formed. If the short leg of the part was drawn down into the vee die, the part will be locked into the vee opening and be very difficult to release. In this case, hitting the flange of the front leg might allow the part to be pulled up out of the die.

The second operation needed to close the acute angle into a hemmed shape is accomplished by installing a set of **flattening dies** into the press brake. Flattening dies are bars made flat on the bottom forming surface, with a conventional tongue on the other side to mount into the ram or the filler block (Fig. 13-2). The die set will always consist of an upper flattening die and a similar size lower flattening die; no gaging is used. The acute angle that has been formed is placed under the surface of the upper flattening die by locating the part on top of the lower flattening die. The operator locates the open hem on top of the lower die so that the entire hem length can be contacted by the upper die when the machine is cycled. Prior to closing the first part, the ram setting of the upper die should be set so that the ram will reverse when it is two metal thicknesses away from the lower die. The initial setting does not allow for potential machine deflection or tonnage adjustments to obtain the needed shape.

When the hem is being closed, the part will move. The common belief is that the short leg will simply fold down to touch the longer leg resting on the bottoming flattening die, but that is not what happens. The short leg will try to hold onto some tiny machining mark on the upper flattening die. Even when that surface has been polished, somehow the corner of the short leg holds on to the top die as the ram is lowered. The long leg is pulled toward the rear of the lower die as the part rotates around the top corner of the short leg, and the closing action takes place. At the same time, the long leg will move upward until the top surface of the part hits the front edge of the upper flattening die. As the stroke continues downward, the short leg will move until it is about parallel to the longer leg. If the stroke is stopped at that point, and the part is inspected from the end of the die set, it can be seen that when the two legs are almost parallel, the long leg will still be slightly up in the front, still touching the front edge of the top flattening die. The inside of the part will have closed to a one metal thickness inside radius or a distance of two metal thicknesses between the two legs. If the ram is then moved downward slightly, possibly one-half of a metal thickness, the bottom (long) leg will move down to its original position, flat on the lower die. The short leg will roll around on the radius, and the end of the short leg will leave its contact point with the upper die and move downward toward the top surface of the long leg of the part. As the ram continues to the bottom of its original stroke setting, the one metal thickness inside radius observed above will have the material in that radiused arc forced down into a new shape. The force required to crush the arc builds up very rapidly. The tonnage required to obtain an acceptable hem depends on how the part will be used.

After the hem has been formed, and the hemmed dimension is measured, it will be discovered that the outside of the radiused end will probably be more than two metal thicknesses. At the point where the short leg is touching the long leg, the measurement will be close to two metal thicknesses. If a difference in the two measurements can be seen visually, the part is said to have a **teardrop** shape. Some teardrop is normal for a hemmed part; but the question will always be how much is acceptable? The only way that the legs can be made closer to parallel is to apply more tonnage. This may be limited due to the capacity of the press brake. In most cases, when parts that are 18 gauge and thinner are flattened with a tonnage of 10 to 15 tons/ft, the teardrop is minimal and the shape is acceptable.

Most hems are designed to be in the 0.375" to 0.5" long. This minimizes the problem of the material being difficult to remove from the first operation die. It allows the use of a more conventional vee die opening (although probably smaller than recommended). It is important to control the forming speed of the long front flange to protect against any backbend or distortion. The larger the vee opening, the slower the part flanges will move upward. Engineers try to minimize the hem size in order to reduce material cost and the weight of the part.

Some parts actually are designed with a longer than normal hemmed section. When the normal $^3/_8$" to $^1/_2$" hemmed leg is closed during the flattening operation, the short leg actually rolls around the one metal thickness inside radius and forms to about the same place on the top of the longer leg. If a normal length hem is being formed, the edge of the short leg will stay in contact with the surface of the longer leg even as it moves slightly while the hem is being closed. If the hemmed leg is abnormally long, however, it will touch the longer material about $^3/_8$" to $^1/_2$" from the radiused area, and then often open up as the radius area is being flattened. When the flattening stroke is completed, it may be found that the edge of the hemmed leg will not be in contact with the long leg, but will be in contact halfway back into the hem length. Restriking the part with additional tonnage, or making a new part with added tonnage, will not correct the shape. Apparently, each material thickness has a normal point of contact that will allow the short leg to touch the long leg as it is flattened and provide a consistent shape. If the part has too long a hem, it can create the problem of the hem being open at the edge. Finding a way to eliminate the distortion cannot be guaranteed.

Some part designs require a hem near other previously formed bends. Occasionally, the previous bends in the part restrict the area to be hemmed from being completely under the upper flattening die. As a result, the edge of the upper flattening die can only contact the sloping edge of the part in its downward stroke. The upper die will not grip the sloping surface, and the part will be pushed out to the front of the press brake toward the operator. No closing action will occur. The only way to close the hem is to hold the long leg in position with a fixed stop. The pressure that kicks the part out of the die area is very great, so the front stop must be strong enough to withstand the kicking pressure. The stop must also be adjustable to determine the best location for the upper closing die to contact the sloping surface of leg to be hemmed. Even using front stops, the long leg will still have a tendency to try to move upward, and the operator must take care that he is not injured by sharp edges of the part during the closing cycle.

Hems or seams should not be designed in an attempt to increase part stiffness along the hemmed length. When the operator picks up a flat blank, he will normally lift it in the center of the piece. As the part is lifted, if it has enough length (e.g., 6' long), each end of the part will sag downward. In some cases, the part will be difficult to load on the top of the lower die, especially if the sag is greater than the stroke opening between the upper and lower die. When the initial acute angle bend is made, the part will be very stiff with no apparent sag or flexing. After the closing operation, when the hem is removed from the flattening die, the operator will find that the part will again sag or flex almost the same amount as it did prior to the hemming operation (this assumes that the only bends in the part at this time are the bends required to make the hem).

Since many parts that require a hemmed length are not too long, it may be desirable to locate a 30° set of dies on one side of the press brake, and to locate a set of flattening dies on the other. This will cut down on material handling and speed of production. If many parts are required, it may be possible to have an operator at one station and a helper at the other station, allowing one hemmed part to be made each time the ram is stroked. Before this can be considered, the 30° dies and the flattening dies must have been machined to run at the same die height. This machining can be completed when the tooling is originally purchased, or the dies can be modified later as production is increased to justify the tooling expense. In either case, the operator will have to shim the dies, since the forming load for the first operation is much less than the tonnage required for completing the hem. This tonnage difference will cause unique deflections between the bed and ram if both stations are used at the same time. Even if each station is used separately, the difference in loads will require die shimming. When two operators are used to increase productivity, a careful review of what may be required to ensure the safety of both operators is mandatory. A light beam sensing device, dual foot switches, or another suitable method should be considered and provided prior to production.

Remember, if the closing operation is located on the left or right side of the centerline of the press brake, the required closing tonnage may not be attained due to the fact that each housing can only carry one-half of the maximum machine tonnage. The position of the closing die must be located to ensure that the loading on one side will not cause a hydraulic press brake to reverse the ram. This can happen because the housing on that side has been overloaded, even though the total closing tonnage does not approach the machine's capacity.

2) Hems Formed Using Special Press Brake Tooling

When parts are longer than one-half of the length of the press brake, and the production is great enough, it may be wise to consider a special one-handling hemming die. There are a number of different designs, and each one has limitations. The most common special hem die has a common 30° upper die used for normal acute angle bending (Fig. 13-3). The lower die is made up of a combination of an acute angle vee die and flattening station. In the back portion of that die is a rectangular slot that holds a spring loaded "L" shaped die section. This makes up the lower acute vee opening as well as the area that is used for the closing operation. The flat part is laid across the top of the vee opening of the moving section. The upper die is lowered, hits the metal, and as the stroke continues downward, the middle section is pushed down until it contacts the bottom of the rectangular slot. As the stroke continues down, the part is formed to an acute angle of approximately 35° or less. When the upper die returns to the top of its stroke, the springs in the middle section lift it up to the top of its preset travel (often about 0.5" to 0.75"). The front of the upside down "L" shaped section creates an opening between the flat surface of the lower die holder that is in the front of the rectangular slot in the lower die and the bottom of the "L" section that extends toward the operator. The part is removed from the vee opening and laid down with the long flat leg of the part resting on the top of the lower die holder. The outside radius of the acute angle part leg that is to be closed is located up against the vertical leg of the "L" that moves up and down. Ideally, the tip of the short leg is completely under the flat of the "L" section. When the ram is again cycled, the 30° die now pushes the vee section of the "L" section downward, which allows the horizontal leg of the "L" to close the hem.

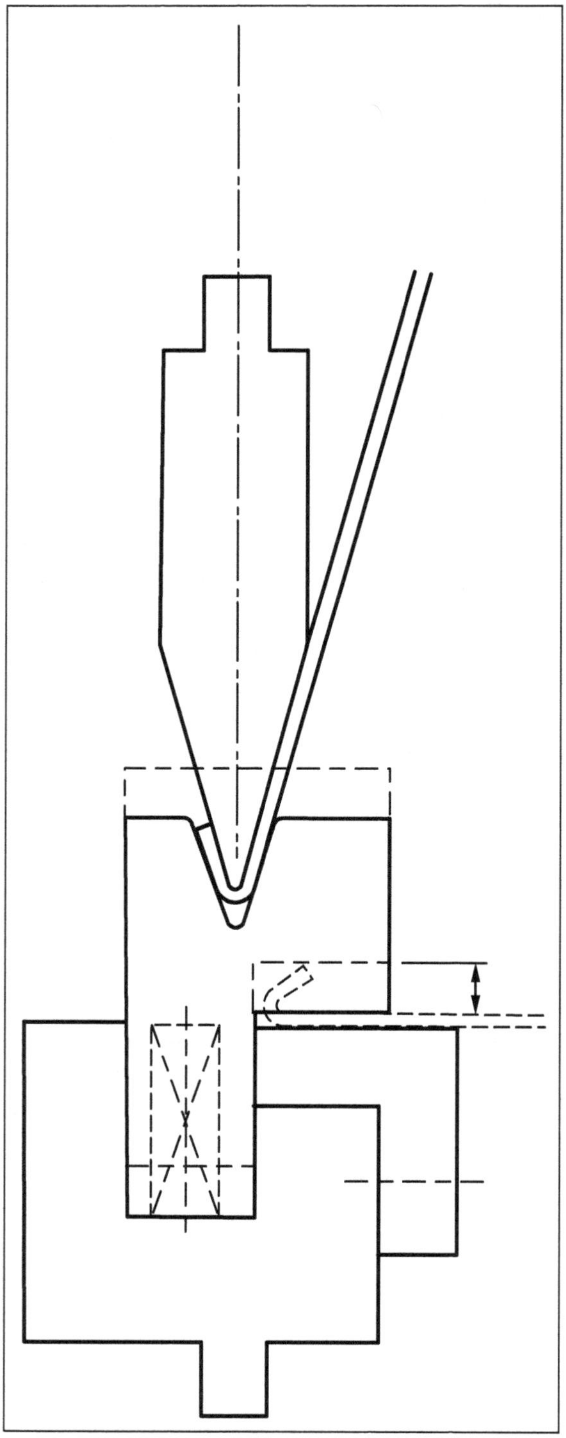

Fig. 13-3. Double deck hemming die.

In practice, the closing operation takes place if the short leg remains under the moving flattening area. The part, however, does not stay in its original position. The closing operation acts the same as if the legs were being hemmed under simple flattening dies. As soon as the upper die touches the short leg, the longer leg will rotate upward until it contacts the front corner of the moving die section. As the stroke continues downward, the short leg will be closed. The longer leg will be pushed downward so that it is almost back to its original horizontal position at the bottom of the stroke. Because the upper die does not have the same forming area pushing down as a conventional flattening die, and cannot build up similar flattening tonnage without risking splitting the “L” shaped section, the resultant hem will always have a greater “teardrop” appearance than if the same hem had been closed with a simple flattening die. The savings in handling time will normally justify this appearance problem since it will seldom affect how the part will be used.

Problems are present in any special die usage. In the double-deck die described above, the vee opening is normally kept on or near the centerline of the bed and ram, which provides the most consistent repeatability and forming pressure. This is a basic press brake design limitation. If the load is moved toward the operator side of the ram nose, but the pressure from the upper die is applied at the centerline (as is the case with the 30° upper and vee opening in the top of the upside down “L” shaped moving section), the middle section will try to open up similarly to a “C” clamp when it is under load. Since the nose of the 30° die will normally have a relatively small radius (possibly $^{1}/_{32}$" or $^{1}/_{16}$"), the radius at the bottom of the vee opening may not be an exact fit with the nose radius of the upper die. This will allow some gap or clearance to be present on the sloping sides of the vee opening. When the load is applied to close the hem, the “L” section may gap open. Although the open angle is very slight, the closing surface of the “L” section and the closing surface on the top of the lower die are no longer parallel.

If the part has a short leg that remains completely under the upper closing die, the bottom leg will move up and down, but the hem will be closed to the normal teardrop shape. As noted previously, if the leg to be hemmed extends past the corner of the moving “L” shaped part, when the upper die moves downward, touching the sloping leg, the short leg will not close, and the entire part will be pushed out of the closing area toward the operator. This problem is often present with all special hemming dies. The distance the middle section can travel is often limited due to the bar sizes of the die required to guide the moving spring-loaded section. Only a specific distance is available to locate the 35° part, which limits the length of leg that will fit in the opening. Some parts require other bends to be formed on the part prior to the hem being closed. Those bends interfere with the moving section if the 35° angle is pushed too far into the hemming area. In either case, a front stop must be used to keep the part from being “kicked out” of the closing area during the hemming stroke. The gapping of the “L” section during the closing stroke helps to magnify the kicking problem.

When a special die is used, normally the die will only form the hem to a specific preset dimension. The problem is that the material thickness of the part will change from lot size to lot size, and sometimes from part to part. A good die design will provide an extra bar in the die that will allow that distance to be shimmed up and down to allow the setup to be adjusted if required. In hemming operations, the distance between the die closing surfaces is critical to obtaining sufficient tonnage to minimize the teardrop dimension.

There are several types of special die designs shown in Fig. 13-4. Each is designed to form a specific thickness of material. Each has a limitation as to hem size and tonnage that is required; and, of course, the cost of the different dies varies due to design, type, length, and requirements of the user.

3) Making a Bend in the Center of a Hem

Occasionally, a part is designed with an abnormally long hem (perhaps 1.5" long). A secondary operation is required to make a bend angle in the center of the long hemmed section. Two major problems are present in the part design. First, as pointed out above, long hems are very difficult to make. The end of the hemmed section often opens up slightly; although it may only be 0.020", it will be very visible. The second concern is that when the

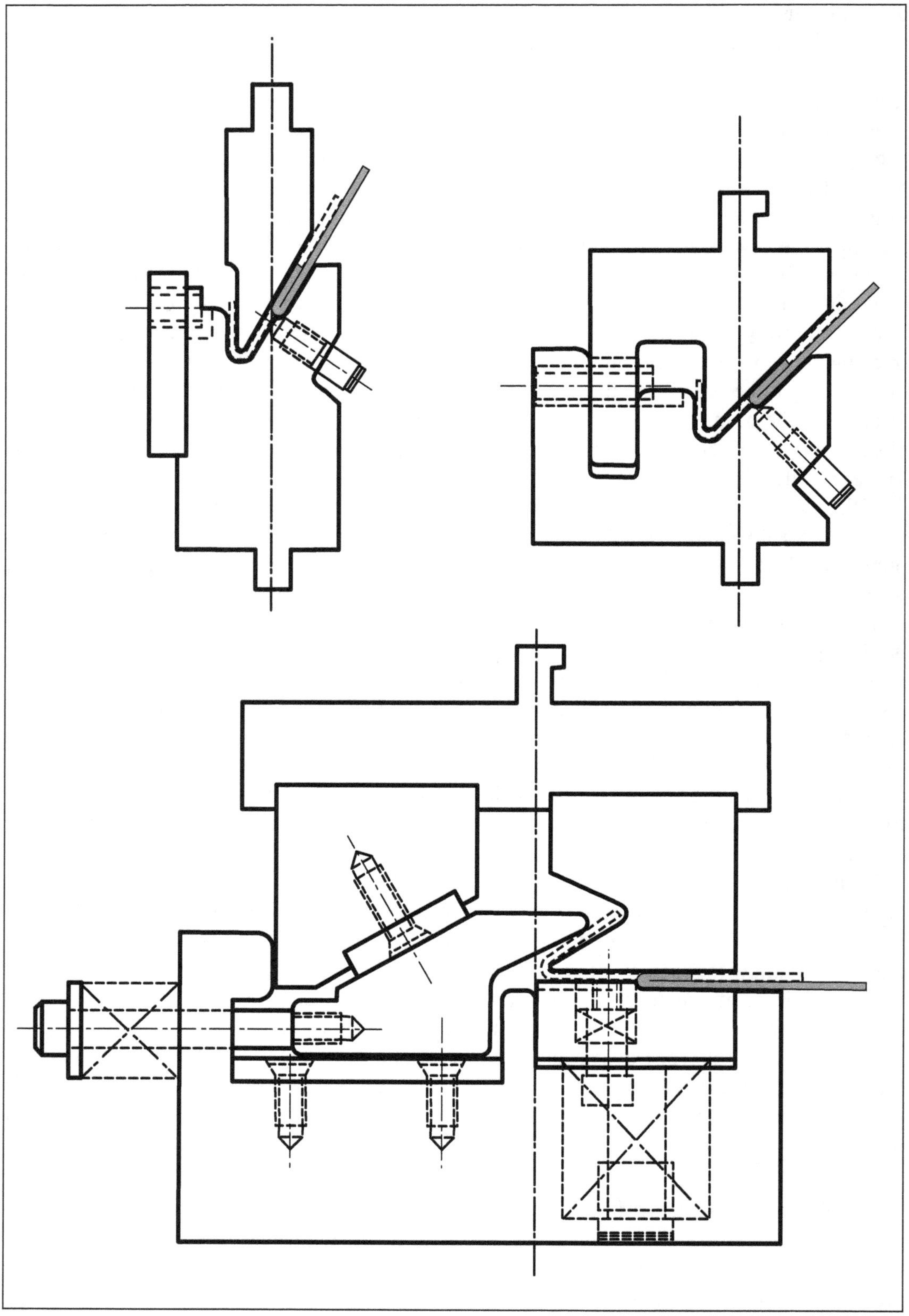

Fig. 13-4. Other designs of special hemming dies.

angle is formed into a hem, the two thicknesses of material provide two different neutral axes. Each will spring back differently, based on the radius formed into the material. The inside leg will have a smaller inside radius than the material at the bottom of the bend. As a result, the closed end of the hem will gap away from the long flange by a noticeable distance. It is not practical to try to close the gap in a press brake operation. The open leg usually has to be clamped to the longer leg, and welded in another operation.

B) Seams

There are various designs that incorporate a seam type shape. The most common is the simple hem shaped part, where the short leg is not closed but left open a certain distance. Other designs have a simple seam on one end of the part and a matching seam on the other end. Next to the second seam is an offset (one metal thickness joggle), which provides for the assembly of an identical part using simple seams to interlock. This allows the outside of the part to remain in the same plane. The common name for this assembly is a lockseam (Fig. 13-5). There are different similar seam designs used to provide a means of connecting two separate parts. A third type of design has a 90° bend opposite the edge of the open leg of the seam; this is called an open standing seam.

A simple seam is initially bent exactly the same as if a hem were being formed. A simple 30° to 35° angle is formed in the part. The angle must now be closed. An upper flattening, similar to the one used to complete a hemming operation, is selected. The lower die must be designed to allow a mandrel or spacer material to be located between the long leg of the part and the short leg that is to be closed (Fig. 13-6). The thickness of the mandrel must match the desired seam opening. When the short leg has the mandrel in place, the closing stroke is made. When the pressure is released, the mandrel is grasped tightly by the short leg of the seamed part. Somehow, the part has to be "stripped" from the mandrel (or vice versa).

The same action that takes place during a hemming operation takes place while closing the seam. The first operation will form the acute angle with an inside radius slightly larger than one metal thickness. When the closing operation forces the short leg downward to the point where it is parallel, the seam distance is two times the radius, or an opening of about two metal thicknesses. Since that is seldom the needed dimension, as the top die is lowered, the end of the short flange rolls down until it hits a solid part. In the case of the seam, that part is the mandrel. The radius of the end of the part is reformed into a shape that allows the two legs to form a seam shape, but the edge of the part will be pinched on the mandrel. Operators use different methods to minimize this problem. Often by adjusting the bottom of the stroke position several thousandths of an inch, the shape of the reformed radius causes the short leg to open slightly at the edge of the mandrel, and reduces the point of contact on the mandrel to one place. Once that pressure point has been released, the part can easily be slipped off the mandrel.

Some designs allow a special stripper to be used to release the part while the upper die is being returned to the top of the stroke. Other designs allow the operator to manually push the part off the mandrel. It is not uncommon for the closing operation to require the part to be turned over prior to closing, and have the short leg hooked over a mandrel mounted on the lower die. This assumes that the mandrel mounting has sufficient strength to withstand the pushing and pulling needed to release the closed seam. The closing tonnage should never be higher than 2.5 times the recommended air bend tonnage. For example, if 16 gauge mild steel was air bent using the recommended 0.5" vee die, the tonnage needed would be approximately 3.8 tons/ft. To close a seam should take about 3.8 × 2.5, or 9.5 tons/ft. If the part is hit too hard,

LOCKSEAMS

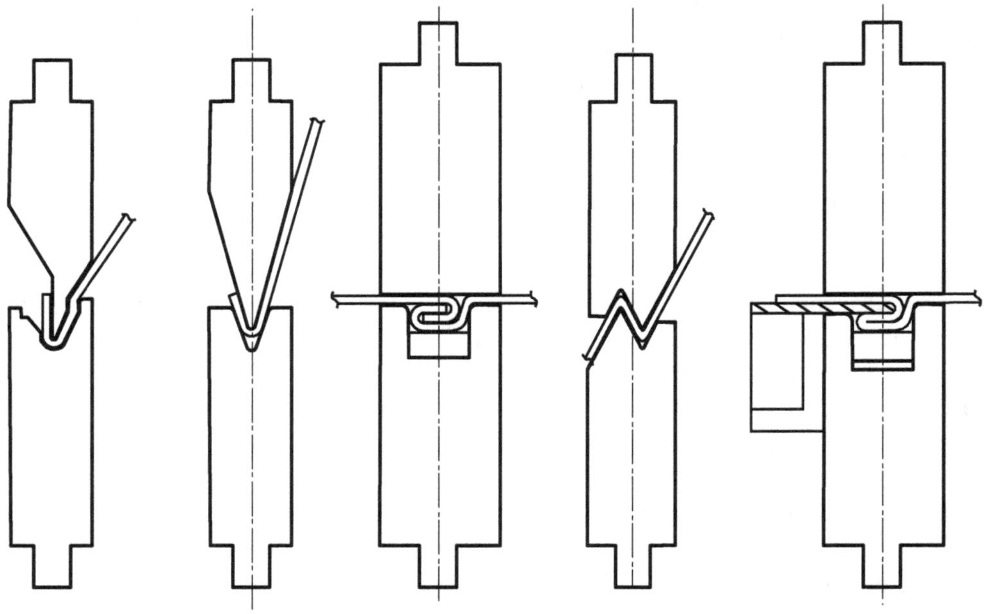

Lockseams may be made in a number of ways depending upon the style of lockseam desired. However, the single lockseam and the Pittsburgh lockseam are the most common. The three dies shown at the left, above, are used for the single lockseam. By shimming the insert in the flattening die, various gauges of metal can be seamed. The two dies shown at the right, above, are used for the Pittsburgh lockseam. This is used for joining two sheets at right angles as in duct work. The flattening mandrel and shim under the insert are changed for forming seams in various gauges of metal. The seam is finished in the field by sliding the joining sheet into the opening and hammering down the projecting edge as shown lower left.

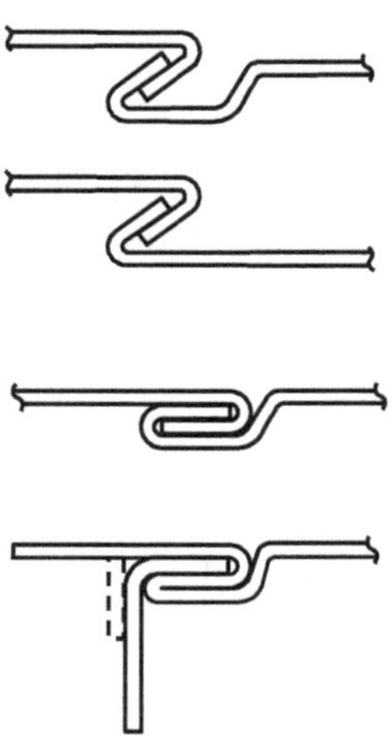

Another type of open standing seam can be made using the first two dies illustrated at the right.

Dies for forming one type of connector strip are shown at the far right. This strip is formed in two operations.

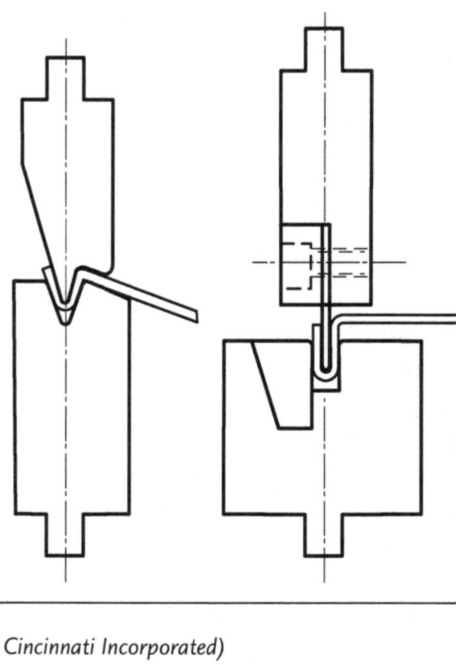

Fig. 13-5. Lockseam and standing seam dies. (*Courtesy of Cincinnati Incorporated*)

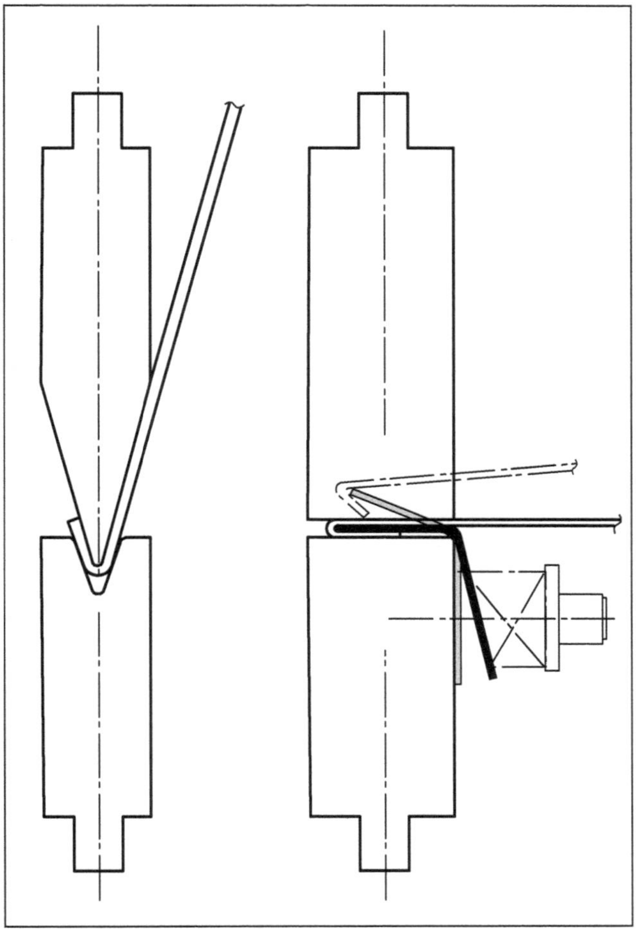

Fig. 13-6. Simple seam dies running side-by-side. Note the mandrel required to obtain the proper seam opening.

the mandrel may stick to the inside of the seam to the point that the part may be damaged in order to remove the mandrel. In this case, hitting the part harder to improve perceived quality does not work.

It is good practice, if special tooling is purchased to make special shapes, to have part development completed by the tooling supplier. Samples should be submitted for approval prior to the dies arriving in your facility. This assures that production will stay on schedule. As with special hemming dies, different types of seam tools have limitations. Fig. 13-7 shows methods of forming various seams; this may help you get an idea of the number of handlings that may be needed to form your part. In some designs, the forming tonnages to make preliminary bends prior to forming the seam (e.g., offsets) need to be identified to ensure that a large enough press brake is available.

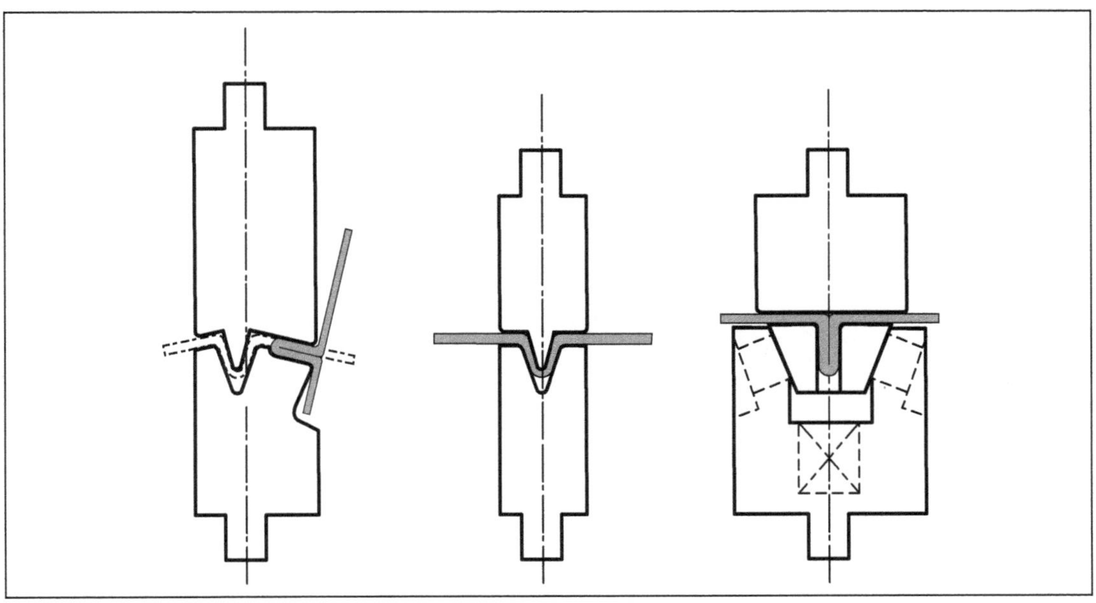

Fig. 13-7. Other types of seams and possible tooling to form them.

Chapter 14
Radius Dies

Many parts that are formed on a press brake require a "radius" shape (Fig. 14-1). The radius can have many uses including: enhancing the appearance of an assembly, providing an area in an assembly that does not collect dirt (or can easily be cleaned), creating a chute to remove parts, making a tube shape, or even making scraper blades for a bulldozer. The material may range from thin sheet metal to heavy plate. It will be impossible to review all of the many configurations that may be used. The discussion that follows will provide the basic information required to understand what can be made and what practical limitations should be expected.

A) Evaluating How to Form Parts that Have a Radius

If a simple radius up to 1" in size is to be made in mild steel or stainless, a special set of radius dies will normally be needed. One basic consideration that should always be remembered is "the larger the radius, the greater the springback." It may be necessary to review Chapter 5-B concerning springback. Simple 90° angles formed by air bending mild

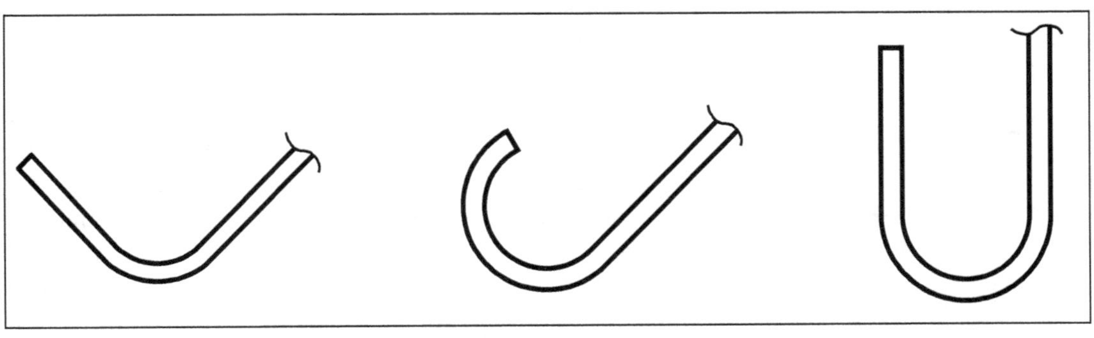

Fig. 14-1. Common radius shapes.

steel sheet metal will have springback of 2° to 4°. A 1" radius in a 90° angle in the same mild steel, if formed in one stroke, will probably spring back as much as 10°. The springback will cause the inside radius to become slightly larger. Most radius bends show an inside radius on the engineering drawing. The main exception occurs when the part has to fit a mating part and match the outside radii.

Assuming the 1" inside radius is the one to be measured, the probability will be that the top die radius should be $^{31}/_{32}$", and the angle of the die cut to an included angle of 80°. If the lower die is made with the same radius and angle to match the upper die, there is a good chance that a sheet metal part formed using that die set can be formed to a good 90° angle having an inside radius very close to 1". The springback of the material, being approximately10°, allows the inside radius to increase about $^{1}/_{32}$". As a result, an acceptable part may be formed. However, the quality of the 1" part radius might be questioned. One of the rules given in arriving at an inside radius, when making a simple angle using air bend style vee dies, is that the inside radius is a function of the vee die opening (Chapter 3-A). It was pointed out that the approximate inside radius could be calculated by multiplying the vee die opening by 0.156. If the vee die opening is made to provide an 80° included angle arc, plus the material flat needed to control the bend, the distance across the opening of the lower die will measure somewhere between 2.25 and 2.5 inches. Multiplying 2.5 × 0.156 would result in the inside radius being a little greater than $^{3}/_{8}$". What will happen when the part is actually formed?

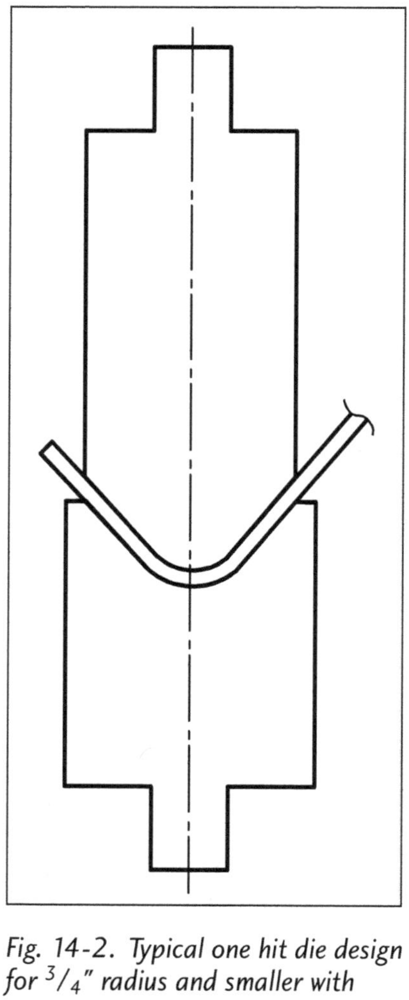

Fig. 14-2. Typical one hit die design for $^{3}/_{4}$" radius and smaller with springback allowance included.

When a part having an inside radius of less than 1" is made, the part will try to break away from the upper die and form its own natural radius. The restriction of the friction occurring on the outside surface of the material as it moves over the corners of the vee opening during the forming stroke usually prevents the part from breaking away from the nose of the die and forming to the natural $^{3}/_{8}$" inside radius (estimated above). When the inside radius is less than 1", the inside radius will normally be close to the radius machined on the upper die plus the increase in part radius that will occur from the springback in the part when the forming pressure is released. This works because the tension on the outside of the material caused by friction tends to hold the material tight enough so that the material will wrap around the nose of the die while it is being formed (Fig. 14-2).

As the required die opening becomes larger when a larger radius is to be formed, the point of contact by the nose of a larger radius die is in the center of the die. Only a small area is contacted by the upper die, which allows a much larger unsupported distance to be available between the nose of the upper die and the corners of the lower die. As a result, the restrictive load or friction on the outside of the material at the lower die corners is greatly reduced. This allows the natural inside radius, which would normally be formed when a larger vee die opening is used, to be formed into the material at the start of the bend.

Assume that a part, having a radius larger than 1", is being formed to a 90° angle in one stroke. If the stroke of the press brake is stopped when it is only halfway down into the die cavity, and if the operator looks at the end of the part where it is partially formed, it would be seen that the material has actually moved away from the nose of the radius die in the center (Fig. 14-3). The only upper die contact will be in a straight line from the corners of the lower die to the point of contact with the die. The inside leg of the part will continue (in some unknown arc shape) to the inside radius of the material that has broken away and is no longer in contact with the upper die. The only way that the material will again regain full contact with the upper die radius is to have a mating lower die so that, when the material reaches the bottom of the stroke, the lower die forces the "breakaway" area of the part to conform to the true radius machined on the upper die. Many dies have been made over the years that have the radius of the lower die made one metal thickness larger than the radius machined on the upper die. The thought is that by bottoming the radius dies with the material in place, the "perfect fit" will form a perfect part with minimal springback (Fig. 14-4). Unfortunately, the concept does not work in a production setting. Why? Someone may have tested the matched dies with short lengths of material of the same type that would be used in production and made acceptable test parts. However, it was quickly found, that bottoming a full length part throughout its entire arc

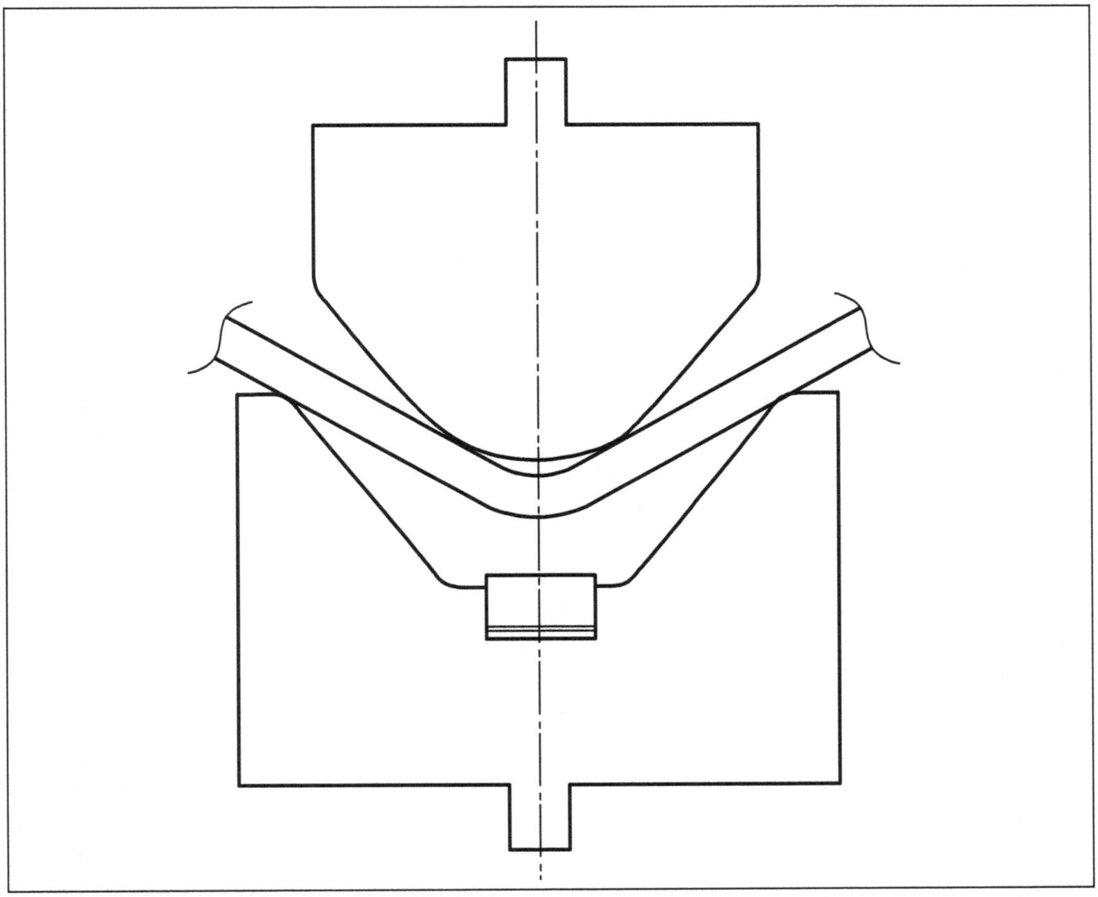

Fig. 14-3. Possible die design for a larger one hit radius die. Note a shimmable pad at the base of the lower die to help push any "breakaway" of the material from the upper die back against the forming radius at the bottom of the stroke.

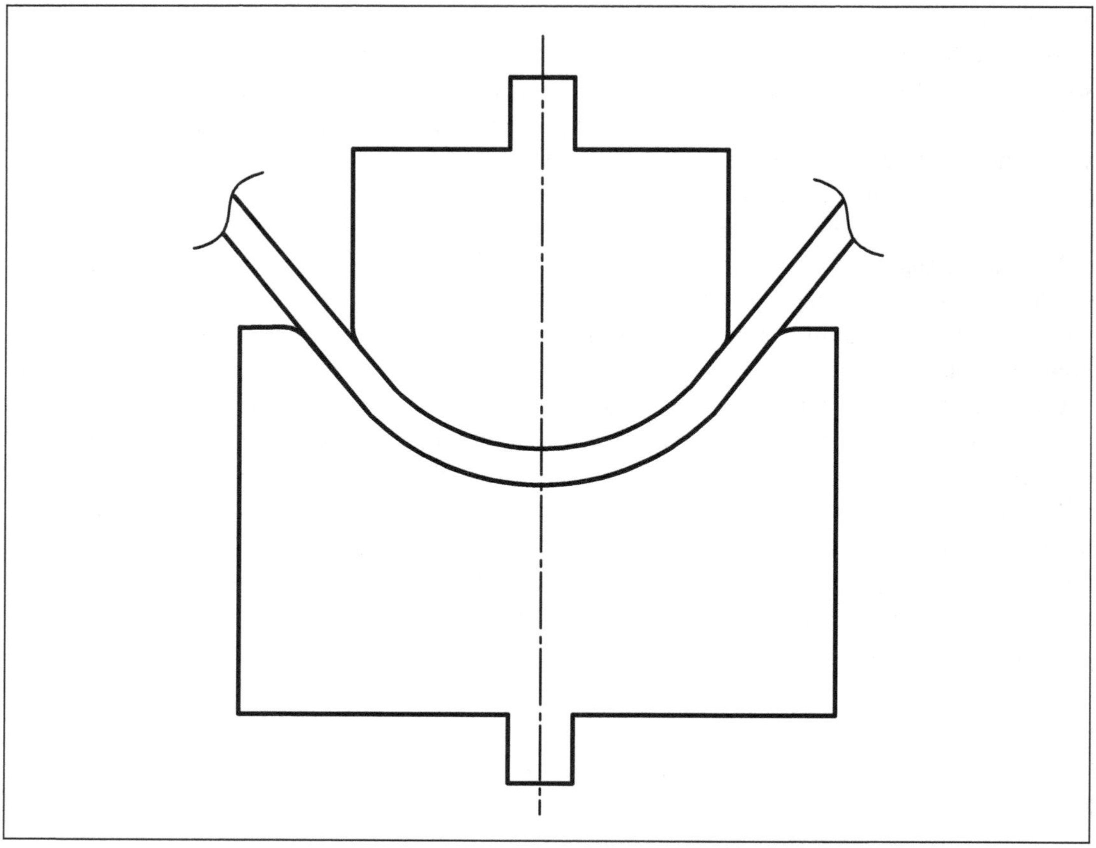

Fig. 14-4. Die shows a radius die to form a larger radius with the lower die made one metal thickness larger than the upper die radius. This design is not recommended.

is impractical. To form a radiused part and to overcome any springback that may result, the entire area that is being formed must have enough pressure applied that the compressive stress of that area is exceeded. In the majority of steels, the compressive strength of steel will equal its tensile strength. This means that if the tensile strength of the steel is 60,000 PSI, when that figure is converted to tons/in.^{2} by dividing it by 2000 pounds, the resultant answer is 30 tons/in.^{2}. If the arc of the radius die is only 1" wide, in one foot, the area would be 1" times the 12" length, or 12 in.^{2}. Multiplying the 12 in.^{2} times the 30 tons/in.^{2} compressive strength requirement to eliminate the possibility of springback, the bottoming tonnage for the radius would be 360 tons/ft. This type of forming tonnage can be justified in very few cases.

Other considerations must also be taken into account, as follows.

1) Even if the upper and lower dies were machined to perfect dimensions to allow a one nominal metal thickness fit, the actual material thickness cannot be guaranteed by the steel supplier. If the material would be thicker than nominal (but still within the tolerance range), the material would lock up on the sides of the lower die, allowing clearance to occur at the bottom between the part and the lower die (Fig. 14-5). If the material ran thinner than nominal, the upper and lower dies would touch at the bottom of the radius, but some "gap" would be present on each side of the upper die radius and the material being formed. This causes the part angle to remain slightly open (Fig. 14-6). Bottoming tonnage may or may not be sufficient to overcome the material variations.

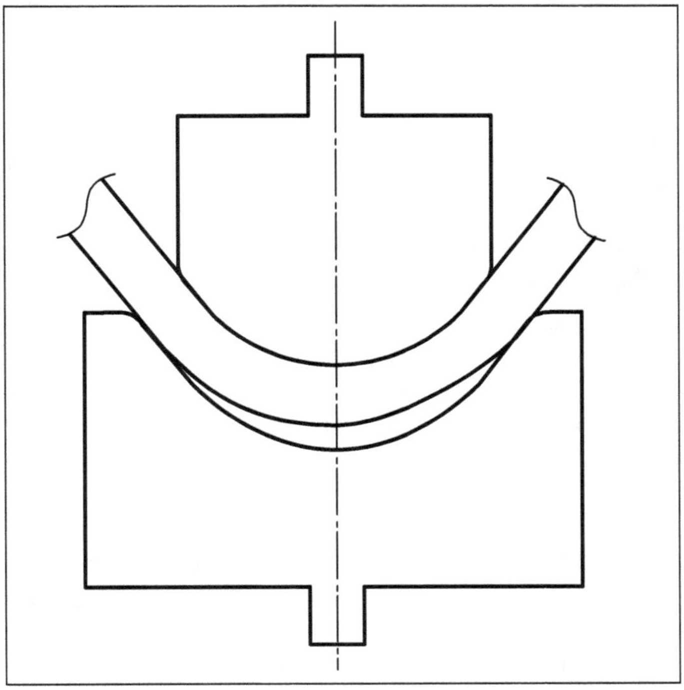

Fig. 14-5. If the die shown in Fig. 14-4 were used to form material thicker than the nominal metal thickness clearance shown, it can be seen that the thicker material will keep the upper die from penetrating into the lower die sufficiently to form a proper angle. This can occur if the material varies within its normal tolerance range.

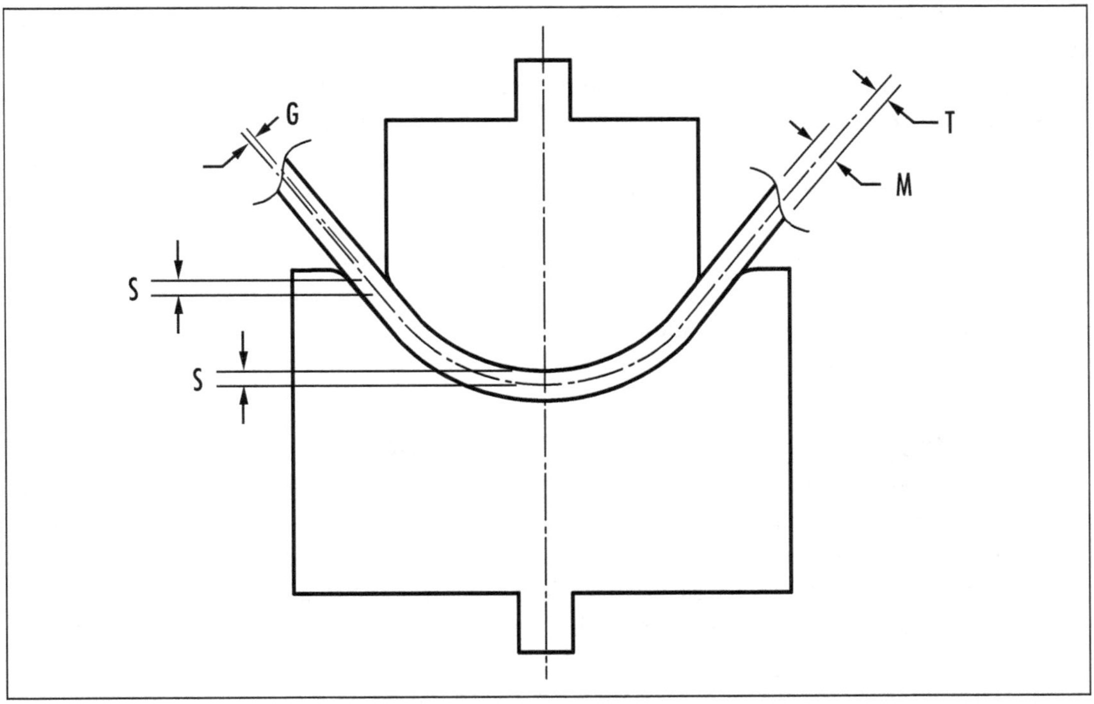

Fig. 14-6. If the dies shown in Fig. 14-4 were used to form material thinner than the nominal metal thickness clearance shown, it can be seen that the upper die could strike the material on the bottom centerline of the stroke but there would be excess clearance on the sides which would allow too much springback and angular variation. The amount of penetration to have the upper die push the thinner material to the bottom of the lower die is shown with the letter S. *The thinner material is shown with letter* T. *The original material thickness is shown with the letter* M. *The amount of gap on the sloping sides of the die created by the thinner material is shown with the letter* G.

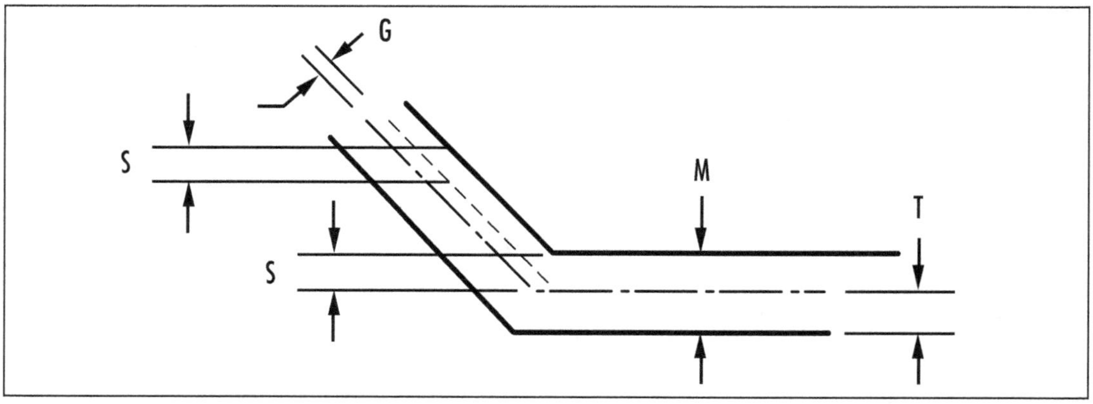

Fig. 14-6a. Sketch showing the same problem with "side clearance" when forming thinner-than-normal material, but using straight lines instead of a radius shape.

2) Another concern is overcoming the effect of the material **breakaway** that will be magnified with a larger radius and a wider lower die opening. This distortion causes too much material to be drawn into the lower die cavity. If the part is fully bottomed, the extra material remains in the die and works against providing acceptable parts.

3) If adequate tonnage is not available to bottom the radius, the upper and lower dies must provide sufficient springback allowance to assure the correct final angle can be formed. When the springback needed to form the part is not as much as the die allowance, the angle is held by raising the ram setting. However, any breakaway distortion will still be present in the part, since the lower die is kept from pushing the part back against the upper die.

4) If too much breakaway has occurred, even if it can be flattened out at the bottom of the stroke, the results of the distortion may still be visible even though the part and radius conform to part tolerances. This is the same problem that occurs if a piece of paper is creased and an effort is made to flatten out the paper. The paper can be made to again lay flat but the location of the crease will remain visible. If the formed part had the breakaway flattened, and the breakaway area can still be seen after it has been placed into an assembly (e.g., stainless steel or painted stock), the part will be judged to be unacceptable. In some areas of the country, this breakaway distortion is called **fluting**.

B) Die Design for Sheet Metal Parts Having a Radius

Dies can be divided into several groups. The first group covers tooling that forms a specific radius in one thickness from an inside radius up to approximately 0.75". The dies should be made for one specific material thickness and one unique radius. The vee opening must be wide enough to allow for a 2.5 to 3 times metal thickness flat past the machined die radius in order to control the bend. This is the same rule that has been mentioned several times previously when discussing different types of bends.

The nose radius of the upper die should be made to the radius required, unless more than 10° of springback is expected. In that case, an allowance for any increase in the inside radius that may result from springback must be machined into the upper die. A 10° allowance for springback should work for almost all mild steel radii unless the thickness is less than 20 gauge. Stainless materials, and some aluminum types that have a "T" designation, will require more than the 10° allowance. Dies to form the stainless steel should work satisfactorily if a 15° provision is used. Thin aluminum aircraft material often has strange hardness requirements that increase the springback greatly. Testing the dies to determine what angle is needed is almost mandatory. The angle cut on the upper and lower dies should be the same. This will minimize any problem with setup and die alignment. If the radius at the bottom of the lower die is made the same as the radius of the upper die (or reasonably close), any potential material breakaway that may occur will be rounded when the part reaches the bottom of the stroke. The final outside radius will then appear to be the correct radius.

Parts with a radius 3/4" to a maximum of a 4" can be satisfactorily formed to a 90°angle in one stroke using proper tooling. The same method of radius allowance and springback allowance that is estimated for the smaller radius parts is still used. However, increasing the springback allowance to as much as 15° or more may be needed. The big difference is the addition of a square or rectangular piece of mild steel bar stock at the bottom of the lower die. The lower die is machined with a slot on the centerline at the bottom of the forming arc. A piece of cold rolled steel bar stock is placed in the slot so that the top of the bar is tangent to the radius that is machined into the lower die. If the part being formed requires the maximum penetration into the lower die in order to use all of the springback allowance, it would touch the bar only when the outside radius reaches the bottom of the stroke (Fig. 14-7).

In most cases, the part penetration will be much less since the material springback is less. The problem is that the material breakaway will bring extra material into the die cavity. To correct the appearance problem, and to provide consistency in the angle, the pad can be

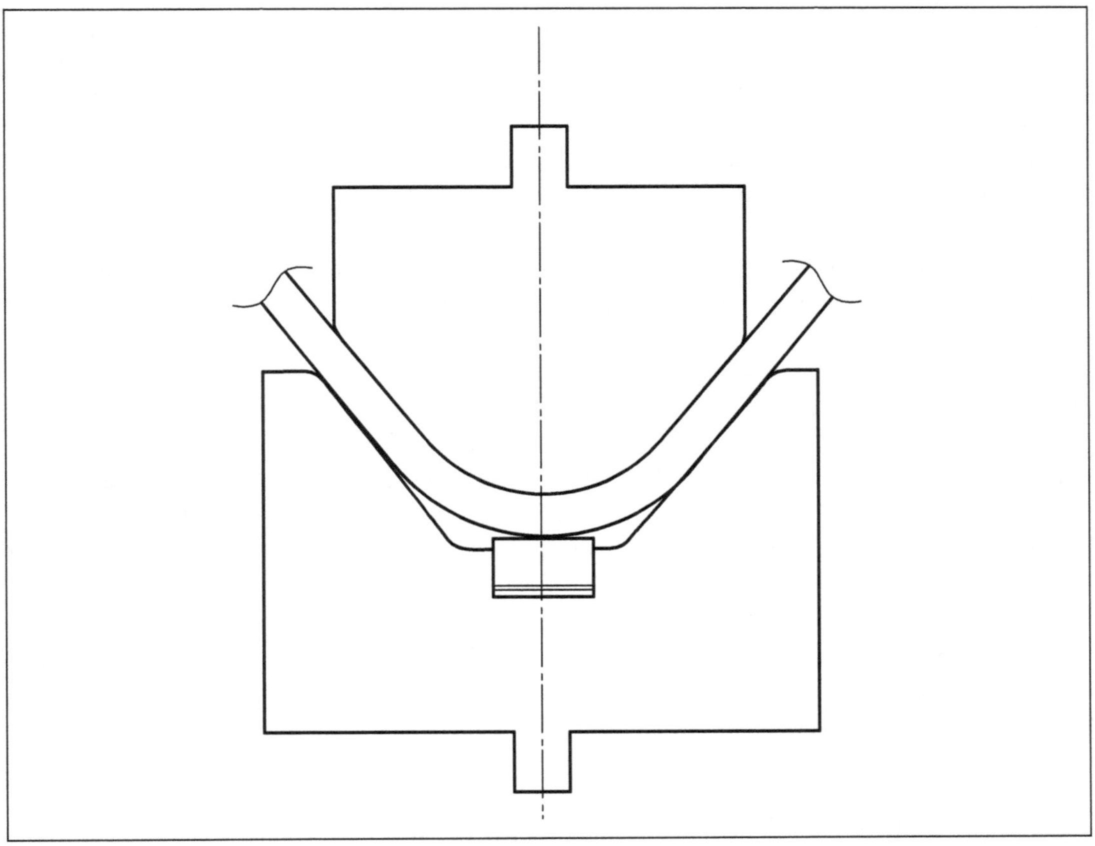

Fig. 14-7. One hit radius die forming a larger radius with the lower die having a shimmable lower pad to compensate for material breakaway. The excess material from the breakaway is pushed into the open space in the bottom of the lower die. Any change in the required part radius is minimal.

shimmed up to a position where the outside of the part radius will touch the top of the pad at the bottom of the required forming stroke. When the contact is made, the part will move away from the upper die. As the stroke moves down the final distance created by the breakaway gap, the pad will push the part back against the upper die at the bottom of the stroke. The extra material present in the gapped area will mushroom toward the front and rear of the pad in the space created when the pad was shimmed upward. The shape of the radius is changed very slightly and is difficult to identify when the part is measured.

To form larger radii to as much as a 90° angle, a one hit type approach has high risk. The larger die bars may extend to the front and rear of the bed and ram and not be properly supported. Unsupported tooling can deflect inconsistently and can rock under load, causing potential damage to the press brake. A multiple hit approach is strongly recommended using tooling that will work in the press brake size and tonnage available. A multiple hit die or bumping die (so named when larger radii or plate forming is to be done) should also be designed with a bumping pad option in the lower die. This provides three points of contact on the outside of the part, and helps remove material variation as a variable in forming a specific radius.

The dies should always be tried out prior to releasing the tooling for manufacturing, since the radius may have to be changed. In most cases this requires machining. Occasionally, the upper die can be made from a round piece of cold rolled steel bar stock. This assumes that the part radius will be acceptable after it springs back. If the radius is less than $^3/_4$", the round bar stock is a good option.

In order to determine the correct radius to machine on the upper die—to obtain a good final inside radius—the following ideas might be considered. As pointed out previously, if the inside radius is $^3/_4$" or less, and the material to be formed is mild steel, the upper die radius can be made to the radius on the drawing. If the inside radius is 1", a nose radius of $^{31}/_{32}$" seems to work in the majority of cases. A 2" inside radius might form to an acceptable inside radius using an upper die radius of 1 $^{15}/_{16}$". Larger inside radii need careful analysis. Machining a larger radius often requires using a larger bar of steel. The radius often needs a special template, used to properly machine the radius. The larger bar takes more time to machine, and the cost of making the dies becomes a big consideration.

To determine the amount of springback, and what type of radius is needed when larger radii are required in sheet metal parts, the following method has been successful. Shear or cut several strips of the material type and thickness that will be used to make the radius. The strips should be about 1" wide and 12" long. Care must be taken that the strips lay flat and do not have any distortions such as "bow" or "twist" which are caused by shearing narrow strips. Some factories have an inventory of different diameter bar stock. Bar sizes may range from 4" to 8" in diameter. If the needed round bar inventory is not available, it will be necessary to use thick plywood or other wood that approaches 1" thickness. The wood can then be saw cut to various radii that have a 180° arc but have a flat diameter. The wooden pieces should look like a semicircle. Using a piece of large blank paper, draw a series of full size circles that start with the required radius and includes at least sixteen larger and smaller circles in $^1/_8$" increments. Test pieces are then laid on top of the paper circles to determine what radius has actually been formed.

If bar stock of various diameters is available, the test can be done at the storage location with the bars left in the racks. If the test is to be done using the wooden semicircle pieces, the wood will have to be held in a conventional vise. The testing is done in the following manner. At the end of each bar, or on the top of each semicircular wood section, make a mark using an ink stick, at the "12 o'clock" position. Make additional marks at 10:30 and 2:30 to show the equivalent of a 90° arc at the top of each template. Take a strip of the 1" × 12" steel (cut from the same material that will be used to make the final part). Using a "C" clamp, fasten the center of the strip to the top of the bar or wooden template so that the contact point will be located at the 12:00 mark. Take each of your hands and form each side of the strip down the arc of the bar or wooden semicircle until the marks at 10:30 and 2:30 are reached. The strip should have been formed to a 90° arc (Fig. 14-8). Carefully release the strip from the "C" clamp. Lay the formed strip on the paper circle template and match the part to the closest radius that fits the inside of the arc. Record the radius. Next, measure the formed angle. The original radius is known, the radius after the strip has been formed is known, and the amount of springback from bending the part to 90° is now known. Another test must now be made in a similar manner to determine what radius will be required to obtain a final radius closer to the part drawing radius. In most testing, the number of bars available, or the number of wooden templates that have been cut, will not provide an exact answer. If three tests are made using radii that are in the area of what is anticipated as the final die radius, the tests will give the person designing the tooling enough information for a close estimate for the initial machining of the dies. The machined dies should form a part very close to not only the radius required, but also have sufficient springback allowance that the dies can make acceptable test parts. Minimum remachining should be expected. This testing concept is provided since

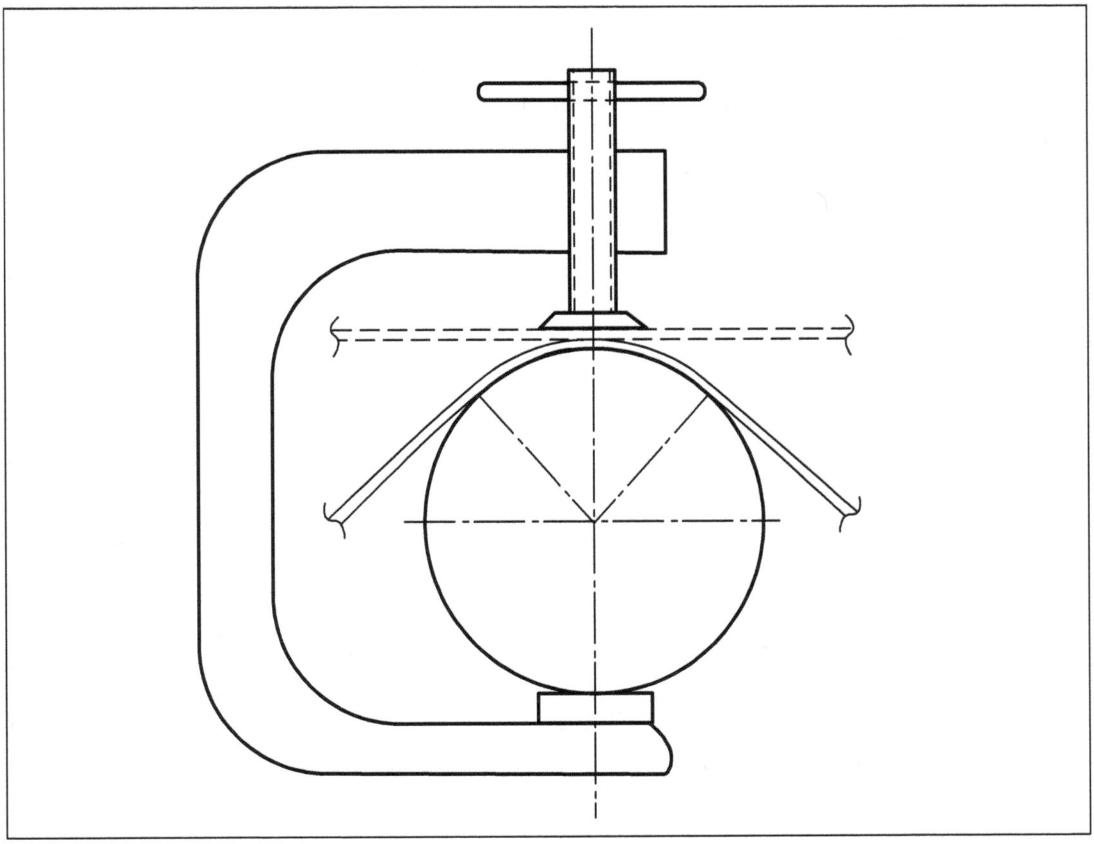

Fig. 14-8. Clamping a test strip to a known diameter bar of steel in order to hand bend the material to a 90° bend, let it springback, and measure the strip for springback and the resultant inside radius.

many companies actually make their own upper dies, and form small lot sizes using rubber or polyurethane pads. Refer to Section G in this chapter for suggestions for making "bar and stem" upper dies.

C) Radius Bends in Plate

If the parts having a radiused shape are to be made from plate, the radius is usually larger. If the material and radius is compatible with available bar sizes (such as using a 6" × 6" die steel bar as a lower die used to make a 2" inside radius in $^{1}/_{4}$" plate), the same method used to form sheet metal parts can be used. The lower die must have the bumping pad to control part breakaway that also occurs in heavier materials. The majority of plate, however, will be formed using a bump forming method. The upper die will be machined from a wider bar so that the higher forming load required for bending plate will be carried safely by the ram nose. The lower die should be made wide enough to hold front and rear forming inserts that can be adjusted back and forth to allow the vee die opening to be changed.

The lower die should have a special location for an adjustable bumping pad (a simple square or rectangular cold rolled steel bar) so that when the part is formed down into the shallow vee die opening, the part will touch the top of the pad and create three points of die contact with the outside radius of the part (Fig. 14-9). Since most press brakes used to form plate are hydraulic, they often can be set to reverse at a specific tonnage. This option allows the outside of the material to penetrate into the vee die until it strikes the "bumping pad." The upper die will reverse its travel when the tonnage set by the operator has been reached. This concept keeps the outside of the material consistent, while allowing the reversing position of the upper die to change when any differences in material thickness would cause the bottom stroke setting to vary.

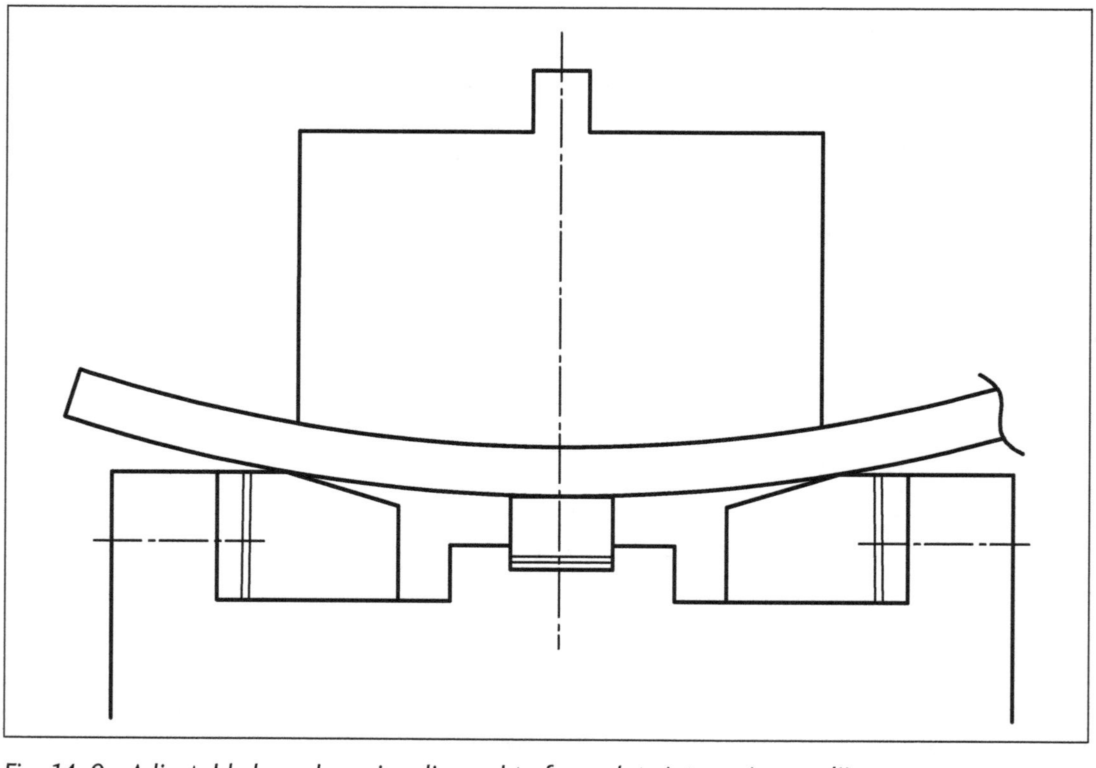

Fig. 14-9. Adjustable lower bumping die used to form plate into various radii.

To form an acceptable radius in plate, it is best to make some tests first. Most plate bending is very low volume. If the initial bends are made in new parts, chances are that the parts may have to be scrapped. The first item needed is a template that has the correct radius on one side. Each company has its own preference. The template can be made from anything, including sheet metal, plywood, or cardboard. The height of the template is restricted, since the operator must have room to locate the template on the top of the formed part between the part and the nose of the upper die. The length of the template is also limited by the stroke distance that is being used (if it is available).When the stroke is up, the operator must be able to use the template to measure the formed radius at each end of the part. The best templates are no longer than 6" to 12". Longer templates make it very difficult to handle, plus the operator cannot see the radius that has been formed in the rear of the dies since the stroke distance is normally not sufficient to perform a full inspection.

The second concern is tonnage. Does the press brake have sufficient tonnage to bump form this part? Review the vee die opening of the lower die that will be used to bump form the radius. Compare the tonnage needed to air bend that thickness of material using that vee die opening to the tonnage chart. Add any bending factors that must be considered for the material that is being formed. It is possible that the vee opening selected may have to be opened in order to reduce the estimated tonnage. To bump form the part and possibly reverse the ram by having the part contact the bumping pad in the bottom of the vee die, the estimated air bend tonnage above must be increased by about 20% to 40%. Many operators actually use much more than the minimum tonnage needed to bump form the part, since the vee opening available or the die design may not allow any adjustments.

To form the part, multiple strokes are required. The part is indexed forward about one-half of the vee die opening after each stroke. The first stroke requires that the edge of the material rest across the vee opening and be supported on the back surface of the vee opening by at least one metal thickness of material. As a result, the part will have a noticeable flat at the edge. The distance of the flat will be measured from the tangent point of the inside radius to the edge of the material Depending on the opening of the vee die selected, the flat can easily be 2 to 4 additional metal thickness in length. In some cases, the operator actually pulls the part edge backwards after the first several bends so that the edge of the material rests close to the bumping pad surface. By rehitting the part in that location, it is hoped that some of the edge flat will be formed into a radius shape and minimize any problems caused by a noticeably unformed leg. The edge flat can be a major problem if the material is formed into a tubular shape or is to be rolled into a pipe. The edge flat, which may have been formed on both edges of the blank, will cause the part to have a peak at the joint or seam. The seam is defined as the location where the two edges of the material join together to create a round tube shape—this is not the seam that was described in the hemming and seam chapter.

D) Locating the Bends When Bumping a Radius

At this point, it is assumed that the dies have been properly set up to make a good radius, the forming tonnage is within the press brake limits, and the part will require multiple hits to obtain the arc required to make the part. Note that if the radius tooling allows the part to be completed in one stroke, simple back gaging (or, in some cases, front gaging) would normally be used. Multiple hits are required when: the forming die would have to be made so wide that it could not be safely mounted into the press brake, the cost of the tooling would be excessive, or attempting to form the part in one stroke would limit the ability of any operator to correct for possible material variations.

After the first bend is made in a piece of plate, or if the material is sheet metal that requires a large radius, the next bend becomes a challenge. If the part is moved over the lower die opening so that the material is laying flat across the opening and supported on both sides, when the next stroke is made, the resultant shape will look like the part has a radiused shape, a flat in the material, and then another radius. This method will not make an acceptable part. The closer together the arcs produced by each stroke are kept to each other, the better the appearance. The additional strokes increase the amount of handling. Additional strokes also add to the total time to make the part. Therefore, the cost of manufacturing must be considered.

If the increments are about one-half of the vee die opening, the radius produced on many parts will be acceptable. The concern is, when the part is positioned to make the second hit, the front (longer) leg of the blank is tilted up in the air at some angle. This occurs since the material in the rear of the die is radiused from the previous bend, while the unformed material in the front of the die is still straight. If the leg is not too long or too heavy, the operator may be able to hold it in position in order to make the next bend. If the part is too heavy (e.g., plate), the front leg must either rest on support arms that have provisions to locate the approximate angle required to allow the next bend to be made, or in some cases the operator can use a crane to hold the material at the approximate angle needed .

Some operators do not take the time or effort to create a support arm that allows the part to be tilted when it is indexed into the lower die to allow part contact with the front and rear corners of the vee die. They simply leave the material flat, not tilted, and still index the part one-half of the vee die opening (or less) while attempting to form a smooth radius. Laying the front leg flat with the arc of the first bend only partially across the vee die opening with no material contact at the rear of the lower die will not work very well. As soon as the stroke is made, the straight leg extending from the die centerline to the front corner of the vee die is forced down into the die cavity, hopefully allowing the additional radius forming to take place. Since there was no initial contact in the rear, the part will move toward the operator as the upper die tends to force the material back toward the previously formed radius. Often, one end of the part will move slightly differently from the other, causing a spiral shape to be formed. Holding consistent gaging from stroke to stroke under those circumstances would be more luck than expectation.

Assuming that the problem of how to index the part forward consistently to make an acceptable radius has been solved, the next concern is the increments needed to give the best radius in the least number of strokes. This requires testing. A simple method is to obtain four pieces of the material approximately 12" long and wide enough to allow the operator to locate the increments safely during the testing. Lay the first blank on the top of the lower die with the edge of the part positioned in such a way that it will not slip into the die during the forming stroke. Make an inkstick or some other type of mark on each end of the blank directly over the front radius on the operators' side of the lower. Move the blank forward one half of the vee die opening. Make another mark on each end of the blank located on the top of the same radius. Continue the indexing and marking of the top of the blank until there are more marks on the part than the distance required to form the radiused shape. By using the first blank and duplicating the marking that has been made on the second blank, two identical pieces with gage locations are now available.

Take the first test piece, and locate it in the initial position by carefully aligning the inkstick marks for the first bend over the lead-in radius of the lower die. Form the initial bend. Index the part forward. Locate the part in such a manner that the front leg will be slightly up, and the initial arc that was just formed is touching the back area of the die cavity at some point. Carefully align the second inkstick mark at each end of the blank with the front lead-in radius. Make the second stroke. Continue the indexing, aligning the next mark for each bend until the approximate angle designated on the drawing has been reached. Do not attempt to form the test piece to an arc of more than 120°. If the arc is too great, there is a chance that the leg in the rear of the arc that has just been formed will hit the upper die during the forming stroke and distort the part.

Now, take the initial part out of the press brake and check for radius, appearance, and angle. The angle will almost always have to be corrected. Often the appearance of the radius will be close but not close enough. It will be determined that the gage locations should be changed slightly. The second test blank, already marked with the locations used to form the first piece, should be used to make the corrections for the second part. It is always best to use a different color marker so that there is no confusion when locating the new gage points. Now mark each of the test blanks with numbers from 1 through 4. The third blank should be marked with the new locations so there is always a flat blank left with the last gage positions identified. The second part is now bent and inspected. It is normally much better but needs some fine tuning. The third blank will be marked with any new gage positions. All of the last locations are now made on blank number 4. Blank number 3 is formed and is inspected. It is usually acceptable. If not, additional blanks will have to be made. When a good part is made, one flat blank with the final gage locations should always remain.

A simple template can be made from a piece of cardboard, approximately 4" wide and the exact width of the actual blank that is to be used to make parts. Notch the cardboard with little vee cutouts on one side with one notch located at each one of the gage locations marked on the flat blank. Using the template, each of the parts that require a radius and angle similar to the one that has just been tested can be marked on each end of the flat blank prior to forming, which will provide consistent parts each time the radius has to be formed. The fastest way to mark the parts is to lay the template on one edge of the blank and spray paint down the edge on the top of the template. Gage marks will be made at each notch identifying the proper gage location. The template is then turned over and located at the other the end of the flat part. The notches in the template are again spray painted, providing a symmetrical gage pattern on the part. The gage template life will depend on the volume of parts that have to be made and the type of material used to make the template.

E) Tube Forming

Tube forming is done on the same type of lower die that is used for the radius bends described above. The upper die nose will often be made from a bar that will match the width of the vee die opening that has been selected. The upper die will have a relatively thin stem mounted on the top of the radiused nose bar in order to provide maximum clearance so the tube can close without being distorted during the final stroke. After the last stroke needed to form the tube, the part is pulled off the upper die endwise (Fig. 14-10). High tube dies normally take up a great deal of die space on a hydraulic press brake. Care must be taken to ensure that sufficient stroke remains to load the part, gage it, form each increment, and remove the final shape. It probably will be necessary to remove the filler block to obtain additional space. Provision must be made to fasten the lower die to the bed top if the die holder is taken out.

If the tube is larger than 4", the part can have one edge bump formed to an arc of 120°. The piece is then turned 180°, and the other edge bumped to a similar arc. The part is then nested around the upper die, moved back and forth using increments that allow each additional stroke

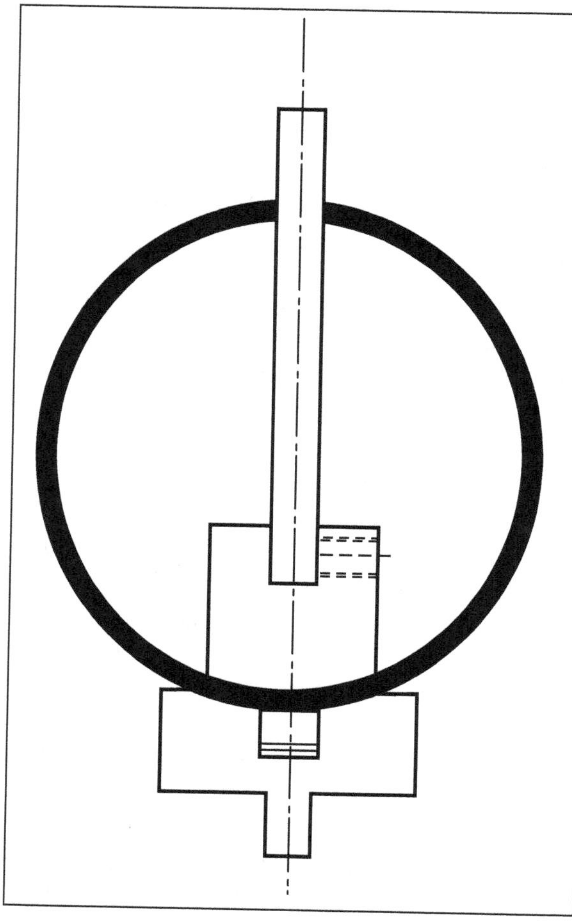

Fig. 14-10. Common die design for bumping thinner material into a tube shape. Note the thin stem on the upper die to allow minimal gap when the final hit is made. The part must be removed endwise.

to gradually close the part around the stem of the upper die. The last hit, which closes the tube, must be directly opposite the opening of the tube. Often the ram setting is raised slightly to prevent the final hit from locking the two edges at the seam of the tube tightly on the thin stem of the upper die. If the part is pinched too much, not only may the part be distorted, but it may be next to impossible to remove without using special pulling devices. It is best to leave the gap open slightly and close it at the time the seam is welded in final assembly.

Smaller diameters can be formed in a similar manner. Often each side is formed to an arc of approximately 90°. The lower die is changed to one that has a die cavity somewhat less than 180°. The third stroke to complete the tube, using the original upper die and the new lower die, wraps the blank around the upper die until it touches the thin stem of the upper die. The upward stroke causes the upper die to lift the tube up out of the lower die. It is removed by pulling it endwise off the upper die. The shape of the tube will intentionally be open to allow for removal. It must be closed after removal in a separate operation prior to welding the seam.

If the tube must be closed at its seam, and be held to a tight tolerance as far as diameter or roundness, a special closing die can be designed. The tube is made in the three hits required to make an open tube. A special sizing die is needed to hold the exact diameter of the outside dimensions of the final tube when the dies are in a closed position. The part that has been partially closed is placed on the top of the lower sizing die. It cannot fit down onto the die since it is still slightly open. The mating upper die moves down and forces the part down into the lower cavity and forces the open upper section of the tube to roll around the radius in the upper die into a closed tube shape. At the bottom of the stroke, tonnage must be built up to permanently "set" the material. This is done by applying enough pressure so that the compressive stress of the steel has been exceeded (Fig. 14-11).

The sizing or "setting" pressure is obtained by calculating the area that is being compressed. If the tube was cut across its diameter, the exposed material would be one metal thickness on either side of the tube (or a total of two metal thicknesses of the part thickness). If the area under compression is calculated for 12" of a part that is made from 16 gauge mild steel, the area would be 2 × 0.060" (the nominal thickness of 16 gauge) × 12", which equals 1.44 in.^{2}.

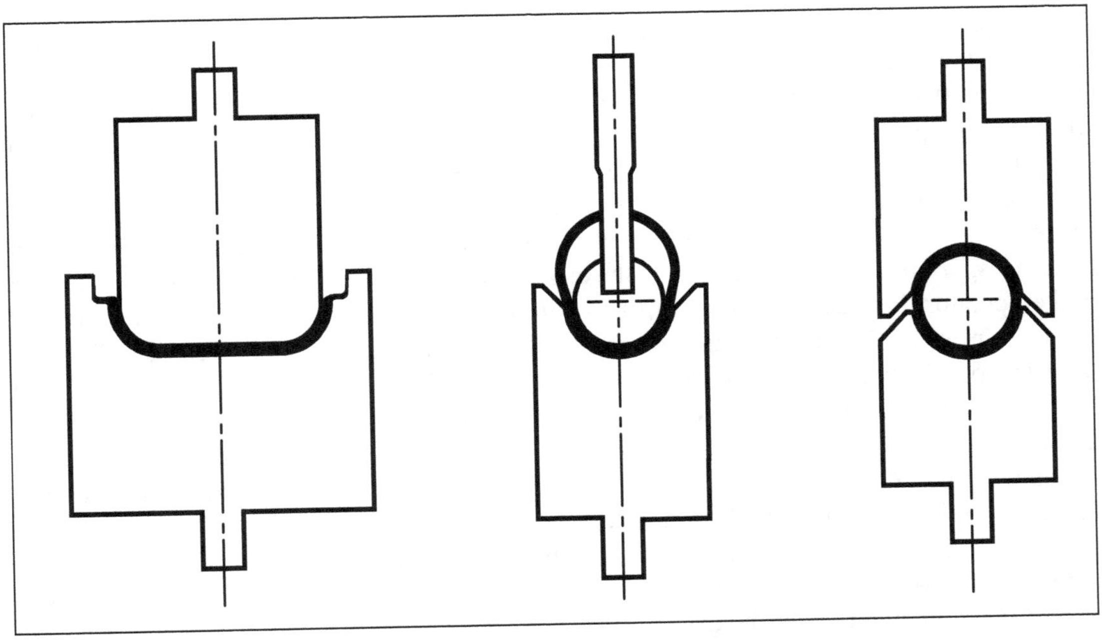

Fig. 14-11. Special three stroke dies to form thin material (14 gauge or thinner) into close tolerance tubes. The sizing used for the final closing requires high tonnage which may limit the length of the parts that can be made.

The compressive strength of mild steel is the same as its tensile strength. That means that if the tensile strength is 60,000 PSI (30 tons/in.^{2}), the estimated load requires to size the tube would be 30 tons/in.^{2} × 1.44 in.^{2}, or 43 tons/ft. It can be seen that sizing dies not only require a great deal of tonnage, but the tooling will be very expensive to manufacture and test to optimize the final part. In some cases, removal of the part from the sizing die is difficult and may have to be reviewed to ensure operator safety. It should be pointed out that very few parts are made in this manner, and they are generally limited to diameters of 4"or smaller and relatively short lengths.

F) Tolerances of a Part Formed by a Radius Die

Engineers have a habit of dimensioning sheet metal, and even parts made from plate, as if they have been machined to their final dimensions. As a result, what is called for on the drawing and what will be made in the press brake area is usually not even close. Engineering drawings have been reviewed calling for a 4" inside radius bent to a 90° angle in 16 gauge mild steel, held to a tolerance of ±$^1/_{32}$". If the engineer would overlay three pencil drawings of 90° angles made to radii of 4", 4 $^1/_{32}$", and 3 $^{31}/_{32}$", those drawings would be so close to the same pencil line that it would be difficult to tell the difference. Most radii formed in sheet metal are measured using a radius gage that measures the inside radius to see how close the part has been bent to the required radius. If the part radius is too small, the radius gage will not fit very well, and will show what appears to be a large gap between the gage radius and the inside radius of the part. If the part radius is too large, the radius gage only touches the inside radius of the part in one location, and it will roll around enough that the person doing the measuring may have a difficult time being sure that the formed angle is close to 90°. Another problem occurs if the part is handled in such a manner that the formed part is dropped. Dropping a part often causes the radius and angle of the bend to open slightly. As a result, a part that was judged to be acceptable when it was formed may be rejected by the time it reaches the assembly station.

It is best if some thought is given as to how the part, having a specific radius that will be formed using press brake tooling, is going to be used. The operator must be instructed as to how much variation from the radius gage is acceptable. The difference in daylight under the radius gage may not affect the use of the part. If the true shape was somehow identified, the shape actually may be a compilation of several different radii with a range of ±1". Unrealistic tolerances, printed in the little boxes on the engineering drawing, will not be held by the press brake operator. If management attempts to enforce them, the cost of the part will increase to the point it will be impractical to process. Clear directions should be given that allow the operator to quickly steady the formed part, locate the radius gage in the arc, and visually see if the daylight under the gage is acceptable, or if some handwork is needed to make the part acceptable. This concept is important with thin materials; it becomes critical when a radius is being formed in plate. A piece of plate is too heavy to take out of the machine, measure the radius, relocate it in the tooling, and repeat the process every few strokes. As pointed out previously, the operator must have a simple template between 6" to 12" long that can be used at the press brake to determine if the radius being formed will be acceptable. Instructions as to how much daylight under the gage will be allowable should be in writing to minimize any second guessing.

The reason that the tolerances for radius forming usually specify daylight under the template is that the space under the template means that the actual radius that has been formed is smaller than specified. Once the part is formed and is moved out of the fabricating area, the part is often handled, stored, dropped, and processed through the assembly area. All of these activities will cause the original radius formed on the press brake to open up or become slightly larger. A formed part can always have the radius made larger, either intentionally or by accident. The part will never have the radius made smaller unless an additional operation to reform the radius is required, or if the part is accidentally damaged.

In many cases, the tolerances of the radius and final angle placed by engineering have to be modified. The standard plus or minus (±) tolerance sounds good, but if the angle is made open or closed, often the part cannot easily be assembled. It is always best to make the parts with the correct angle specified, but in some cases it may be best if all of the angular tolerances were either all open or all closed. Each part that has a specific use should be reviewed to determine how the part is to be used and how the tolerances should be controlled.

G) Making Your Own Upper Radius Dies

Many times, it is required to form a radius in a part but the needed radius die is not available to meet the production timetable. The first step in order to determine if a simple solution is possible is to make a basic full size sketch of the bend angle formed to an 80° included angle. Measure the straight legs of the part to locate the minimum width of the lower die. Check the current tooling inventory to determine if a possible lower die is available. In many instances, only an upper die must be made. To make an upper die, purchase a round cold rolled steel bar of the proper length. The diameter will be limited to what is available from the steel service center supplying the bar. Select a diameter that will provide the needed radius minus $^1/_{32}$" per inch of radius. For example, to form a 1" radius in mild steel, a 1 $^{15}/_{16}$" diameter bar (if available) will provide a $^{31}/_{32}$" radius. The smaller radius will allow the material to spring back to about a 1" inside radius. In addition, a $^1/_2$" wide × 4" to 6" high cold rolled bar will be needed to provide a mounting stem for the bar. The height of the 1/2" bar depends on the available die space when a filler block, the lower die, and the round bar and stem are mounted in the machine.

When the bar and stem are ready to be assembled, place the lower vee die in the filler block and tighten the set screws. Install the 1/2" stem in the ram nose and clamp it as would be done with any simple upper die. Raise the ram to allow sufficient room for the round bar to be located in the vee opening, and place the round bar into the vee die opening. Slowly lower the ram so that the lower surface of the 1/2" stem contacts the round bar and is pushed completely into the ram clamping slot; do not build up any tonnage. Raise the ram slightly. Align the round bar by moving the lower die and filler block front to back so that the round bar is centered on the centerline of the 1/2" bar. Lower the ram so that the two bars are touching. Tack weld each end of the round bar to the 1/2" bar, making a 1/2" long weld in the front and rear of each bar, plus a small tack in the center of the bar length. Raise the ram enough to remove the now-welded upper radius die. At a separate location, add additional tack welds about 1" long in the front and rear of the new die. The welds should be spaced about 12" apart along the die length, with the weld locations the same on both sides of the bar.

This type of die will often make 500 to 1000 hits before the forming load will deform the point contact of the round bar with the 1/2" bar surface and the welds will start to crack. Attempting to weld the bars full length will cause the bars to bow due to the heat of the welding, and make the die difficult if not impossible to use.

If a lower die must be made, the problem is more complicated. For a 90° bend, the upper die must penetrate about 40% of the vee die opening. It will be important to make a sketch showing both the upper and lower dies, since the die height of the "home made" dies will be important. The lower die will probably be wider than the filler block and may have to be mounted on the bed top. This means the filler block would be removed, and the stem of the upper die made higher to ensure that the die space created will allow the upper and lower die to come together and form the part.

The lower die can be made from two of the same diameter bars used to make the bar of the upper die. The two bars must be spaced apart so that the point of contact of the formed material on each of the lower round bars is at the same place, so that a machined lower radius die, the one sketched originally to determine the size of the vee opening, would contact the part. A rectangular plate, made of cold rolled steel at least 1" thick, is used for the base of the lower die. The width of the plate is sufficient to support the two round bars plus some wedge supports mounted on the outside of each round bar to help keep the bars in a fixed location. The wedges should be spaced 12" to 18" apart, depending on the diameter of the round bars and the thickness of the material being formed. Care must be taken not to put too much heat into the material when all of the bars and wedges are being welded to create the lower die body.

The main limitation of this die will be the length of legs of the part required. In order to span the round bars making up the lower die, and not slip past the contact points of those bars into the opening at the bottom of the die, the edge of the part must extend past the rear die bar and must be at least 0.5". It is also pointed out that a method of clamping the lower die to the bed top must also be designed. Often, drilling holes through the lower plate and using the clamping method used to hold the filler block secure is easiest. This die is certainly not for high production, but may allow enough parts to be formed to meet the immediate need.

H) Forming a Radius if the Part has a 90° Flange at Each End

Some part designs require that the radius be formed after each end of the blank has been flanged (or some other shape made) in order to provide a place for mounting the part, or simply to give rigidity after the piece is installed. If parts with this design are formed using special press brake tooling, the radius will not be consistent along the bend line.

The lower die must be made with end plates to help contain the previously formed flanges. When the bend is made, the radiused bend area will spring back the amount normally expected for a radius of the size and shape for the material being used. The problem is that the radius next to the flanges will not be able to spring back because the flanges at each end have introduced a great deal of stiffness into the part. When the flanged leg is bent, assuming it did not distort during the forming cycle, its springback will be much less than the radius shape along the remaining length of the bend line. As a result, the bend line will have a closed angle at the flanged area gradually opening up to the angle expected if the material had normal springback along the entire bend line.

The higher or stiffer the side flanges, the more variation will occur. If the side flanges can be made shorter, allowing more springback in the flanged areas, the angular difference will be reduced. Some designs have been modified so that the flanged area has been eliminated or reduced sufficiently in the area of the radius that the formed difference along the bend line was considered usable.

Occasionally, additional stiffening ribs can be added to the part. They will be located parallel to the side flanges and spaced in the radiused area to reduce the springback in the part length, and hold the formed angle consistent enough along the bend line that the part can be used. In most designs, the side flange distorts or wrinkles excessively during the forming stroke, so that the part has to be redesigned to not only minimize that distortion but also to correct the angular springback problem. In all cases, the part will require a great deal of time and money to develop an acceptable shape.

Chapter 15
Rib Shapes

To stiffen a flat sheet in order to keep the part from flexing, it is often necessary to place a "rib" configuration the full length of the part. Rib designs vary widely depending on the purpose and how the rib may be used in the part after all of the forming is completed. The three basic designs are: the simple vee shaped rib, which makes a "V" groove in the flat surface of the part; the round or semicircular rib shape, which does the same thing as the "V" shaped rib but provides a different appearance; and the rectangular or "hat" shaped rib. The rib shapes are often seen in products that range from the sides of truck trailers to the roofs of commercial buildings (that product is called roof decking). Rib shapes have been designed to provide room and strength to lift pallets of metal skids, provide a method to drain liquid from one surface to another, and provide strength and stiffness in a variety of panels. Rib shaped parts are a common product made by press brake users. The discussion in this chapter cannot review every shape, but will provide information that may help minimize problems that will occur when a new rib design is to be manufactured in your plant.

In press brake work, it is important to remember that a press brake is not a press. In some instances, press type tooling can be used in a press brake. When rib shapes are formed using a press brake, the material is pulled in from each side of the blank. As a result, the width of the blank will be reduced by the amount of material needed to form the rib shape. Some parts have a design where the rib does not extend the full length of the part, but is stopped a certain distance in from each end of the length of the blank. When the part is formed, the metal in the area where the rib is formed is drawn down into the lower die cavity, but each end of the blank that is not being formed is suppose to remain flat. In a press, a heavily loaded pressure pad will hold the flat blank around the perimeter while the rib shape is "drawn" into the material. In some cases, the material is actually stretched or elongated enough that the ribbed areas of the part plus the unformed ends remain almost flat. In most cases, the material around the perimeter of the oversized blank that is outside of the pressure pad or clamping plate (normally called a draw-ring) will wrinkle and have to be trimmed off in a later operation. In press brake work, if the dies can be designed to form an acceptable rib shape, and still flatten out each end of the part, then when the pressure is released, the part will pop up into a twisted shape and no longer lay flat. This is called an **oil can** in the part. The name is derived from the distorted piece of metal mounted in the base of a simple old-fashioned oil can. The design allowed the user to turn the can over and, by pushing down on the distorted base, oil could be pumped out of the can through a flexible stem and nozzle.

When looking at the new design of a rib, if it is to be made using a press brake, it should be a full length rib. Since it is not usually possible to make a bend, turn the part over to make a second bend, and turn the part over again to complete even a simple rib shape in most parts, the rib will require a special set of dies designed to form the section in one stroke. To make an acceptable rib shape will require the cost of the special tooling plus a press brake with enough forming tonnage. Inspection of the drawing will usually confirm that the radii called for at each bend line of the rib is going to be difficult if not impossible to hold. Often, the depth of the rib is measured from the top surface of the material to the outside of the rib shape. This can be a problem if there is material variation, or if the rib design requires a great deal of springback after the forming pressure is released.

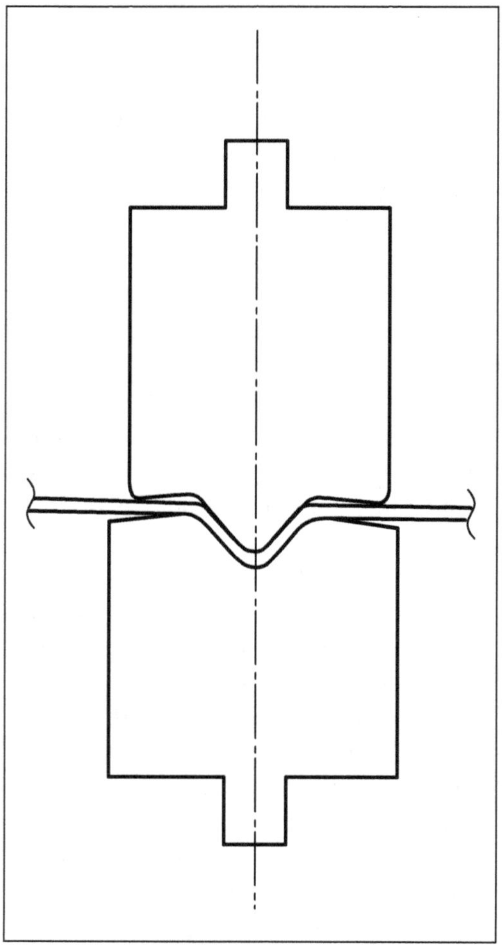

Fig. 15-1. Tooling used to make a simple V shaped rib.

A) Basic Vee Shaped Rib

The upper die used to form a vee shaped rib looks at first glance like a flattening die with a vee shaped nose sticking out in the center of the flat surface. The lower die looks like a mate to the upper die. But looks can be deceiving (Fig. 15-1).

The upper die nose is made to air bend the 90° rib shape. The initial nose radius of the die is made approximately one metal thickness (or larger if a larger radius is needed on the section). The upper die bar must be wide enough not only to include the width of the 90° rib shape, but also wide enough to push the horizontal legs back down without causing abnormal marking on the top of the part. The lower die width will be similar to the upper die. The vee section should have the same configuration as the upper die, as far as angles, to allow visual die alignment during the tooling setup. Visual alignment allows very close adjustments to be made quickly with minimum risk of error.

A major decision is determining the best radius to machine on the corners of the vee die opening to create the top of the rib shape. The part drawing often shows about a one-metal inside radius for all three bend angles of the rib shape. The actual vee die opening will produce a normal inside radius (remember the 0.156 × the vee opening rule) in the bottom of the rib shape during the initial forming of the part. As the metal starts to lock up at the outside corners of the upper die and the corners of the vee opening when the material is drawn down into the lower die, the bottom inside radius is reformed into an elliptical shape. The part that will form the top two angles of the rib shape will initially form to an inside radius that is a function of a theoretical distance between each outside corner of the upper die and the nose of the upper die. If this distance is multiplied by the 0.156 times the vee die opening formed, the result will always be much larger than one-metal thickness. If each lead-in radius of the lower die has a radius that is too sharp, the larger radius made on the part during the forming cycle will loop around the sharp radius, overbend, and then spring back erratically. The problem is magnified due to the fact that the nose of the upper die is normally forming the part to a 90° angle while the top flanges are being formed to a 135° angle. The springback between the two angles will not be the same.

To allow the special tooling to produce a good part at a consistent forming tonnage, it is best to have the drawing modified to allow a larger inside radius to be made at each of the 135° angles. This will provide a starting point for part consistency. The tooling manufacturer should initially machine the lead-in radii at least two times the metal thickness. Testing may require that the radii be increased.

When the pressure is released, each angle will spring back. Hopefully, the top surface of the rib shape will be flat. More than likely, the top flanges will **wing up** (slope up) or **wing down** (slope down). After inspection of the part, checking part shape and making several parts at different ram settings within an acceptable tonnage range, the decision will be made to remachine the top (flat) surfaces of each die to adjust for the **winging**. Often this means that top surfaces of the lower die are machined either up or down to provide for the springback. The upper die is remachined to match the angles on the lower die. When the dies are retested, the top surfaces will be very close to flat. If the ram adjustment is changed slightly to compensate for normal material variations, the top flatness can be controlled, but the depth of the rib will be slightly changed. Normal allowable linear tolerances should allow for the possible depth fluctuation.

The deeper the depth of the rib, and the sharper the lead-in radii that have been machined on the lower die, the higher the risk of wire draw becoming a problem (wire draw is discussed in Chapter 11-C). The distortion will magnify any springback problems. This concern is very real when a rib has a sharper

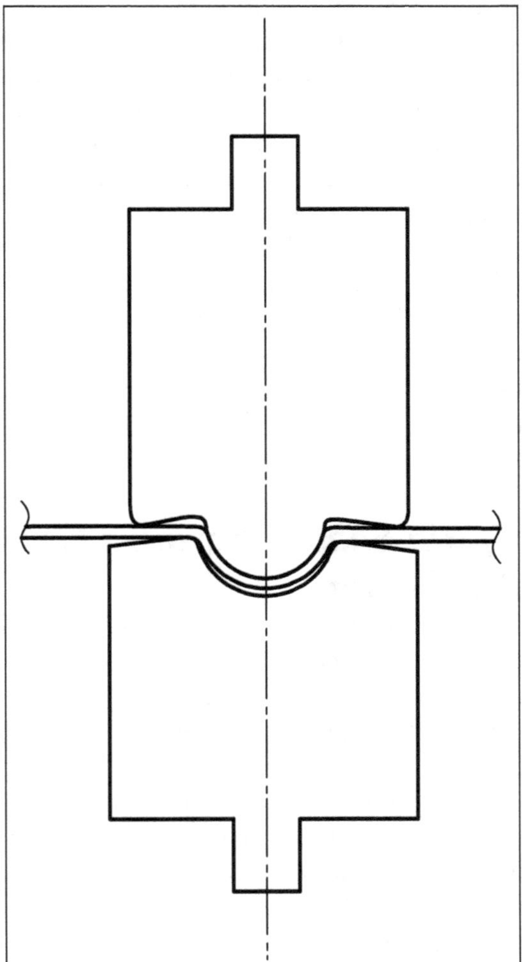

Fig. 15-2. The tooling required to form a radius rib shape must have sufficient springback provision to compensate for normal material variation.

inside angle than 90°. If the rib angle required must be as sharp as 45°, the depth of the rib requires a great deal of material to be moved over the lower die corner radii—making those radii very critical.

B) Radiused Rib Shapes

Instead of a simple vee shape at the bottom of the rib shape, some designs call for the rib to look like a 180° arc (Fig. 15-2). What is drawn on paper may not be easy to make on the press brake. A radius shape will spring back after forming more than a simple straight line bend in the material. Depending on the design of the rib, it is not uncommon for minimum allowance to be made for the top radii of the rib blending into the flat surfaces of the part. When the engineering drawing is actually made showing the outside radius blending into the top flanges, the radius of the rib is no longer 180°, but only radiused to the point where the top radius on each side of the rib is tangent with the inside radius shown on the drawing. Since the actual rib is seldom formed with the corner radii required at the top flange, some type of compromise shape will actually be formed.

The springback will be greater than a V shaped rib because of the large radius rib radius and the radii of the top flanges. While the large radius of the rib will spring back more than the top flanges, it will appear that the allowance for the flange springback is actually allowing the flanges to form upward, like a bird with its wings up. The larger the radius required in the ribbed area, the larger the radius on the top of the lower die opening (trying to match the natural inside radius forming at the flanges). Holding the top flanges "flat" with a radiused rib shape takes a great deal of practice and skill on the part of the press brake operator. As a result, this shape is not selected too often, unless the design requires its use. Occasionally a very deep rib with a full 180° radius at the bottom of the rib is needed. This design may require two hits to complete. The final shape will probably not be like the drawing, and part removal from the lower die after forming may be limited by the amount of die space available in the press brake that will used to make the part. This type of shape often needs a great deal of development to arrive at the best method to make good parts.

C) Rib Shapes Having a Flat Bottom or Four Bends in the Rib Design

Rib shapes that require four bends in its design fall into two categories. The first category is a rib design that is open at the top, has sloping sides down to a flat bottom, and a bottom made parallel to the top surface of the part. The second category is the rib that has vertical sides, and is often called a **hat shaped** rib. The opening of the part is almost the same at the top as at the bottom of the part. In either design, the forming tonnage will be much greater than that required for the simple vee rib. The corner radii at the opening of the rib blending the vertical or sloping legs to the top flanges will have the same considerations as the vee rib. Each inside radius that blends into the flat surface at the top of the rib opening will be much larger than metal thickness. Again, since this is a known problem, the engineering drawing must be evaluated to determine what radius will work.

1) Open Rib Tooling

The open rib shape that has open or sloping sides is often made in tooling that appears fairly simple. If the rib is relatively small, so that an available size die bar can be used, the upper die can be made in one piece. The same consideration is applied to the lower die. If the rib shape is larger, the upper or lower die may have to be constructed from several different size bars. The sloping sides of the rib should allow for some springback. Since the upper and lower dies are normally made to matching shapes, when the ram setting is raised or lowered to compensate for material variations, the sloping sides of the part will not accidentally bottom out if the material is slightly thinner. Fitted dies may cause the material to hit solidly on the sloping sides if the material is thicker than expected. Modifications to the die shape will be needed to provide for springback in all four bends. In order to obtain a flat top surface, the radii of the upper die nose may be made smaller. In order to help hold the bottom of the rib flat, or to minimize **belly** in that surface, it has been found that adding a rectangular slot the width of the bottom flat of the part in the lower die and filling the slot with a piece of cold rolled steel bar stock will help. This provides the operator the ability to shim the bar up or down to control part quality. The ability to shim the bar so that it touches the outside flat surface of the bottom of the rib when the forming stroke is nearly completed minimizes the breakaway that will occur as the part is formed into the lower die (Fig. 15-3).

Some parts have a rib shape located near one edge of the blank. When the stroke is made, the front of the blank is forced up into the air until it touches the top corner of the upper die. The maximum amount and angle of movement can be measured by taking a straight edge and placing it to touch the front edge of the upper die and the corner of the radius at the forming nose at the flat of the rib shape. If this angle forces the material up too quickly or too far, when the part is forced back down to the original horizontal position at the bottom of the stroke, the movement can be faster than the operator can control. It can cause part distortion, a loud noise when the part is forced back down on to any support arms, and it may even strike the operator's hands if he does not remove them quickly enough. In order to minimize this concern, the flats on either side of the forming nose are often made slightly wider. This reduces the maximum angle the part can move upward. The larger distance might increase the radius at the top of the open rib slightly, but not enough to be readily seen. If the press brake is a hydraulic, some of the machines have helpful options that slow the downward speed during the forming cycle.

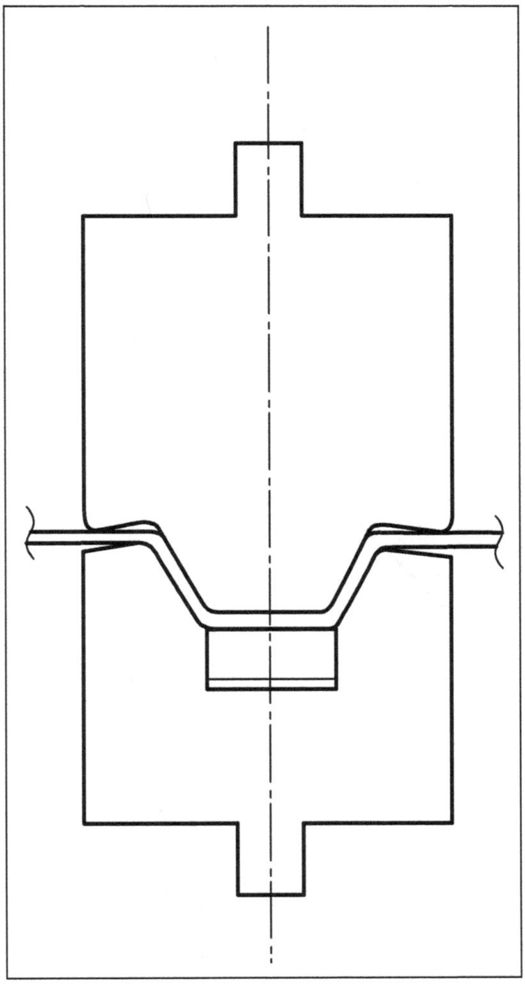

Fig. 15-3. A shimmable pad is used in the bottom of the lower die to hold acceptable clearances and to help flatten the bottom of the rib when any breakaway occurs.

2) Hat Shaped Rib (or Channel) Tooling

When the sides of the rib shaped part are vertical, there is a good chance that the drawing will call for all of the radii to be the same. In some cases, it will be decided to make the part in two hits using offset dies. The upper offset die may have to be tall enough to allow relief to be cut into the front surface to provide clearance for the first hit offset. After the first bend is made, the part is turned 180°, gaged to provide proper spacing between the offsets, and the second stroke completes the hat shape. The advantage to the two stroke method is that all of the inside radii are the same, the top flanges might be easier to keep flat, the die cost is less than a one hit die, and the forming tonnage may be lower. The main disadvantage is two handlings are needed, and two gage positions always add to the possibility of error (Fig. 15-4).

If the hat shaped part is to be formed in one stroke, the two radii in the closed part of the hat will have inside radii of about one metal thickness. The top two radii will be much larger. The problem with wire draw is very common in this shaped part if the top corner radii are made too sharp. The lower die must have a pressure pad to help hold the bottom of the part flat and lift the part up out of the lower die after the forming cycle. A side insert should be provided in the lower die to allow clearances to be adjusted for material variations. Depending on the size of the lower die, a simple method of mounting on a wider filler block or on the bed top of the machine must be provided. The upper die must be wide enough not only to form the hat shaped part, but also to minimize the motion of the part as discussed earlier. Another concern will be that the vertical legs of the hat shaped section may overbend slightly at the bottom of the stroke. When the part is raised up out of the lower die, the operator finds the formed part is sticking to the upper die. In some cases it would be difficult to pull the part off by hand, which is not recommended since it may be a safety concern. To remove the part, a simple hook type stripper might work. Some parts have very short top flanges that do not extend much further than the width of the upper die. In that case, a special release wedge may have to be added, which will substantially increase the cost of the die (Fig. 15-5).

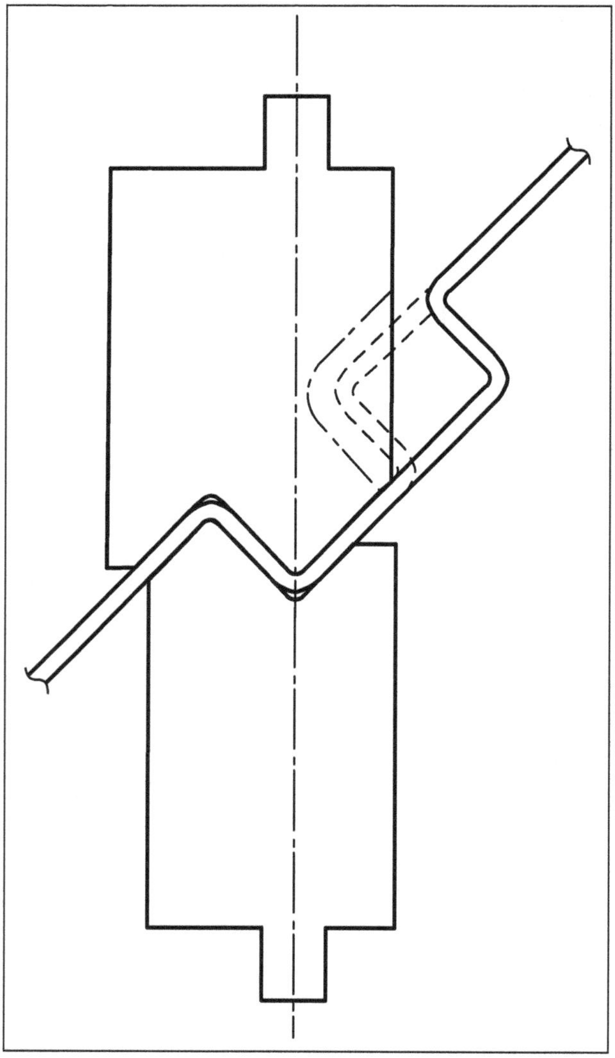

Fig. 15-4. Offset dies can be used to make some hat shaped rib parts. Part handling and gage limitations often limit their use.

3) Panels

The two shapes described above in this Section are often repeated in panels. If the top and bottom flats are the same, and are symmetrical in appearance, they will be called

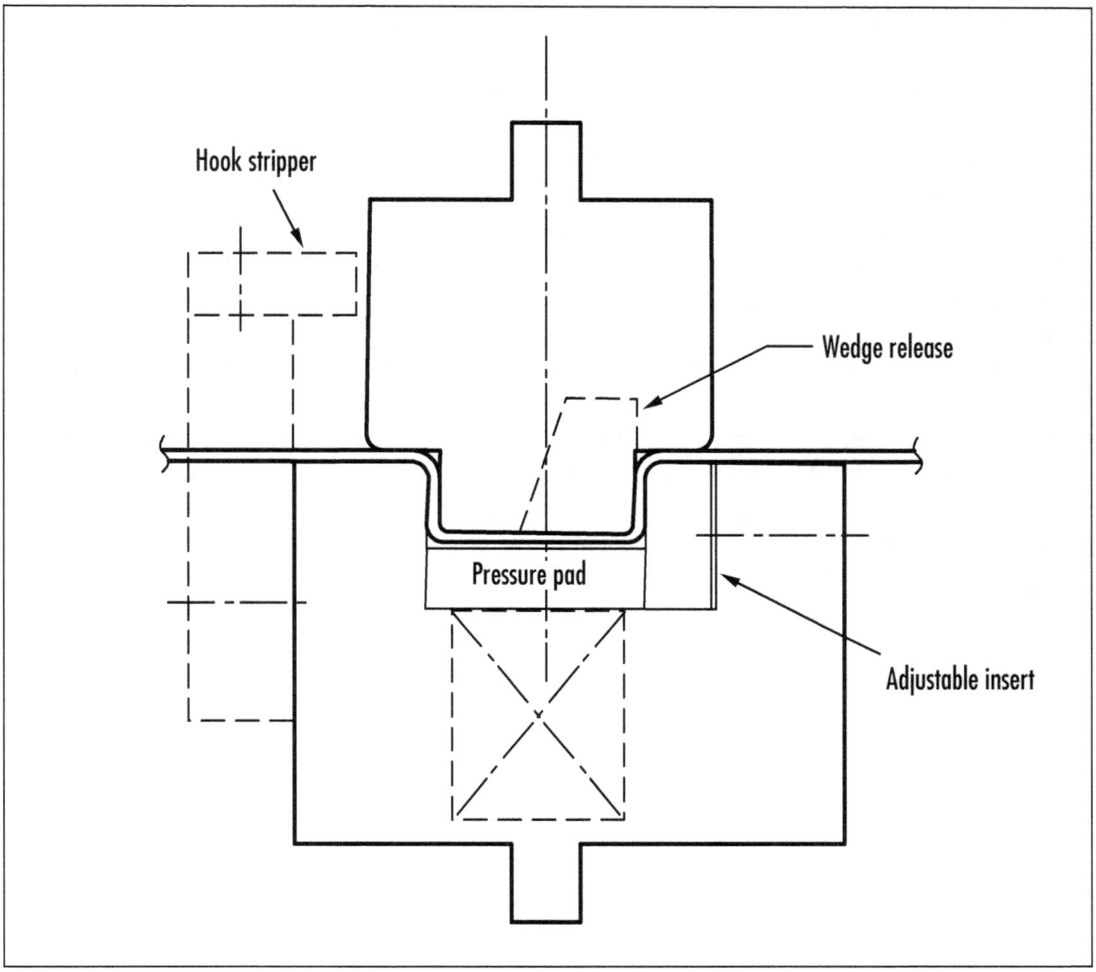

Fig. 15-5. A special hat shaped channel die is used for most panels, or if the hat shape rib is repeated in the part. Note the spring loaded pressure pad and adjustable side insert needed to provide quality and compensate for material variations. If the part sticks to the upper die after the forming stroke, either a hook-type stripper or a wedge release may be needed.

a corrugation. Corrugating dies are discussed in a later chapter. If the rib shape is simply repeated randomly in a panel, it may be desirable to front gage the part using "flip finger" gages (refer to Chapter 7 G-5). The support arm not only holds the unformed sheet at die level, the front gages can easily be adjusted by the operator to compensate for the material draw required to make each rib. A set of support arms in the rear of the bed provides support for the sheet as it is moved through the die after each rib is made. It is very poor practice to push the unformed part to the rear of the press brake and try to use an NC-type back gage to push the part forward toward the operator. The first concern is operator safety. It is always cause for concern to have an operator push a deep flange or wide sheet against a relatively small gage finger. Very few operators take the time to set up sheet support arms in the rear of the press brake that will allow a part with a deep flange length to be pushed over the lower die and slid back to contact the back gage. Some NC back gages may not be designed to allow sheet supports to be mounted on the rear, or may have limitations regarding their location. The concern is: how confident will the operator be that he has actually contacted both gage positions if he pushes a wide part back through the forming die set until only a very short flange remains in the front?

The second concern is that few, if any, NC back gage manufacturers recommend that their back gage be used for pushing parts forward. Most designs do not provide for the rigors of pushing parts of different sizes and weights. In addition, when the part is formed, the long flange that is now in the rear of the machine will move up and down during the forming stroke. Depending on how the operator has programmed the gage, the gage may start to move to the next gage position at the bottom of the stroke, and move into the same location needed by the flange that is in the process of being formed. Often this can cause damage to the NC back gage.

Although some companies have successfully used the back gage to locate and index panels from back to front, the potential concerns above make this a high-risk method. Prior to selecting this type of forming, it is strongly recommended that this type of NC gaging be attempted only after approval from company management. It has been pointed out that most companies that manufacture NC back gages do not design them to push material toward the operator. If any damage occurred to the back gage when it was being used for this type of operation, it is highly probable that any warranty remaining on the back gage would be voided.

Chapter 16
Corrugations

Corrugations are parts that have a repeated design. The shape of the part would theoretically look the same if the part were turned over and laid on top of a previously formed part. The design of a corrugation can be a series of symmetrical radii such as those found in a corrugated pipe, or it could look like a series of hat shaped channels located next to each other to produce a pattern that looks the same from the top or from the bottom (Fig. 16-1). On occasion, the corrugations will look like a repeated series of up and down lines. Upon inspection they are the same as offset bends located next to each other (Fig. 16-2).

The various shapes are often formed on a press brake using special tooling. The main problem is always: "how will the last several bends in the corrugated panel be formed?" The majority of corrugated panels have the shape repeated the entire width of the sheet. When the sheet is pushed into the forming die by the operator to locate the final bends of the corrugated

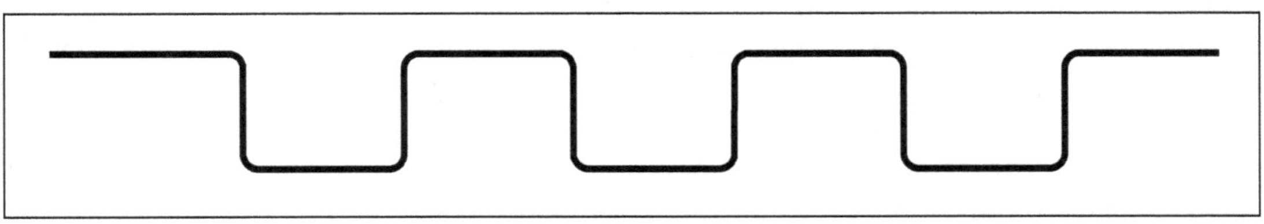

Fig. 16-1. Hat shaped corrugation.

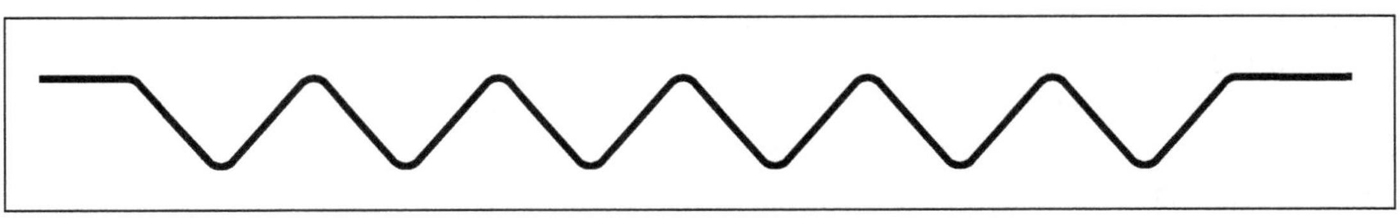

Fig. 16-2. Offset corrugation.

panel, any front gaging that was being used (e.g., flip finger gages) cannot extend into the tooling. The part now has all but the last bends formed, and the operator may have lost the ability to locate the part without creating a safety problem. In many cases, it is necessary for a helper or second person to work in the rear of the machine to locate and remove the completed panel. A major problem will also occur if the press brake being used had been equipped with an NC back gage. It may have to be disassembled or modified, depending on the length of the corrugated piece. If a helper must be stationed in the rear, compliance with state safety codes and/or OSHA safety guidelines must be addressed. Each situation is different, depending on what must be done in order to make the part. Sometimes, creating a safe distance is sufficient; sometimes an operator station may have to be added to the rear of the press brake.

The corrugated panels are normally shorter than the distance between the housings of the press brake. The only exception is if the throat depth allows the panel to be made the full length of the bed and ram without interference. Depending on the width of the panel and the available throat depth, the formed panel should be removed to the rear or removed endwise. Attempting to slide a formed corrugated sheet back toward the operator is usually difficult and time consuming.

Corrugations should not be gaged using an NC back gage for the same reasons discussed in Chapter 15 C-3. Some companies have tried to form a corrugated panel by forming one half of the panel, turning it around, and trying to complete the part by attempting to match the last corrugation in the center of the sheet. This is possible for some shapes, but it is seldom (if ever) successful. The method of forming the different shapes, plus common problems, will be separated into the following four general categories.

A) Hat Shaped Corrugations

The only practical method of forming a hat shaped corrugation (Fig. 16-1) is to form four bends at a time using a hat shape channel type die described in the previous chapter. The same problems that are present with any one hit die still remain. The bottom two inside radii formed in the bottom of the die will be relatively sharp. They may have to be modified in order to hold the opening at the top of the hat shape close to the needed dimension. The radii at the top of the hat shape will be larger to minimize wire draw and control top flatness.

Since the shape of the bends are not like the dimensions shown on the drawing, it will be a problem to hold the width of the panel as additional square shaped ribs are formed into the part. The lower die may have to be relieved in the rear to provide clearance for the previously formed rib. The die must have a pressure pad, which will help hold initial gaging plus lift the formed shape up to the top of the lower die so that the sheet can easily be indexed to the rear. As the part is moved into position for the next bend, the rib that has just been formed will drop down in the back of the lower die as the rib clears the forming surface. It is important to have support arms in front to support the flat unformed sheet. It is also important to have support arms in the rear, located to support the bottom of the formed rib and long enough to support the entire panel after the last rib has been moved to the rear far enough to clear the lower die.

When the part is being formed, the front of the sheet is moving up and down, and often strikes the front support arms with a great deal of force. As the ribs are formed, and each rib moves to the rear of the lower die, the formed sections each move up and down and will strike the rear support arms with sometimes more force than in the front, since the stiffness of each of the formed sections is much greater. Each time the formed rib hits the support arm in the rear, all of the bend angles that were formed in each rib open up slightly. In addition, the flatness of the formed sheet may bow in some manner, either up or down. This will affect the final overall width of the panel.

When the hat shaped rib is being formed, the new rib will pull material in from the front and rear of the flat sheet in order to make the correct shape. If there is more material in the front than in the rear, it takes more force to pull the metal in from the long length in the front than from the short leg in the rear. As a result, the material will not be pulled in exactly the same since the pressure pad in the die will not have sufficient clamping pressure to eliminate the effect of the material weight. As the additional ribs are made in the panel, the pull-in will become balanced for the ribs formed in the center of the sheet. As the part weight shifts to the rear, the movement of the front and rear flanges reverses. The result is that the adjustable gages used to locate each rib must be moved to compensate for the difference in draw needed for each rib. Thus, the top opening produced by the rib die (or hat shaped channel, as many will call the shape) will have a slightly different spacing than the flat formed in the bottom of the channel die. The main goal will be to hold the width of the formed panel to a consistent dimension with the correct number of channels or corrugations.

One of the tricks of the trade is to glue a $^{1}/_{8}$" thick strip of polyurethane to the top of each support arm. This will minimize the effect of the material forcefully striking the arms during each forming stroke, and will reduce the distortion caused when a previously formed rib strikes the rear support arm. The strips will also cut down on the loud noise created from the material hitting the arms. The flatness problem is solved by having provisions in the dies that allow the ram to be raised or lowered slightly to compensate for the distortion. If the ram is adjusted too much, the depth of the rib will be slightly changed, which can then affect the gage positions and possibly the blank size.

B) Offset (or Reverse 90°) Corrugations

This corrugation design (Fig. 16-2) often suggests that the shape can be formed using simple vee dies. After spending hours trying to determine how to gage the part—discovering that it will require the operator to continually remove the part from the die area, turn it over, and make the next bend—most of the panels are made using a special offset die. The exception is if the panel is very narrow. If so, a set of bottoming dies made more than twice the length of the panel width can be cut in half. One half would be placed into the press brake in the same way as any die set. The other half is turned over and installed next to the first die with the vee opening on the top and the normal upper die on the bottom. When the part is formed, one bend would cause each flange to move up the same as any standard bend made using simple vee dies. The part is then moved sideways so the blank is located on what is the forming nose of the upper die—now upside down in the setup. A second bend is made creating an offset shape. Note that the unformed material in the front is kept horizontal prior to being formed. The formed leg in the rear of the lower (or upside down upper die) is sticking up. The upside down vee die must have sufficient clearance on the rear side so that the flange sticking up is not hit by the die as it moves down to make the second stroke (Fig. 16-3). The panel is made by moving the part left to right, from one die to the other, as each stroke is made. Gaging and handling the part safely are the big concerns. Many times, this type of panel is somewhat narrow. It is always best to use some type of squaring arm in the front to ensure that the bend is square to the edge of the material. Gage locations that are close to each other are very sensitive to minor movements.

When an offset die is selected, the part is supported and gaged in the front just as would be done with the hat shaped rib die. The difference is that the formed offsets will be creating what looks like stair steps going up when observed in the rear of the offset die prior to each forming stroke. Unfortunately, the movement of the part while the next offset is being formed does not allow for any supports to be used. As the part is indexed through the dies, the weight of the material becomes very heavy in the rear and will start to sag without any support. Again, management must determine how to safely complete the forming process without endangering the operator on the front or a helper in the rear.

Holding flatness and overall panel dimensions are the main problems to be addressed after the actual difficulties of forming the part have been solved. This type of panel is often designed so that a flat leg extends from each end of the corrugated shape with a 135° bend at each edge of the panel. This magnifies the forming and gaging problems. It will also require a secondary operation to complete the part. Gaging to obtain the correct dimension of the flats may not be too easy since the blank size can change if material varies slightly or the ram setting is adjusted to hold part flatness.

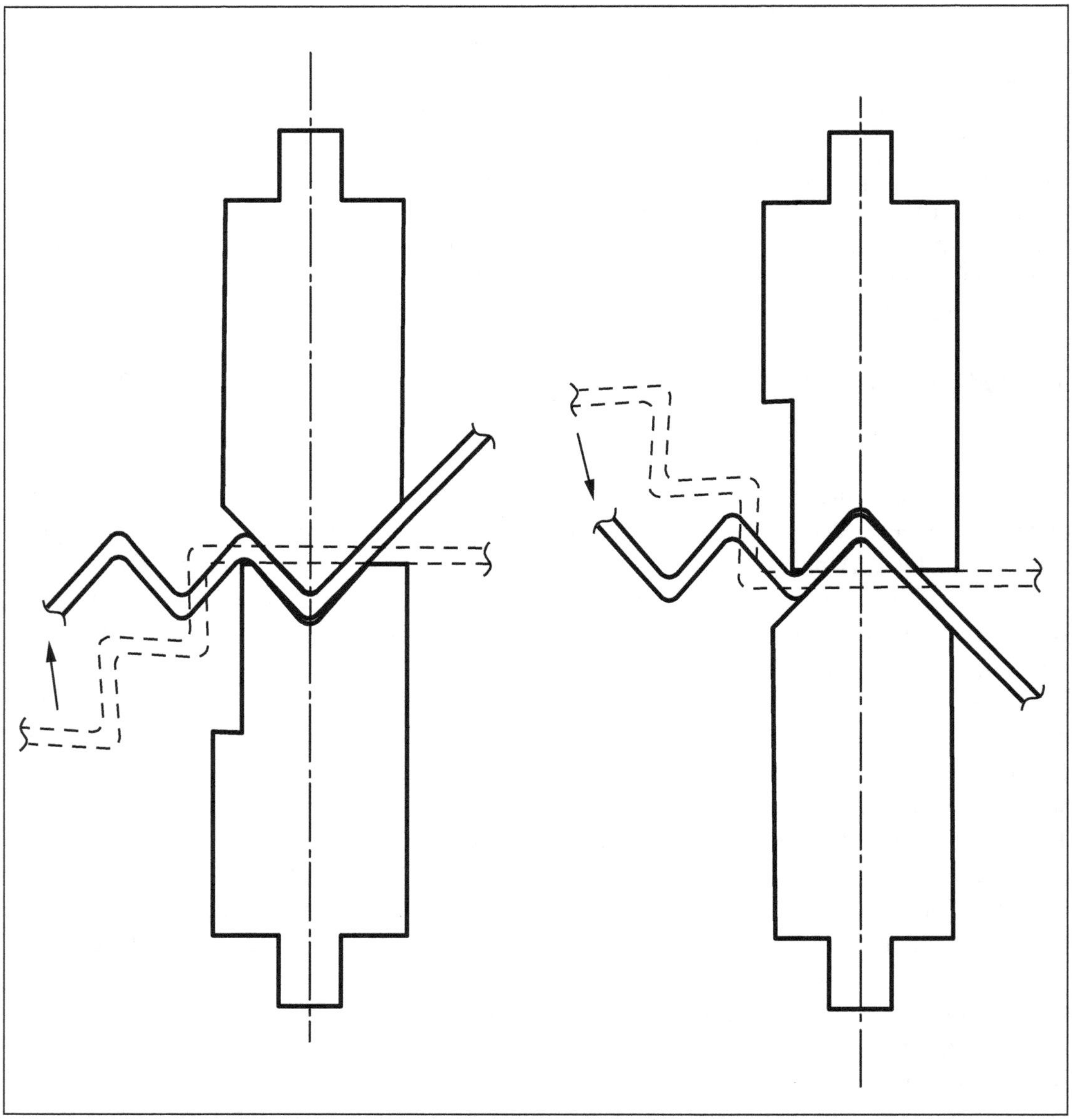

Fig. 16-3. Simple vee dies used to form an offset corrugation, one bend at a time.

C) Radius Shaped Corrugation

The most popular use of a radiused designed corrugation is to make pipe. A flat sheet is made into a corrugated panel and then formed into different diameter pipes. There, a large sectional culvert is made with a 2" depth × 6" pitch corrugation. Smaller sizes include the $^1/_2$" deep × 2 $^2/_3$" pitch, and the $^3/_4$" × 3" pitch corrugations used to make nestable pipe and some types of siding. The 2" deep corrugation is made in material from 12 gauge up to 1 gauge thick, the $^1/_2$" deep corrugation ranges from 16 gauge to 10 gauge in thickness, and the $^3/_4$" deep corrugation is normally thin materials up to 16 gauge thickness. Special dies have been developed to form all of the above corrugations on a production basis. Many other types of corrugations have been made, and all of them require special press brake tooling.

The types of corrugations that are made in production-type dies are made in tooling that has a forming station on the operator side of the centerline of the press brake ram. In the back of the ram centerline is a special pressure pad that clamps a formed corrugation (using a great amount of spring pressure) while the next forming stroke makes the next corrugation. If the material is thin, up to a maximum of 14 gauge, a pressure pad that only holds one formed corrugation will work (Fig. 16-4). If the material is 12 or 10 gauge, the pressure pad is made wide enough to clamp two formed corrugations, which allows additional spring pressure to

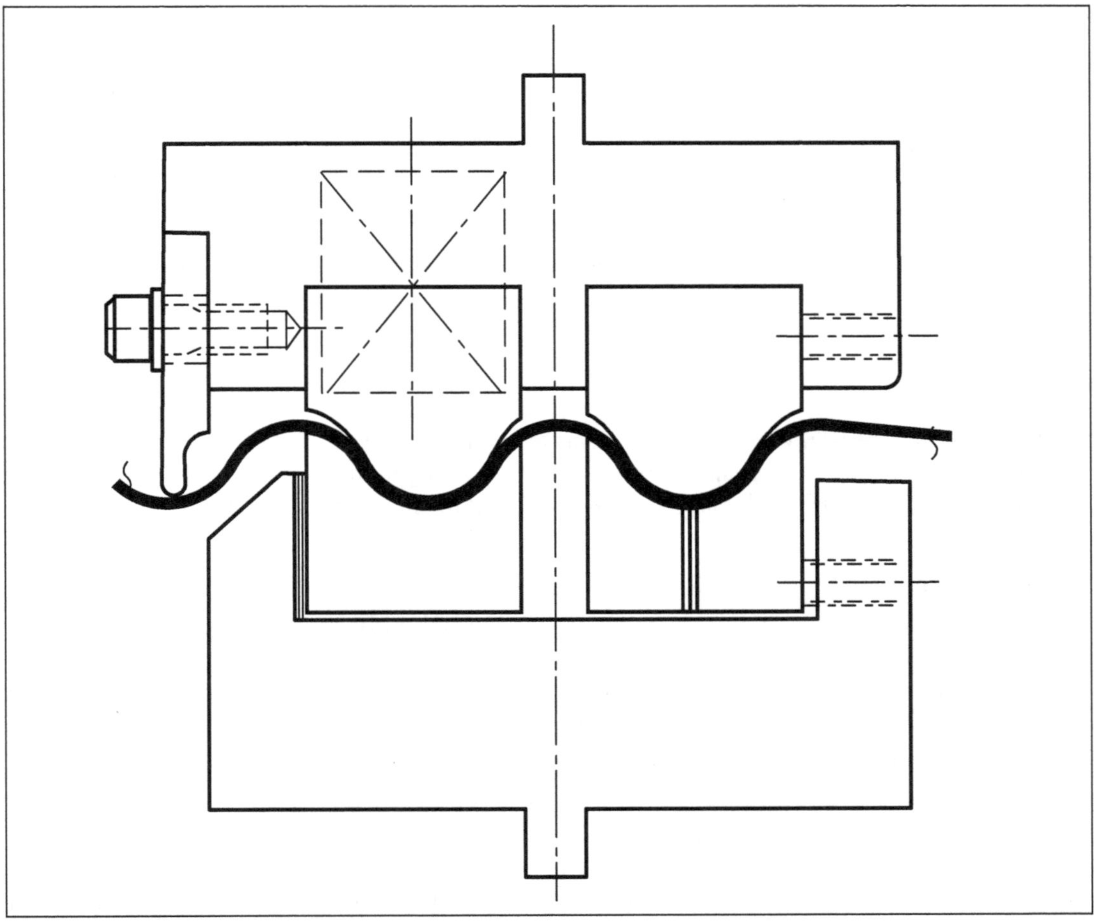

Fig. 16-4. A "hold one-draw one" die design used to form a radiused corrugated shape in materials up to 14 gauge thick. If the material is very thin (20 gauge or thinner), several sheets of material can be formed at the same time.

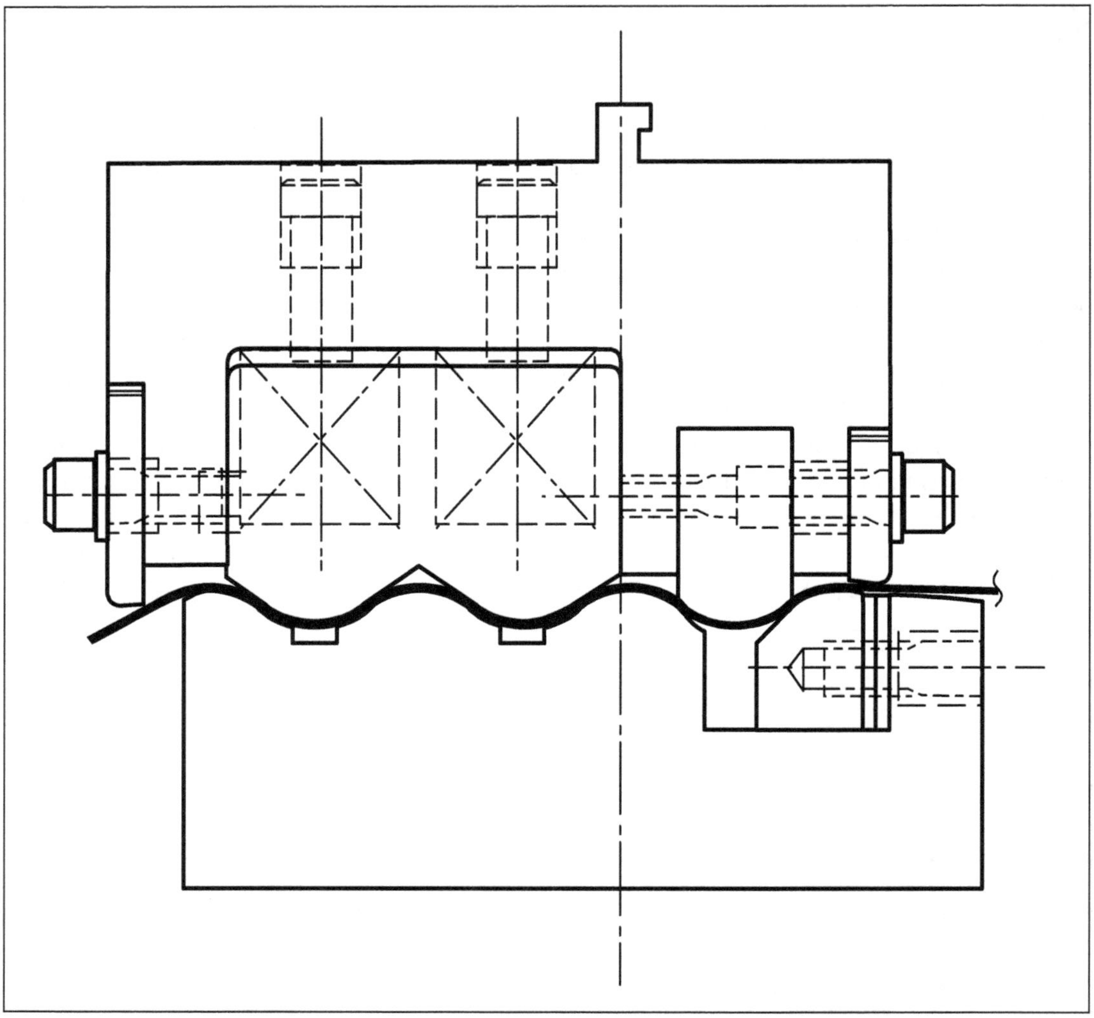

Fig. 16-5. A "hold two-draw one" die used to make corrugations in 12 and 10 gauge mild steel. Thicker materials require a hydraulic pressure pad to keep the previously formed corrugations from slipping backwards toward the forming station and losing the pitch of the corrugation.

be used to clamp the previously formed ribs (Fig. 16-5). If the material is even heavier, for example, the 1 gauge (approximately 0.25" thick material) used to make the sectional culvert panels, it may be necessary to design a hydraulic clamping system that also clamps two previously formed corrugations. The tooling for this type of press brake is very expensive; and a press brake is often dedicated to the manufacture of these parts.

For low production, special dies can be designed to form one corrugation at a time without clamping the previously formed corrugation (Fig. 16-6). In this case, there will be some modification of the drawing dimensions needed when each corrugation is being formed. The previously formed corrugation is located as close as possible to the distance the next unformed corrugation will draw or move toward the die during the forming stroke. The unformed length of the blank is still extended out in the front of the lower die, resting on support arms that are complete with locating gages. When the stroke is made, the blank will move in from the front and from the rear. Hopefully, the corrugation that has just been formed will have its top radius blend in with the top radius of the previously formed corrugation, allowing the pitch of the distance from the top of one corrugation to the top of the next corrugation to be held. The amount of draw will vary from one stroke to the next because the weight of the material in the back and the weight of the material in the front is constantly changing, making gaging difficult.

The second problem with the simple corrugating die is that it does not allow much flexibility in the distance of the flange length that is required by the leading edge and the trailing edge of the part. The leading edge must be long enough to span the lower die, and extend far enough to the rear to allow the upper die to push the edge of the material down to start the first radius. If the leg is originally designed to be shorter, it will have to be modified after the panel has been completed. The same problem often occurs on the last hit. The drawing calls for the trailing edge to pass through the die to end at a certain point. Since the part must complete the last full radius, a trimming operation may be required to obtain the leading or trailing edge dimension.

Once a method has been tested using this type of die, and the result seems to make usable corrugated panels, all gage positions and location of support arms plus normal setup information should be written down. When the part is again made, the next setup will not be as time consuming since all gage locations, any shimming, and any machine settings (such as ram settings or tilting the ram) are already available to the operator.

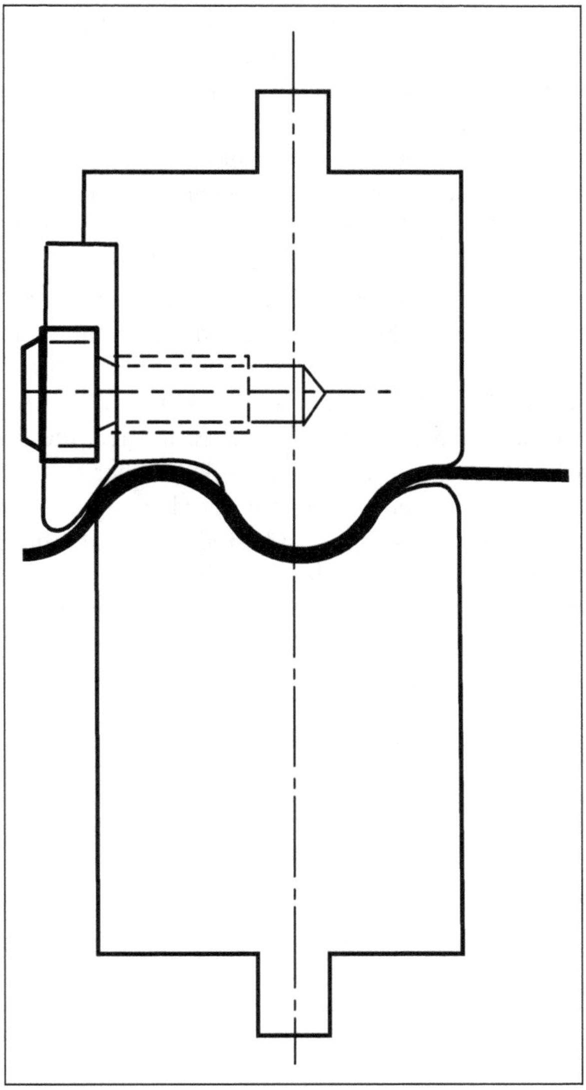

Fig. 16-6. Forming a panel using the simple design shown is very time consuming and requires an experienced operator to form corrugations one rib at the time. The material moves in from the front and rear during the forming stroke, making gaging, as well as holding a consistent pitch or corrugated shape, very difficult. This design is only used for low production work.

D) Inspecting the Finished Panel

It has been pointed out several times that the corrugated panel produced will not look exactly like the drawing. The major concern will be usability and uniformity. The actual corrugation shape must be consistent in its variation. The top radii of the hat shaped corrugation will be different from the bottom radii. The spacing at the top will be different from the bottom. The offset type of corrugation will depend on gage and draw-in variations. The radiused corrugation that is made with a pressure pad holding the previously formed rib will force all of the radiused shaped material being formed to draw-in from one side. This will cause the angles in the corrugation to be different from one side to the other.

After all parties have agreed that the shape is acceptable, the panel should be measured. This is done by using a simple tape measure to measure each side of the panel length. The tolerance will vary depending on the length. A reasonable tolerance for a 24" to 32" wide corrugated panel is $\pm ^{1}/_{8}$" for material up to 10 gauge. Parts made from $^{3}/_{16}$" or thicker metal may have more variation. The next measurement inspects the diagonals of the panel. If the panel is 10' long, the diagonals may have a variation of $\pm ^{1}/_{4}$". The above suggestions will quickly change depending on part usage or on the style of corrugation. The offset type of panel that is narrow but requires many strokes to make a 4' length may have more variation than a hat shaped ribbed panel that only has several rib corrugations. One of the concerns identified by checking the diagonals of the panel is to make sure that the part is not **fan** shaped. This can occur if the material is thicker on one side compared to the other side. The simplest correction is having the ram of the press brake tilted slightly. This forces the thinner material on one side deeper into the die cavity, which compensates for the thicker material variation on the other end.

Because of the differences in the shape of the corrugations, formed sheets have to be stacked in the same direction. If one panel is turned 180° and placed on top of a similar panel, it would be obvious that the two panels don't fit together. When laid on top of each other in the same direction that they were formed, the fit should be almost perfect. This means that when the parts are being installed, either in the plant or in the field, assembly instructions must be provided to ensure proper lapping of the parts. Any problems by customers using the corrugated parts will also be reduced. If the finished panel is to be used separately, such as for the floor of a container, the final shape is not as important as the final outside dimensions. When necessary, any length variations can be corrected by modifying the initial blank size.

Chapter 17
Box Forming

Forming box shaped parts appears to require simply a set of ordinary vee dies. The upper die must be cut to the length of the inside dimension of the box's longest length, and high enough to clear the side flanges of the box. So, what is the big problem? The concern is how well the four sides match at the corners. There are many types and styles of four sided boxes, each requiring review prior to being sent to the press brake forming center. If the box has top flanges, do they bend inward or are they outside the inside dimensions of the box? Is there only one bend in the top flange or a complex shape? How is the blank going to be made? Are the notches to be cut out on a laser, sawed out, or will a special blanking tool be used? What is the best way to gage the part to be able to hold uniform dimensions?

Every combination of box design cannot be covered in this chapter. The basic method of forming boxes, plus the obvious concerns, will be addressed. The majority of boxes are made with simple vee die-type upper and lower dies. Most of the time, it is not one box size that has to be considered, but a family of boxes. An immediate concern is the height of all of the box sides. The last bend of the deepest box often determines the height of the upper die. When the final bend is formed, the previously formed side flanges must clear the front corner of the press brake ram.

In order to calculate the minimum height of the upper die needed to make this final bend, the formula often used is: the upper die height equals the box depth divided by 0.7, plus the width of the ram nose divided by 2. The normal ram width for most press brakes up to 230 tons is 2.5". This calculation requires that bottoming dies be selected, since no provisions for overbending the angle are allowed in the calculation. Since the air bend of the final flanges will cause interference with the previously formed side flanges, and there is the potential possibility of those flanges hitting the ram, all four sided boxes will be made using some type of bottoming die (Fig. 17-1).

The basic calculation should assume that the maximum die space is 12". This often requires that the normal filler block be removed from the bed top of most press brakes. For most mechanical press brakes, the maximum die space is 12", plus the 3" of additional space created when the stroke is at the top. A hydraulic press brake of the same capacity that would be used for making boxes needs 15" of open height. This allows for the same 3" of stroke and 12" of remaining die space found on the mechanical press brake. If 3" is estimated for the forming stroke, and if the lower die is mounted on the bed top and is no higher than 1.25",

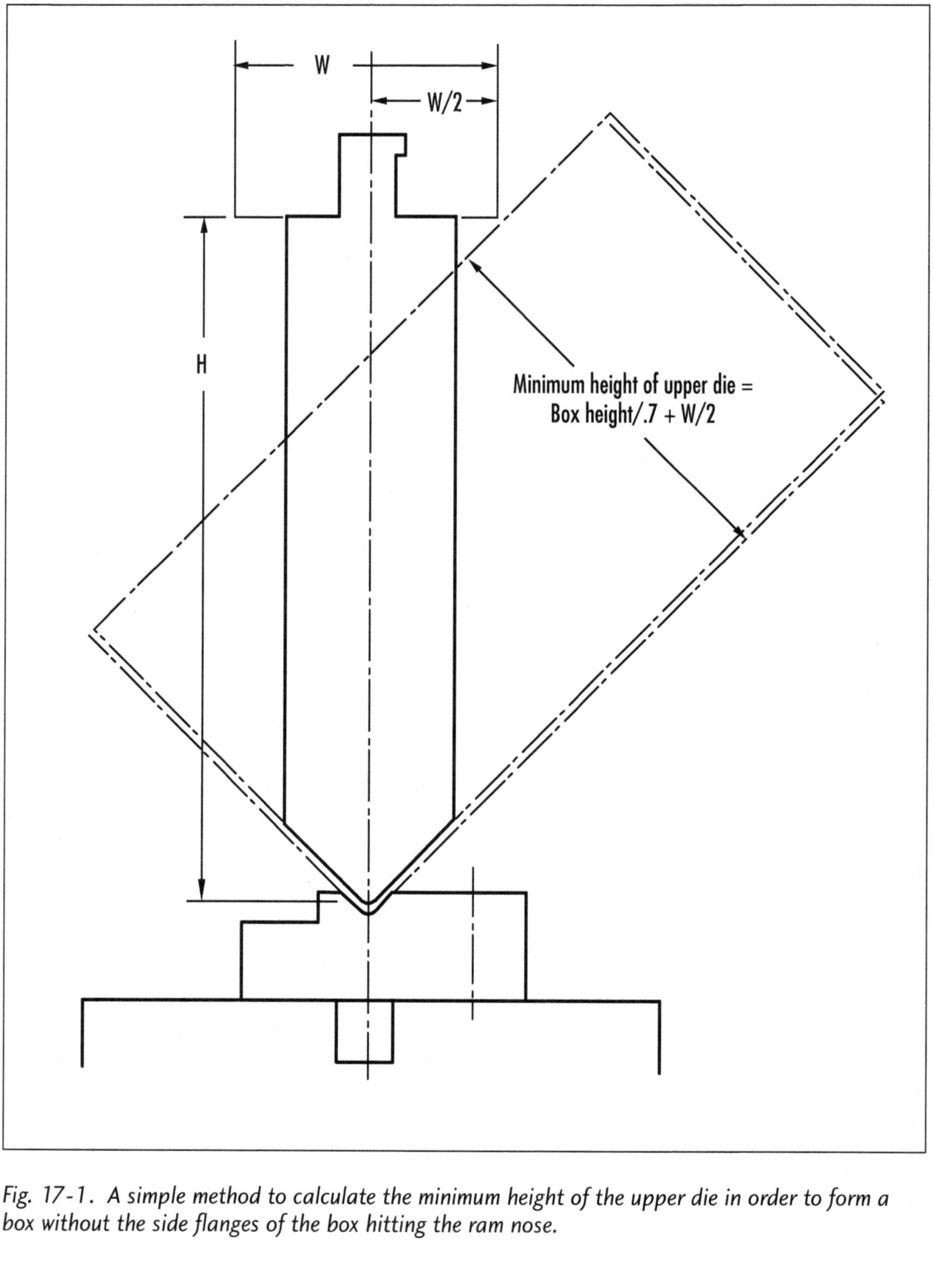

Fig. 17-1. A simple method to calculate the minimum height of the upper die in order to form a box without the side flanges of the box hitting the ram nose.

then the maximum height (depth) of a simple box without return flanges that can be made is 6.5" (Fig. 17-2).

If the box has inside return flanges, the upper die will have to be made to keep the forming nose the full length of the bend line for the last bend. However, immediately above the forming nose, the length of the upper die must be reduced enough to allow the side return flanges of the previously formed sides of the box to clear the upper die body without any interference (Fig. 17-3). If the box has external top flanges, the lower die must be higher than those flanges so that the unformed box blank rests on the die.

The maximum calculated box height must now be reduced, if necessary, by any vertical internal flanges at the top of the box. As was just pointed out, if the lower die height must be increased, the maximum 12" die space distance will limit the height of the upper die. The question is: if the 3" forming stroke is reduced, can the box with external flanges still be easily removed from the machine after the last bend is formed? If the box has internal flanges (often found in N.E.M.A. electrical box designs), the internal box may have to be twisted after the ram has returned to the top of the stroke. This provides a diagonal at the top of the box sufficient for the box to be removed from the full-length forming nose of the upper die. This

Chart To Determine The Height Of The Upper Die For Given Box

Depth Of The Box To Be Formed	Min. Height Of Standard 90° Male Die	Min. Height Of 30° - 60° Male Die
1"	2.75"	2"
1.5"	3.5"	2.6"
2"	4.2"	3.125"
2.5"	5"	3.82"
3"	5.625"	4.25"
3.5"	6.3125"	4.875"
4"	7"	5.4"
4.5"	7.75"	6"
5"	8.4"	6.6"
5.5"	9.125"	7.2"
6"	10"	7.75"
6.5"	10.6"	8.32"
7"	11.25"	8.875"
7.5"	12"	9.5"
8"	12.7"	10.1"
8.5"	13.375"	10.625"
9"	14.1"	11.2"
9.5"	14.8"	11.8"
10"	15.5"	12.325"

The above calculations are based on a press brake with 2.5" thick ram.

Fig. 17-2. Chart showing the minimum upper die height for a variety of simple box sizes.

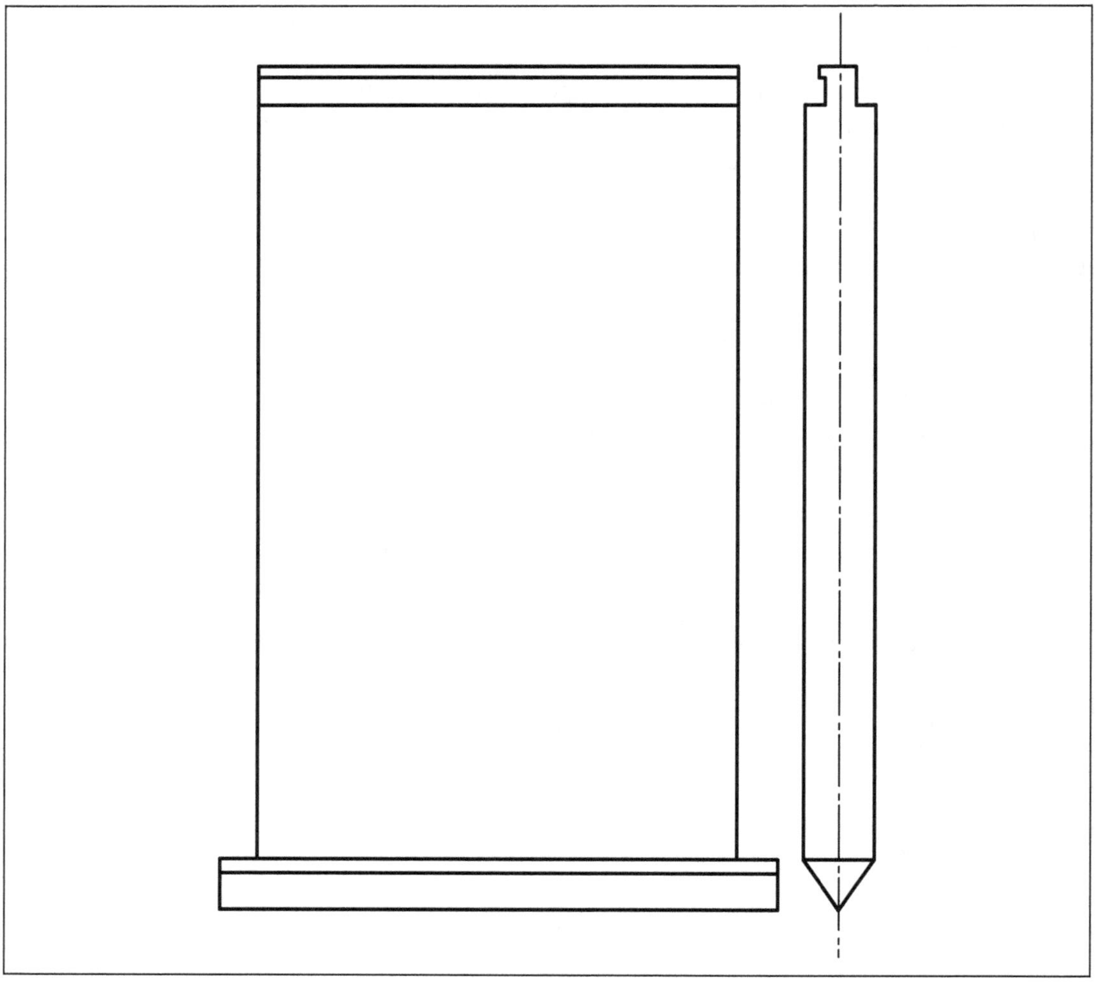

Fig. 17-3. Upper die with "horns" or cutouts that may be needed to clear the inside return flanges found with some box designs.

twisting or turning of the box quickly uses up any stroke opening. A 3" stroke may not be sufficient to remove some of the box sizes.

A) Families of Boxes

Often, box shapes are not limited to one size box that needs only one or two lengths of upper die to complete the forming. In many companies, the boxes must be made in many different widths and depths. If properly designed, the depth of each box will allow sufficient width for the upper die to complete the final bend of the box shape without interfering with the opposite flange (the flange on the operator side of the box). A family of boxes will start with smaller sizes (often rectangular as opposed to square) that also have a smaller flange height. The box size matrixes are normally available in standard 3" or 6" increments. Some companies offer special sizes at additional costs. If a new press brake is purchased to make boxes, it is a good idea to modify the die clamps (on the ram) with the clamping bolts made on 6" centerlines instead of the normal 12" spacing. Short upper die sections mounted in between the normal 12" bolt centerlines can breathe and affect the bend angle. The 6" centerline clamping ensures consistency.

Depending on the box design, the final bends of a family of boxes can be formed with sections of upper and lower dies that have been cut into different lengths to accommodate all of the different dimensions. If possible, the short sides should be formed first. This is recommended because when a part is formed that has side flanges, the additional stiffness created in the blank makes it imperative for the operator to assist in lifting the part during the forming cycle. This prevents the final bends of the box shape from **back bending** over the front corner of the lower die vee opening. The further out to the front the box extends, and the heavier the material, the more difficult it is for the operator to assist in the bend.

Since a number of different die lengths may be needed for making the last bend, the best way to initially review what length dies will be required is to determine if three sides of the box can be made using one standard length of die. This upper die may need to have relief at each end if the boxes have internal top flanges. This thinking will work well on the smaller size boxes. Smaller size boxes are not too big or heavy, but need operator assistance during the forming stroke even when the short sides are extending out toward the front in order to prevent back bend problems. Forming the box is done in this order: one short length of the box is formed, then the part is rotated 90°; then one long side can be formed, and the part rotated another 90° for the final short side to be formed. The part usually has to be moved toward one end of the upper die to allow clearance for the previously formed long flange. The part is then moved over to a second die station, either on the left or right (operator preference), where a specific length of the same size upper and lower die are located to allow the final bend in the box to be made. It may be that two foot switches will have to be selected in order for the operator to move from one station to the other without having to drag the operating station back and forth. Some press brakes have programmable stations that, in some cases, may provide an additional degree of convenience for the operator.

If the final bend is the longest length of the box (as recommended), it is possible to make the upper die in two equal lengths. This is attractive because a family of boxes is often designed

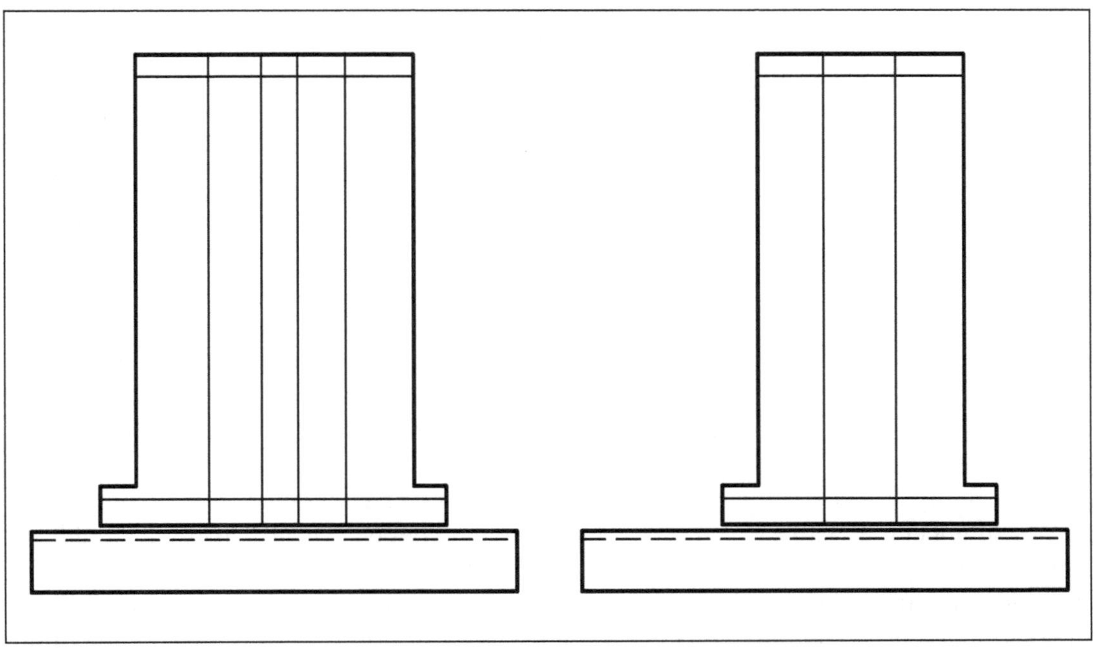

Fig. 17-4. Upper dies showing two stations, and sectioned as required to form a family of different size boxes.

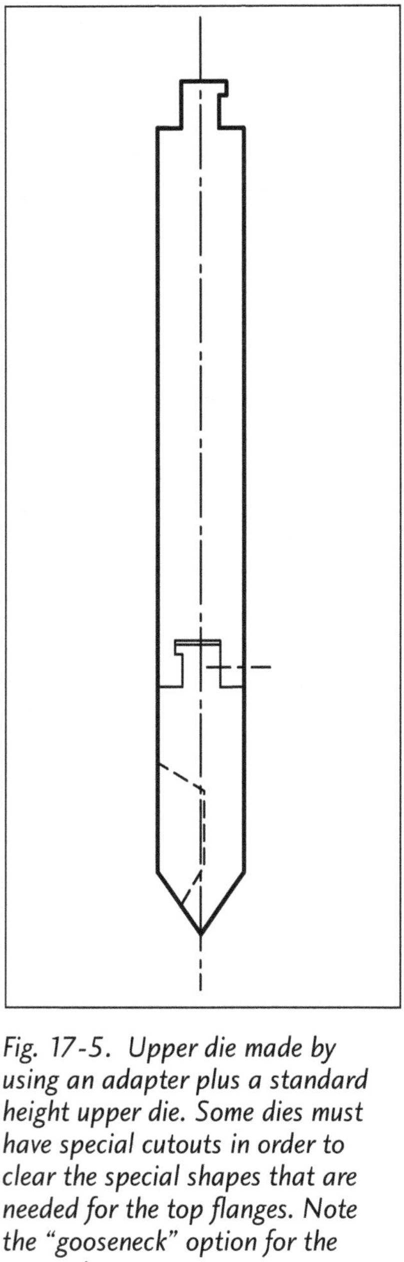

Fig. 17-5. Upper die made by using an adapter plus a standard height upper die. Some dies must have special cutouts in order to clear the special shapes that are needed for the top flanges. Note the "gooseneck" option for the upper die.

with the outside dimensions a conventional length, such as 6" or 12". However, the length of the upper die must be shorter by the two metal thicknesses of the side flanges, plus the distance required for the inside radii of the side flanges, plus about one additional metal thickness. This allows the operator a little distance to nest the part to locate the final bend. The length of the upper die for the smallest box length will be a unique dimension. If the box does not have any internal return flanges at the top, increments are added to the end of the shortest upper length as the length of the boxes increases.

When the box has return flanges at the top that require the opening at the top to be smaller than the box size, the sides of the upper die must have clearance for the return flanges. This relief must be present on the shortest length upper die as well as for the longest. If the smallest upper die length is cut into two equal sections, both the right hand section and the left hand section must have the proper relief. As longer length boxes are to be formed, each section of the upper die for the small box can be spaced apart, and a simple length added between the sections that make up the new size flange length (Fig. 17-4).

Another popular design for the upper die is to make each length in two sections, a top piece and the forming section. The top section is made like a spacer or adapter. It is a sturdy bar (not necessarily die steel) having a conventional tongue on the top to use for mounting the die into the ram. A slot is machined on the other end to hold the tongue of a small upper die that can form all of the box bends. The forming die is held in place by set screws located in the adapter. The advantage is that the adapter can be saw cut into lengths less than the forming length required of the forming die. If the upper box dies require high die bar heights that are not readily available, the total upper die height can be obtained by using an adapter and standard height upper die. The availability of a simple adapter provides added flexibility in the event that special boxes have to be made in a hurry, as new box forming die lengths can be cut from current tooling (Fig. 17-5).

If the family of box sizes includes larger box sizes, the first operation may require several lengths of upper die, with each end section having the relief described. The blank would still be rotated when making the first three sides. Often this is done to allow the use of support arms and simple gaging for the flanges. A second setup would be needed if two forming stations cannot be set up in the press brake at the same time to complete the last side of the box. Always ensure that the box can be removed from the machine after the last hit has been made. Deep boxes will create the most problems.

B) Shape of the Relief on the Upper Box Die

If the box has a return flange at the top, it is a major problem to remove the formed box from the upper die. For many years, the standard horn (relief) was made using a vertical and horizontal cut to remove a rectangular section about 1" above the forming nose of the upper die. In order to remove the formed box, the operator held the box in the formed position when the ram returned to the top of the stroke. The box had to be tilted and rotated to create a diagonal sufficient to clear one top return flange. The box could then be dropped off the die area at that end. It would then be twisted slightly to allow it to be pushed toward the other end of the upper die to clear the relief on that side. The box can then be dropped downward, clearing the upper die. By lifting the base of the box up to clear the lower die, the completed box would be pulled down toward the floor as it was removed from the die area.

In recent years, a new design offered a good alternative method of part removal for some boxes. First, a vertical cut is made from the top of the upper die to a point that would clear the inward flange of the smallest box. A second machining operation on each end of the die is made that slopes from the vertical cut on each side to a point about $^1/_2$" above each end of the forming nose. Often, after the box forming was completed, the operator could hold the box on each side and simply pull it off the upper die. When the sides of the box were pulled over the beveled ends of the upper die, the sloping surfaces forced the sides to spring open slightly (Fig. 17-6). When the part is removed, the sides tend to spring back to their original box shape. This concept works best when there is some depth to the sides of the box (4" or more).

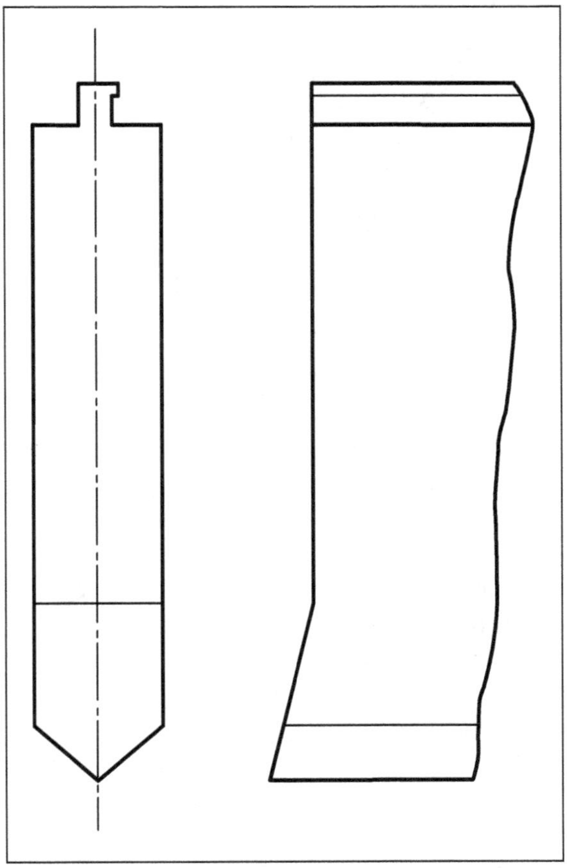

Fig. 17-6. Upper die showing an optional relief. The sloping beveled area may make removal of a formed box easier by allowing the operator to pull the formed box downward, springing the sides open slightly as the box is being pulled off the die.

C) Off Angle Dies

In order to make deeper box sections in a limited die space, the symmetrical 90° (or close to 90°) forming angle of the dies is rotated about 15° toward the operator. The formed angle will only move upward 30° in the front, but the back leg will move upward 60° (Fig. 17-7). This technique is recommended only if the material being formed is thinner than 14 gauge. When off angle dies are being used, the upper die will try to kick toward the front (or operator side). The lower die will try to move toward the rear. The upper die movement is restricted by the ram clamping, the upper die shoulders, and the contact with the material as it is being formed. The lower die movement

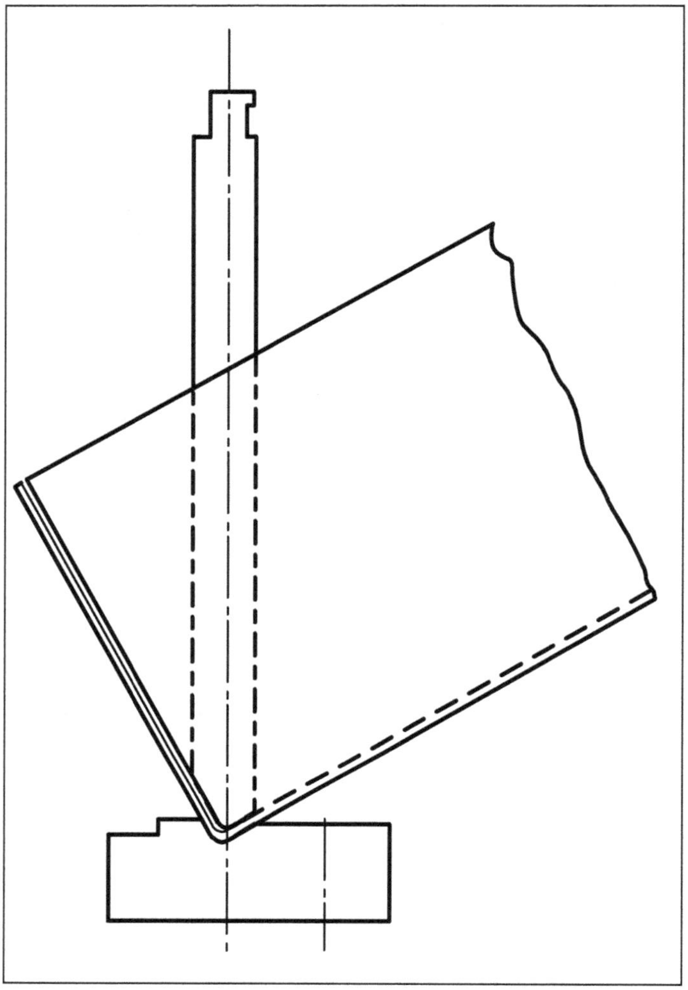

Fig. 17-7. Box being formed using a 30°/60° set of forming dies. This option allows a box with higher sides to be formed in the standard die space available on the press brake.

is restricted by the hold-down bolts in the die and the force of friction created by the forming load. If the material becomes thicker than 14 gauge, special methods to help hold the dies in position, or to minimize the "kicking," must be designed into the tooling.

The advantage of off angle dies, besides being able to form deeper boxes, is that the base of the part being formed does not have to be moved from a horizontal position to a 45° angle. The rotated angle requires the part to move upwards only to a 30° angle. This makes the flimsy blank required for making a box easier to control, and it reduces the risk of damage by back bending. Part removal of the finished box is similar to removal using standard vee dies made from conventional bar sizes. Off angle tools require a larger vee opening since the rear of the lower die must have sufficient depth to provide for the outside radius of the part plus 2.5 to 3 metal thicknesses of material flat, plus room for the lead-in radius of the die. The flat is the same as any recommended vee die opening. A larger vee opening provides a proportionally larger inside radius, which may not be eliminated by using standard bottoming type dies. It is even more difficult with the off angle design. The inside radius that may be on the drawing will have to be modified, and the calculated blank size will definitely be incorrect due to the elliptical shape that should be expected.

The reasons for off angle dies are understandable, but why aren't all dies made this way? The reason is that off angle tooling is very difficult to set up. If the upper die is nested into the lower die without any material between the dies, as the upper die is raised, the clearance between the upper and lower die in the front or the 60° side may be opened to the material thickness that is to be formed. In the rear—the 30° side—the gap between the upper and lower die may only be 40% of the material thickness. If the operator tried to form a part, the upper die would rotate so much that not only can tooling and machine be damaged, but the contact between upper and lower die will not match. As a result, it is difficult to obtain the correct bend angle. Additional forming tonnage will not solve the problem. The proper answer is to align the dies so there is one metal clearance between the upper and lower die prior to clamping the lower die into position. This will be very time consuming since each setup will need the lower die to be incrementally moved back and forth in an attempt to find the proper ram setting. Even when the last ram setting is recorded from a prior setup, it is difficult to adjust some hydraulic press brake rams to a specific ram setting and then try to hold that position. Unfortunately, a great deal of die setup skill is required.

If the dies have been positioned to make a good 90° bend using off angle dies, and if future setups must be made using the same material and ram settings, the operator should have some method to quickly set up the tooling Some users have made special setup blocks that locate the lower die in the proper position in relation to the upper die. It is important that all tools and setup tooling be tagged, stored, and referenced on all routing or setup sheets. Using the off angle dies, the maximum box depth in 12" of die space can often be increased from 6" to about 8".

D) Gaging and Handling

All of the bends of the box, except the last bend, can be made using conventional methods of gaging and handling. Depending on the type of box being formed, an NC back gage may be used if the box has no top flanges. If the box has outside top flanges, an NC back gage must be able to locate all four flanges. When the box is turned over to complete the final box bends, the gage must have a feature that allows the upper die to be lowered, clamps the material, allows the gage to retract, and then has the operator restart the cycle to complete the bend. The gage finger on the back will have moved out of the way of the path of the rear flange with the outside leg that requires a larger arc of movement. The first three sides of the box are made on one set of dies normally located to either side of the special length set of dies used for the last bend.

Before assuming that the NC gage will work, make sure that the lower die is high enough to be used. Most NC gages have a range of vertical adjustment, but in most units the low limitation is about 2" above the bed top. If the lower die is made too low in order to obtain the ability to make a deeper box, some special modifications must be made in the gage fingers, or perhaps the mounting blocks of the entire NC gage unit will have to be changed. If the gage range is modified, the new vertical travel may work with the box forming tooling, but it may limit the gage use for other work that may also have to be done on the press brake. When the box forming requires side-by-side die setups, additional gage fingers will be needed. The operator or setup person must check the tooling setup to ensure that a short leg gage position needed for one die set does not allow the additional gage fingers used for the other die set to strike that lower die—which could cause a wreck.

When the tooling used for the majority of bends (any flanges plus three sides of the box) is set up, often the larger one piece boxes require support arms in the front of the die. The blank size makes it difficult to hold the part square and locate the short lengths of the box against both gage fingers on the NC back gage. Some companies feel that using some type of front gage to locate all of the top flanges places any blank error into the top flange and minimizes the possibility of all of the box flanges not being the same height. In some cases, this could require a third length of the box dies to be added to the dies already in the press brake if sufficient room and tonnage is available.

If the top flanges are outside flanges, when the part is turned over after the top flanges have been bent, but prior to making any bend to form the box, the top flange legs will be facing down. Any support arms must allow clearance for those legs. Another concern is the top

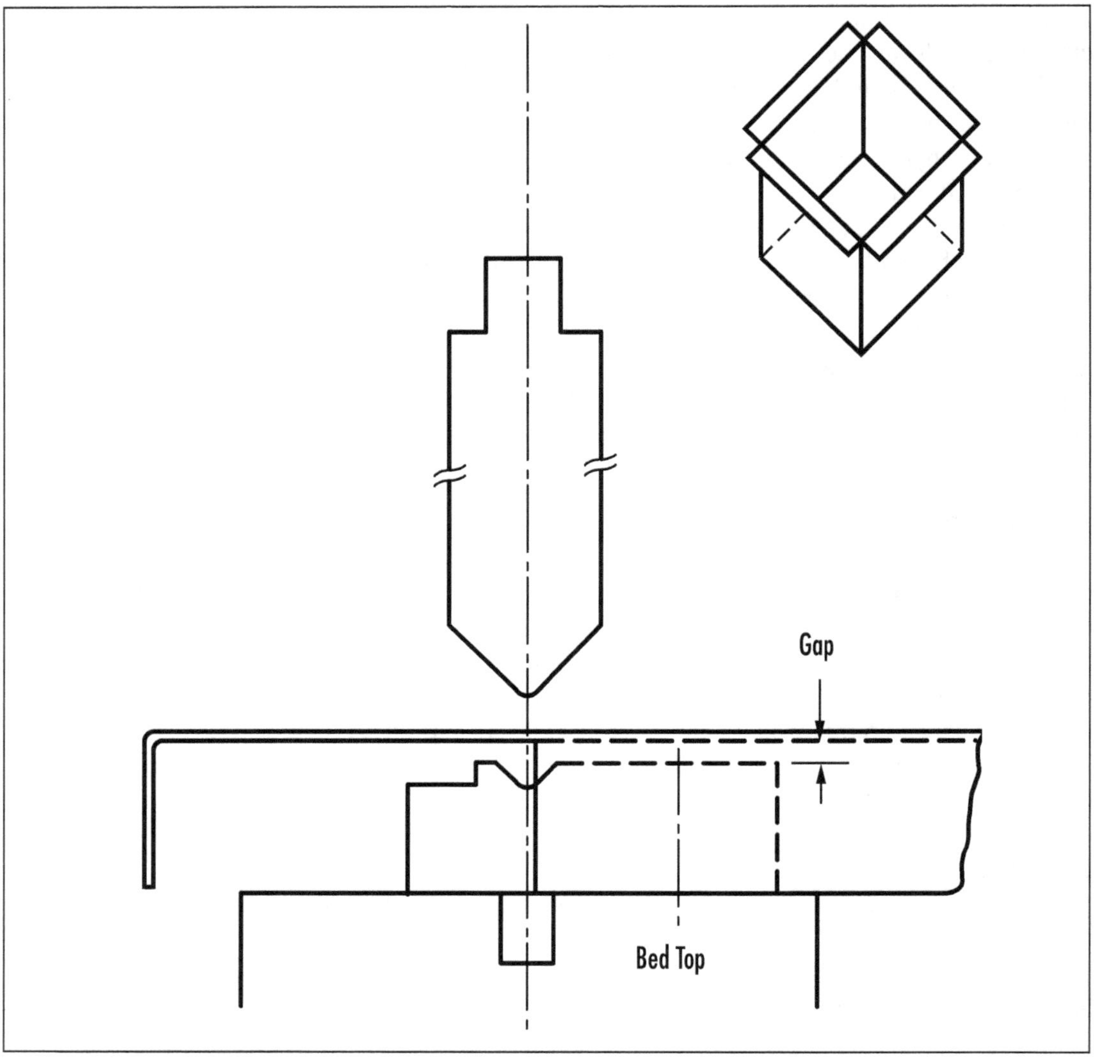

Fig. 17-8. If any previously formed outside flanges are longer than the height of the lower die, they will rest on the bed top prior to the box legs being formed, creating a gap between the body of the box and the top of the lower die. If the gap is not noticed and the bend is made in the leg of the box, the resulting distortion will probably cause the part to be scrapped.

flange dimension. If the lower die has been made lower than the top flange dimension in an effort to provide additional space to make higher boxes, when the blank is moved to the bend line to form a 90° bend at the bottom of the box, the previously formed top flanges of the box will be located over the top of the bed. If the lower die is not high enough, the outside flanges will sit on the bed top and not allow the unformed box blank to rest on the top of the lower die. A bend cannot be made without ruining the entire part. It is best to catch this problem prior to having it discovered by the press brake operator (Fig. 17-8).

After all three sides are formed, the part is moved over to the second die. This die may look exactly like the first die, but it cannot have any front supports or gages. The reason is that after the last bend is made, in order to remove the box from the press brake, the finished box must be held by the operator, lifted out of the vee opening, rotated slightly, pulled forward, and dropped down and rotated upward until the rear flange clears the upper die. If the part has inside flanges, the operator is also trying to pull the inside flange off the upper die. If the box has outside flanges, the operator is lifting the box at the same time it is being rotated to raise the outside flange in the rear over the lower die. In all cases, there cannot be any obstructions in the front of the second die station, since there is a high probability that the box removal action would interfere with the support arms. Some small size boxes—or if the press brake has an exceptionally long stroke providing a large die opening after the final bend is made—are the only known exceptions. Large boxes are often made in three pieces and are welded together to complete the part. Some two piece designs have been developed for some boxes. The decision made by most manufacturers that determines how their boxes are best made is limited by their existing press brake features, the box sizes to be made, available tooling, and the amount of welding needed to finish the box assembly.

Chapter 18
Beads or Curls

There are several common products that fall into the category of beads/curls. An opened, radiused edge on a part may be required to provide stiffness, or to help make a countertop easier to keep clean; this shape is called a **curl**. The rounded shape on the edge of a part (approximately a 330° arc), or a closed circular form, is normally called a **bead**. Small inside diameter beads are often used for hinges. Unfortunately, a rib shape that does not extend the full length of the part (referred to previously as a stopped rib) is also called a bead in some parts of the country.

A) Beads or Hinge Shaped Parts

Engineering drawings will often show a hinge having a full circular shape until the edge of the material touches the flat surface of the part leg. As drawn, the part cannot be made using press brake tooling. Whenever material rests on the backside of the lower die, it must have at least one metal thickness of flat surface contact on a gage ledge to support the material through the initial bending, and to prevent the part from erratically slipping down into the die cavity. Since the material will not allow a perfect radius fit between the upper and lower dies, the first stroke of the three hit forming process used to make most hinge shapes will produce a section of the inside part radius with about a 1.5 times metal thickness flat from the edge of the material to the start of the formed radius. As a result, when a small inside diameter hinge shape is closed, the part will look like a 270° arc with a straight vertical leg touching the longer flat length of the part. As the inside diameters of the different beads are made larger, the actual arc in the part becomes greater, and the edge flat becomes less noticeable (Fig. 18-1).

The type of die required to form a bead or hinge depends on the size of the inside diameter of the part. If the inside diameter is greater than four times the material thickness, the normal method of forming the first bend (because the tooling is the least costly) is to place the leading edge of the part across the radiused lower die opening, and locate the edge on a one metal

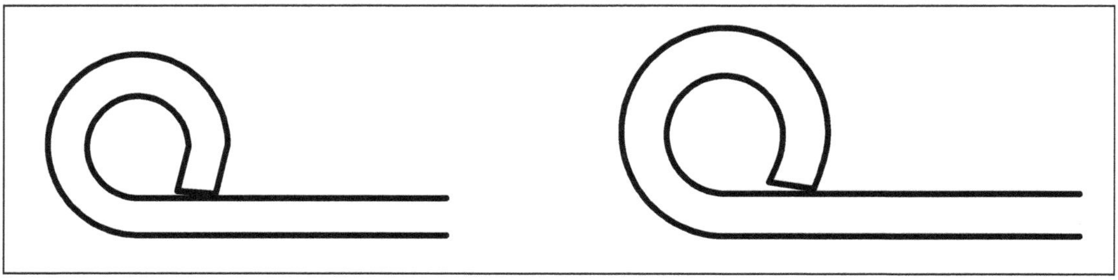

Fig. 18-1. A small bead will have at least a one metal thickness flat at the edge of the material which will easily be visible after the part has been formed. A larger diameter bead has a similar edge flat, but it is not as noticeable in the finished part.

thickness gage ledge in the rear of the die. When the first stroke is made, the part will have about a 90° arc plus the edge flat described. The part is then pushed across the lower die opening with the material spanning not only the lower radius but also the gage ledge used for the first bend. The radiused edge of the part will be located against a back gage on the top rear surface of the lower die. Some small material flat is always required on the rear flat of the upper die for initial support during the second stroke, plus any additional material needed to complete the hinge shape. The second stroke produces an arc somewhat greater than 90° (often 120°). The final stroke closes the shape into the bead or hinge shape. Depending on the part length, the first two strokes can be made on one die set and the closing operation in a second die. Usually, both dies are made to run in the press brake side by side at the same time (Fig. 18-2). If the part is longer than one-half of the bed length, or if the press brake tonnage will not allow the use of side-by-side tooling, all three operations can be designed to work in one die (Fig. 18-3).

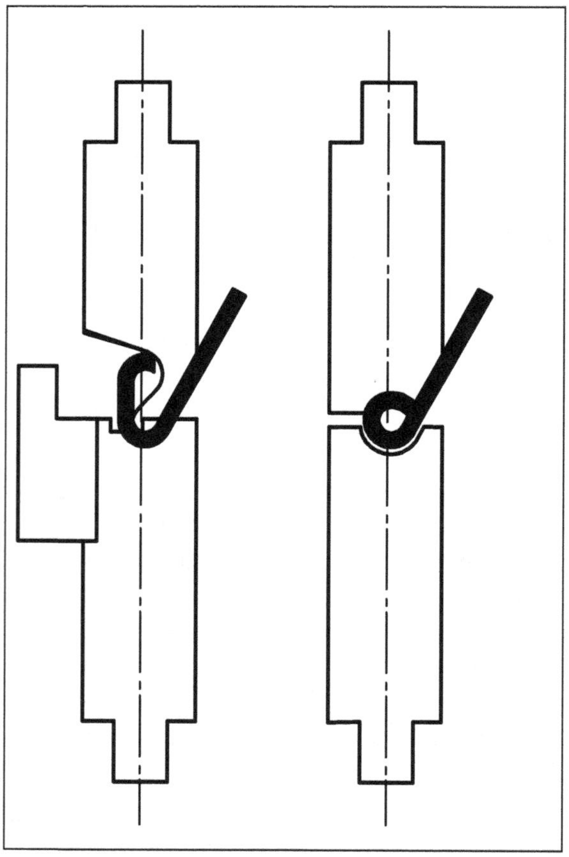

Fig. 18-2. A bead (sometimes called a curl) that has an inside diameter of four metal thickness or more is normally formed using the two sets of dies shown. Note: tooling to make beads or curls are designed for one material thickness. Different gauge materials require a different die set.

If the inside diameter of the part is less than four metal thickness, but at least 2.5 times the metal thickness, a special two station design is required (Fig. 18-4). This is needed because the small increment between the first and second bends used for small inside diameter hinges will not allow the material to span the lower die to make the second bend needed in the normal three stroke forming method. The first die set forms both initial bends of the smaller inside diameter hinge. Sufficient forming tonnage is required to form both arcs and angles on a consistent basis to ensure all of the parts being made will fit into the second station closing die.

The second die is more complex. A special lower die allows the part to be located with the two radiused bends positioned against a vertical section of the lower die. The longer, unformed, leg extends out the front toward the operator with the material sloped up at about a 15° to 20° angle. When the upper die is lowered, a pressure pad in the front of the upper die securely clamps the part. After sufficient clamping pressure is built up, a rounded forming insert forces the 90° arc at the top of the material to start rolling the material into the final bead shape. The upper die is held in position by a "back up" located on the upper die, which slides down the rear of the vertical section of the lower die, and locks the upper and lower die into position while the part is being formed.

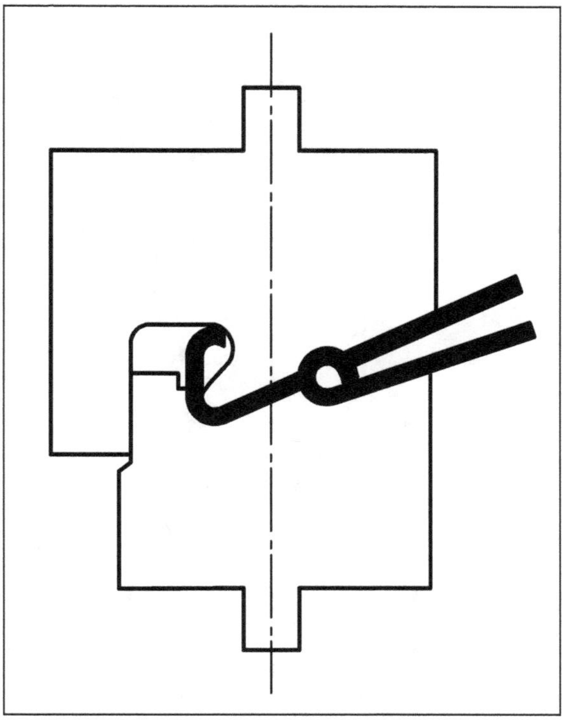

Fig. 18-3. Optional "one" die set that can be used if the part length of the hinge shape exceeds one-half of the machine length, or if a tonnage limitation exists on the press brake.

When a bead or hinge is closed by forcing the metal to roll around, there is a great tendency for the part to try to simply move backwards out of the closing die. As soon as the part does lock into a position where the part will actually curl, the flat leg in the front will try to kick up in the air. If the leg is long enough, the operator often has sufficient leverage to hold it down. Most of the time, the leg will lift up until it strikes the front of the upper die. As the die continues downward, it pushes the raised leg back down until, at the bottom of the stroke, the flat, unformed, leg is back in its original position. Even if the second operation die has a pressure pad in the front holding the unformed leg in place, the lifting force of the front leg can often cause the pad to rock until the stroke forces the leg back down.

Small diameter beads do not curl into a nice round shape. With the edge flat and what might appear to be a kinked radius, the final inside diameter will definitely be out-of-round. The normal tolerance will be $\pm ^1/_{64}$" (or 0.0156") measured on the outside of the formed shape. As the outside dimensions become larger, that appearance of the bead will start to look better, but the tolerance will probably stay the same. If a hinge pin must fit into the inside of the bead section, it is best to have the tooling supplier test the dies to make sure that the hinge pin you have supplied fits easily into the bead without effort and without rocking.

Beading dies only work with one thickness of material. The first operation bends are made with fitted dies made for only one specific material gauge. The closed inside dimensions will be larger or smaller depending on material variations. This is because the metal is closed on the outside radius and all material variations are forced to the inside. Depending on the material thickness and the bead size, the forming tonnages can vary greatly. It is recommended

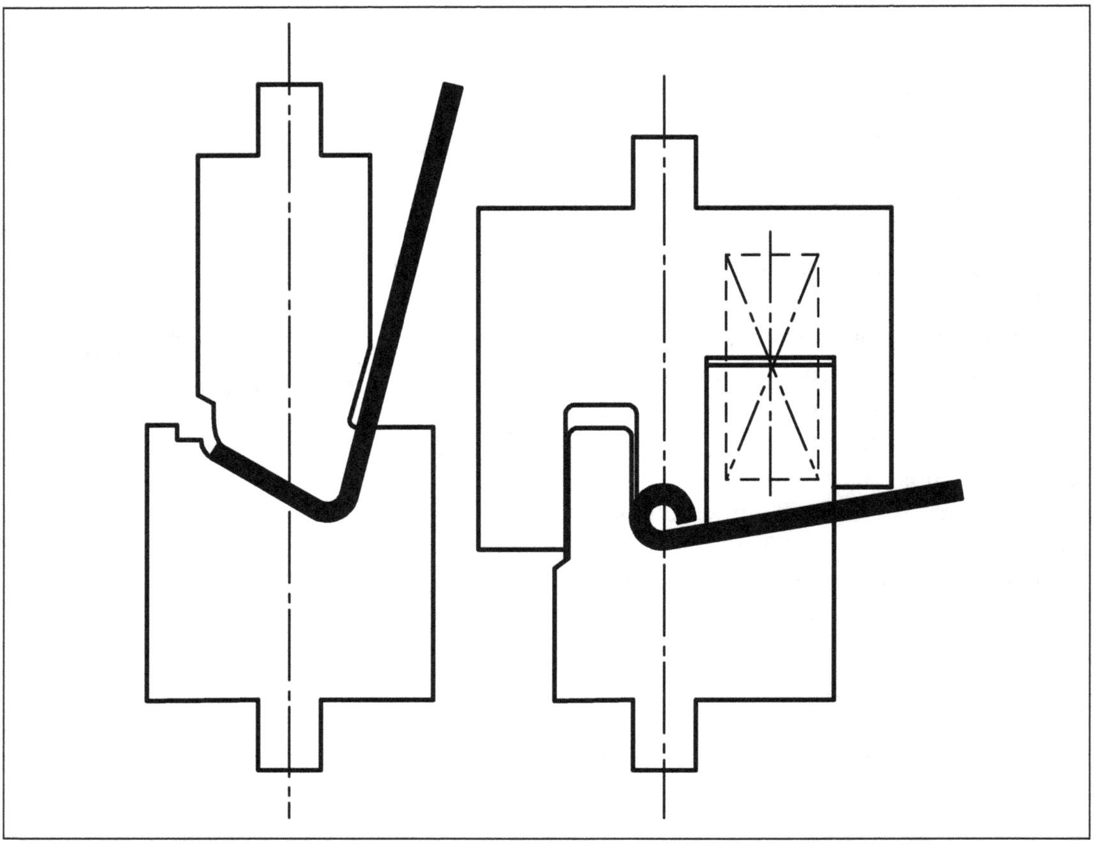

Fig. 18-4. Die design that must be considered if the inside diameter is between 2.5 and 4 metal thicknesses. If the inside diameter is less than 2.5 metal thicknesses, it cannot be formed on conventional press brake tooling.

that, when this type of die is purchased, the press brake die space information plus tonnage available be included on the purchase order with the stipulation that the tooling must be tested and must work on that machine.

B) Curls or Rolls (As They are Called in Some Parts of the Country)

The most popular shape that falls into this category is the sanitary roll used in the restaurant industry. It is formed with a 1.5" outside diameter × 210° of arc. The normal material thickness is 14 gauge stainless steel. This part is made in two strokes, using what looks like a gooseneck upper die having a $^3/_4$" radius as its forming nose. The lower die is a matching die often having a one metal thickness larger radius than the upper die. To form the first 90° radius bend in a roll, the same type of one metal thickness gage ledge that is used to form the first operation in a beading die is machined into the rear of the lower die (Fig. 18-5).

The second and final stroke is the same as the second stroke used to form the bead, except the amount of flat between the bends is reduced so that the arc of the first bend will blend into the second stroke arc. This works since the additional material needed to make a larger radius provides sufficient flat between the bend of the initial arc that will be gaged at the rear of the lower die, and still have enough material to support the start of the second bend before it is drawn down into the lower die cavity. The second bend is located by fixed gage stops mounted on the rear of the lower die. The radius formed in the second bend will have about 120° of arc which will allow the completed **roll** to have about 210° of total wraparound. The maximum amount of wraparound in any curl is determined by the inside diameter size that is to be formed. The curl diameter determines whether sufficient gap remains for the gooseneck upper die to have enough strength to form the radius and still fit into the inside of the completed roll after forming the second bend.

The upper die is often sectioned to allow different lengths of parts to be made. This will allow side flanges to be formed first on different length panels. If possible, the lower die is kept in one length in order to minimize any marking problems. In some instances, the upper die is made higher to provide clearance for previously formed side flanges. One set of dies is required for each material thickness that might be formed, since the fit between the upper and lower dies is important, and the springback between different thicknesses of material may be different.

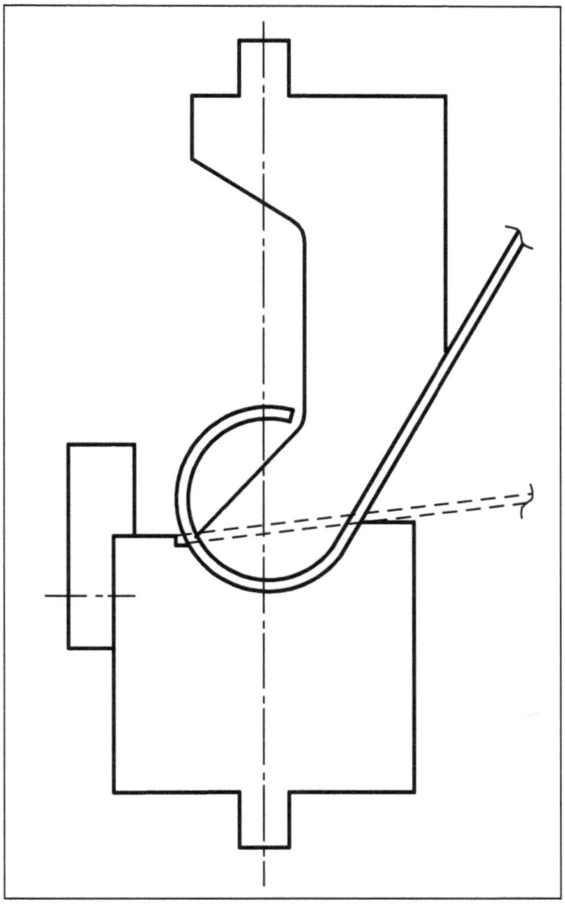

Fig. 18-5. Tooling used to form a standard sanitary roll. Larger or smaller parts of similar shapes can be made using similar dies.

Chapter 19
Wiping Dies

Some rectangular parts having short flanges on all four sides are very difficult to form if they require flanges on each narrow width, and if the long sides of the sheet extend out to the front of the press brake more than 48". The operator cannot easily lift the part during the forming cycle because: 1) it is too heavy to be safely lifted; 2) it is too thin or has holes/cutouts that can cause back bending or distortion as it moves upward to a 45° angle without proper support; and 3) the part may be too long, which can cause the sheet width to **belly** while the bend is being made. This may cause the bend angle to vary.

If the operator assists the process by holding the long but wide part upwards while it is being formed, he/she may hold the part in the center of the part length. In that case, each end will not be supported and will sag down. If the operator has a helper, which thus allows each end of the part to be supported during the forming stroke, the center of the sheet will not be supported. The sagging of the sheet in the center, called a belly effect, can actually cause the resultant angle being formed to vary along the bend line, causing not only angular variations but also flatness problems.

One solution is to form the short 90° flange while the wide sheet remains flat. This type of forming is called **wiping**. The dies can be engineered to allow the part flange to be formed up or designed to form the flange down. Normally, the dies cannot be turned over to allow the operator to choose the direction of the bend. The main tooling limitations are unsupported die surfaces, plus any limitations created in gaging the part. If the part is to have the flange formed down, the main problem occurs after the part is formed. It has to be lifted up in order to move it back to the front of the die. Another difficulty is designing the gaging. Since most of the wiping dies are designed to form the flange "up," the description of the following die considerations will be based on that assumption (Fig. 19-1).

A) Basic Wiping Die Design and Gaging

The lower die must have the front vertical surface of the lower forming corner on the

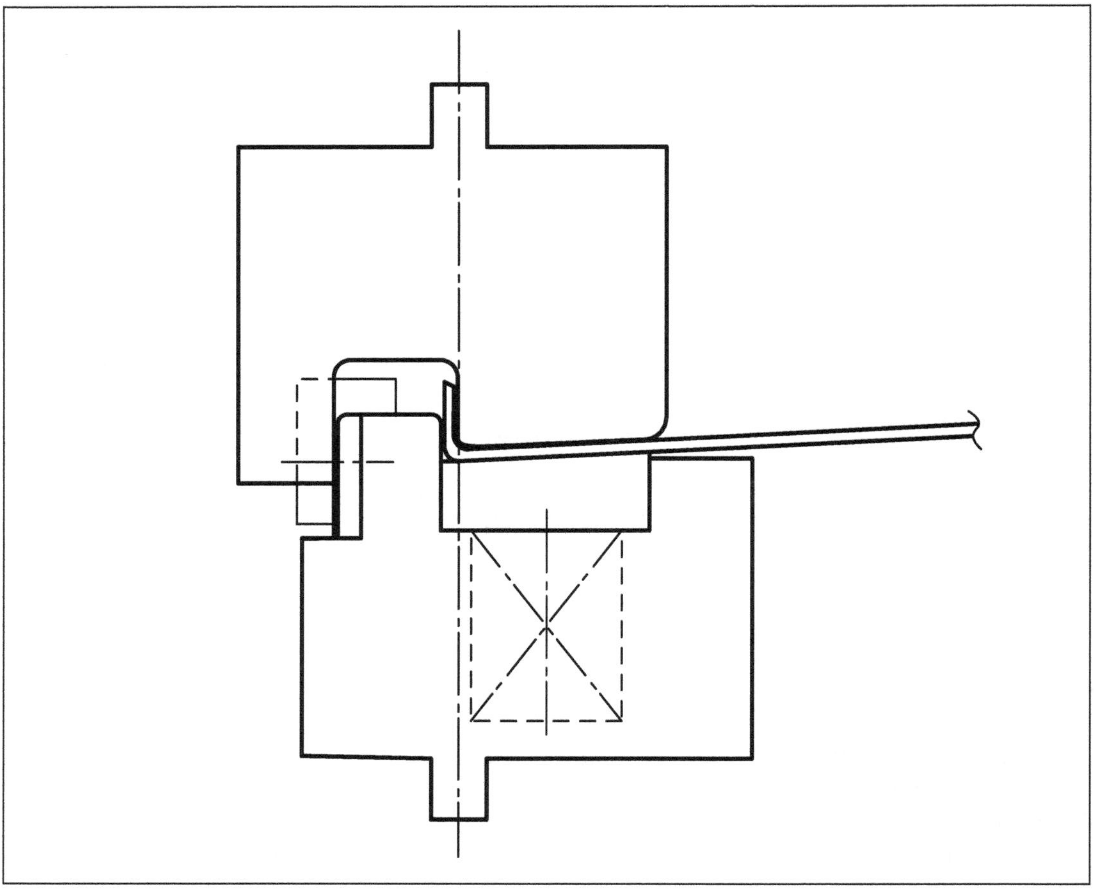

Fig. 19-1. Wiping die used to form a flange on the edge of a wide sheet.

centerline of the press brake. The thickness of the forming insert is located to the rear of the centerline. In the front of the forming insert is a pressure pad that will hold the part against the upper die as it forms the flange. The upper die can be solid in design, except for a **backup** or **heel block** that precedes the forming corner in the stroke, and locks the upper and lower die together. Since a wiping die, by design, will generate a front-to-back separating force, the upper die must lock the lower die forming surface in such a way that the lower die cannot be forced to the back and the upper forming surface wedged to the front. This is why a backup is always needed on a wiping die. A backup can be designed in many ways. It is important that it will be able to keep the upper and lower die together during the forming cycle so that a consistent bend will be formed.

The inside radius formed using a wiping die will be larger than one-metal thickness and not really a true radius. It is usually an elliptical shape. As a result, the part shape will not be "set" like a bottoming operation, but the flange will spring back several degrees after the upper die returns to the top of the stroke. If the lower die is made flat on the top, creating a 90° angle between the top of the die and the vertical forming area, the part will be open 2° to 4°. Sometimes, if the forming radius of the lower die is made slightly sharper, the metal will loop around the corner while it is being formed, and will seem to cut down one or two degrees of the springback. Most wiping dies compensate for the springback by sloping the top surface of the lower die and the top surface of the pressure pad 2° upward toward the operator. This means that any sheet supports in the front have to be spaced far enough to the front to allow the part to be formed, and loop back to the sheet supports without causing any distortion in the part.

Wiping dies are not recommended for material over 12 gauge mild steel. The main problem is that the pressure pad must travel at least six times the metal thickness downward. The forming load will start as soon as the flange begins to wipe upward. As a result, the outside corner of the pressure pad will be tilted down and the springs will start to compress, building up a resisting load. Springs are rated at so many pounds of pressure per $^1/_8$" of compression. The longer the spring length, the lower the spring pressure for each $^1/_8$" of travel. The thicker the material that is being formed, the greater the required spring travel will become. As a result, it becomes very difficult to properly hold thicker material with the pressure pad without the part moving or breathing during the forming stroke. There is also a problem with abnormal die maintenance with the bolts holding the pressure pad. Material as thick as $^1/_2$" plate has been successfully wiped, but the pressure pad was uniquely guided, and hydraulic cylinders replaced the normal springs.

The flange being wiped must be short enough so that it will not interfere with the backup design required to hold the upper and lower dies together. In some cases, the backups can be moved slightly back to accommodate a longer flange. If the material is thin, and the length of the bend less than 48", it may be possible to form a deeper flange by making the die longer and locating appropriate backups at each end. The height of the formed flange is also limited by the amount of open space available to remove the part to the front when the upper die returns to the top of the stroke.

Gaging of the flange is often done by mounting a solid gage block on the top of the lower die behind the forming surface. If the flange depth is too close to the backup bar mounted on the upper die, so that there is no room to mount the gage, the backup on the upper die can be sectioned, and simple hook type gages can be bolted to the rear of the lower die in the space created.

Some parts must have the flanges formed on all four sides. This can require that slots be made in the upper die to provide clearance for previously formed legs as the part is rotated around to complete the part. Care must be taken to ensure the slots are not made too wide, or else the absence of marking caused by the openings in the nose in the upper forming die will show through the material and look like an imperfection. Since there is some risk that the upper die will strike a previously formed flange if it is not properly located at the clearance slot, it is important to use end gages to locate each bend position. Rotating the large rectangular blank around on the support arms can also cause marking on the outside of the material. Some type of covering on the support arms or plastic coating on the blank itself may be required to protect the material surface.

The die must have some method of adjustment available to allow the upper and lower die forming surfaces to be moved forward or backward to adjust for material variations. When this is done, the clearance at the backup area between the dies must also be held. Often this adjustment can be obtained by designing a forming insert similar to one that may be used in a channel die. The insert is held in place by bolts. Shims can be located between the die body and the insert to obtain necessary adjustments.

B) Estimating Tonnage for Wiping Dies

To determine the estimated wiping tonnage, the length of the bend is multiplied by the material thickness to obtain the area of the material being wiped. That area is multiplied by 80% of the tensile strength of the material being formed. The spring pressure from the pressure pad must be added to the calculation, but normally that will only be three or four tons per foot of the entire die length. A simple example would be forming 16 gauge mild steel 48" long but using a 6 foot long wiping die. Multiplying the 0.060" thick material by 48" in length equals 2.88 square inches of formed material. The 60,000 PSI tensile strength of the mild steel is multiplied by 0.8, which equals 48,000 PSI (or, when it is divided by 2000 pounds, equals 24 tons/in.^{2}). Multiplying the 2.88 square inch figure × 24 tons/in.^{2} will equal about 69 tons. Adding the probable spring pressure of perhaps 6 tons/ft × 6 feet, or 36 tons to the 69 tons, will give an estimated forming load of 105 tons. This means that if a hydraulic press brake is to be used for this job, the press brake must have at least a capacity of 105 tons plus a safety factor of 15%, or an additional 16 tons. The press brake probably should have a capacity of 125 tons or more to ensure the wiping die will work. It should be noted that if a full-length part of 6' feet was to be formed, the machine capacity would have to be approximately 160 tons.

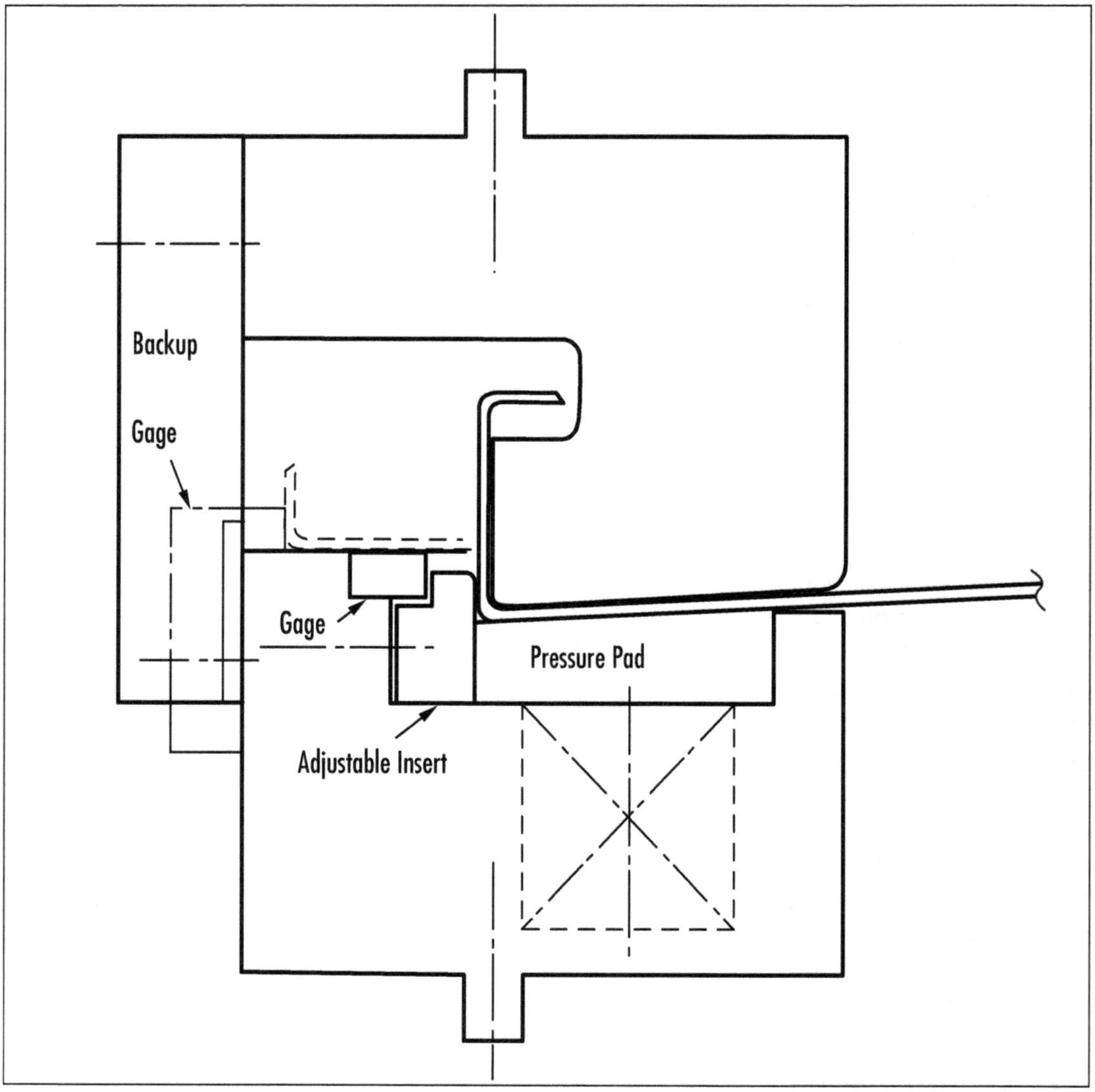

Fig. 19-2. Special wiping die designed to allow two 90° bends to be formed on the edge of a wide sheet in one handling.

A problem can occur when shorter lengths are formed on a die that has a full-length pressure pad, as used in the previous example. The springs under the pad are located to apply uniform pressure along the full length of the pad while a forming stroke is cycled. If a shorter length is formed, the pad will bow near the bottom of the stroke, allowing the ends of the pad to touch the upper die and not provide uniform pressure in the center of the part being formed. Each edge of the part will have high clamping pressure where the pad bow is now contacting the part. As a result, the gaging of the bend may vary or even create a slight bow along the bend line. The easiest solution is to tape a short strip of material, the same thickness as the material being formed, on the top of the pressure pad at each end of the pad not being used to form the short flange. The length of these pieces should allow about 6" of open clearance on either side of the flange being wiped to ensure that the unformed blank is not located accidentally on the top of the filler strips.

There are many combinations of wiping type dies that have been used over the years. Dies have been made to form a large radiused edge instead of the normal metal thickness inside radii that make up most wiped bends. Additional springback allowance is needed in that type of die design. Special wedge action or rocker designs are being used that make two bends in a part in one (wiping) stroke to produce the edge of a panel. Wiping dies with the forming nose made with sectionalized inserts can provide the ability to form families of panel shapes.

It is not uncommon for the edges of a panel to be made in two strokes. One stroke forms the return bend that will be located at the bottom of the panel. The part is then pushed forward sufficiently to allow the second stroke to complete the depth of the panel (Fig. 19-2). The die that locates the inside of the panel corner may have to have clearance notches at each end to provide clearance for side return flanges. Part removal may require some thought and die modification to clear the backups if the panel has to be rotated to clear the forming die inserts, similar to the relief used on box forming dies.

Chapter 20
Plate Forming

Prior to the introduction of the hydraulic press brake, material heavier than $^1/_2$" thick plate was rarely formed on press brakes. Even forming $^1/_2$" material was not too common. The reason is that larger vee openings require a longer forming stroke plus added clearance to remove the part after it has been bent. Mechanical press brakes usually had a 3" total stroke, although some larger tonnage machines had a 4" stroke designed as standard.

The working stroke of the mechanical press brake was limited to one-half of the total available stroke. This was required since the design of any mechanical press brake does not have the energy built up in its flywheel to provide for a bend requiring the full stroke. Prior to 1958, all mechanical press brakes were rated at a maximum tonnage rating starting at the "mid-stroke" or half of the total machine stroke. In Chapter 5-F, it was pointed out that the forming load to make a simple 90° bend builds up to maximum tonnage shortly after the material starts to bend down into the vee opening. The approximate forming stroke needed to make a 90° bend is about 0.4 times the vee die opening. For example, a 4" vee die opening requires a forming stroke of 4" times 0.4 which equals 1.6". This exceeds the maximum 1.5" forming stroke available with a standard 3" press brake. It will work if the press brake has a 4" stroke. As the plate became thicker, mechanical press brakes were not often considered to be the best machine selection to form the parts (Fig. 20-1).

Hydraulic press brakes, with their ability to provide full machine tonnage throughout the forming stroke, quickly became the viable option to form plate. Although the first hydraulic press brakes were introduced after World War II, they did not achieve acceptance until late in the 1950s. Initial problems included ram repeatability and keeping the ram parallel to the bed. Newer hydraulic pumps, seals, and electrical designs, plus specialized "leveling" systems, eventually made the press brakes very attractive for companies interested in forming heavier material.

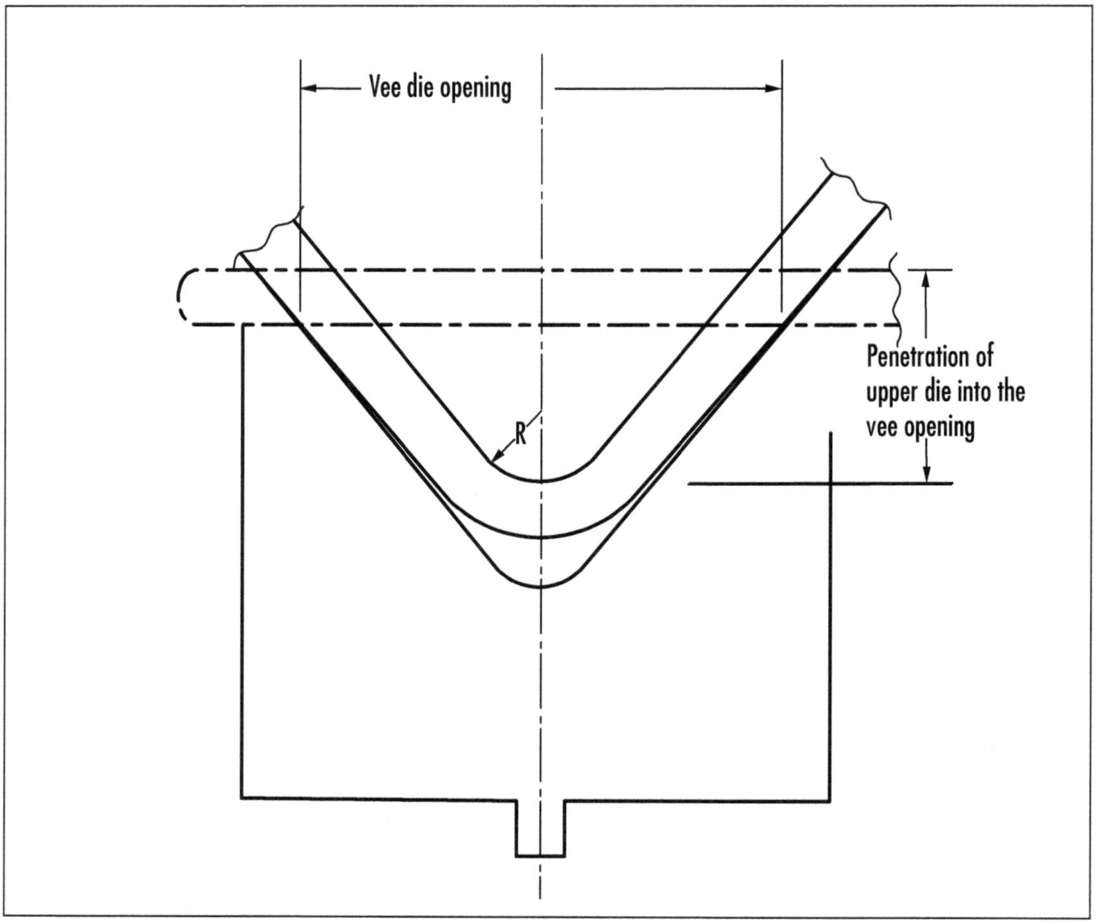

Fig. 20-1. Distance that the material penetrates into an air bend vee die to form a 90° bend.

A) Modification of the Vee Die

The thicker the plate that is to be formed, the more forming pressure is required. It was found that if thicker plate, starting with $^1/_2$" thick material, is formed using the standard eight times metal thickness die opening, there is often some risk of the material developing small hairline cracks along the outside radius of the bend angle. The recommended vee die opening found on most air bend charts will recommend both a 4" and a 5" opening. If the bend has short flanges (12" or less) on one or both legs, the 4" vee die will normally allow the bend to be made without any cracking. Longer flanges can sag downward during the bending cycle, causing some additional friction between the outside of the material and the lead-in radius of the die opening. Most plate is hot rolled and has **scale** on both sides, and the scale can flake off in the radiused area of the bend.

It was discovered over the years that after the vee opening reached 4" wide or greater, it was advantageous to make the lower die angle 80° instead of 85°, which had always been used for conventional air bend tooling. The reason is that the scale from bending plate would drop off the outside radius of the part and collect in the bottom of the vee opening. If a second bend in the part had to be made, the part would be pushed forward, located, and the second forming stroke made. Since the operator could not see down into the die opening prior to the second bend, or if the vee die opening was not cleaned of the scale from previous bends, at some point a bend would be made and the outside radius of the part would "bottom out" on the scale, never reaching the bottom of the stroke and not allowing the correct bend angle to be formed. Changing the lower die angle to 80° provided more room for the scale, thereby minimizing any potential forming problems when forming one piece. The scale would still have to be removed after each part had been formed.

B) Larger Inside Radii

Increasing the vee die opening also increased the inside radius of the formed part. A 4" vee die opening will provide an inside radius of about $^5/_8$" in $^1/_2$" plate. A 5" vee opening forms an inside radius of about $^3/_4$" (Fig. 20-2). When the inside radius is measured, it will become obvious that the inside radius is somewhat elliptical and not a true radius. This problem was first pointed out in the section on bend allowance (Chapter 8). If a simple radius gage is placed inside the bend radius, the fit will look good for about 60° of the arc, but the sides will seem to open outwards slightly and then blend into the vertical legs of the flanges. This gap is the visual result of placing a perfect radius into an elliptical arc. The next measurement is made of the outside radius using a radius gage one metal thickness larger than the one used on the inside radius. The radius gage will appear to fit fairly well, except the legs of most radius gages are not long enough to measure very far up onto the outside legs of the part. The reason for the better fit is that the outside radius flattens out slightly, as described in the example of the rubber eraser in Chapter 8.

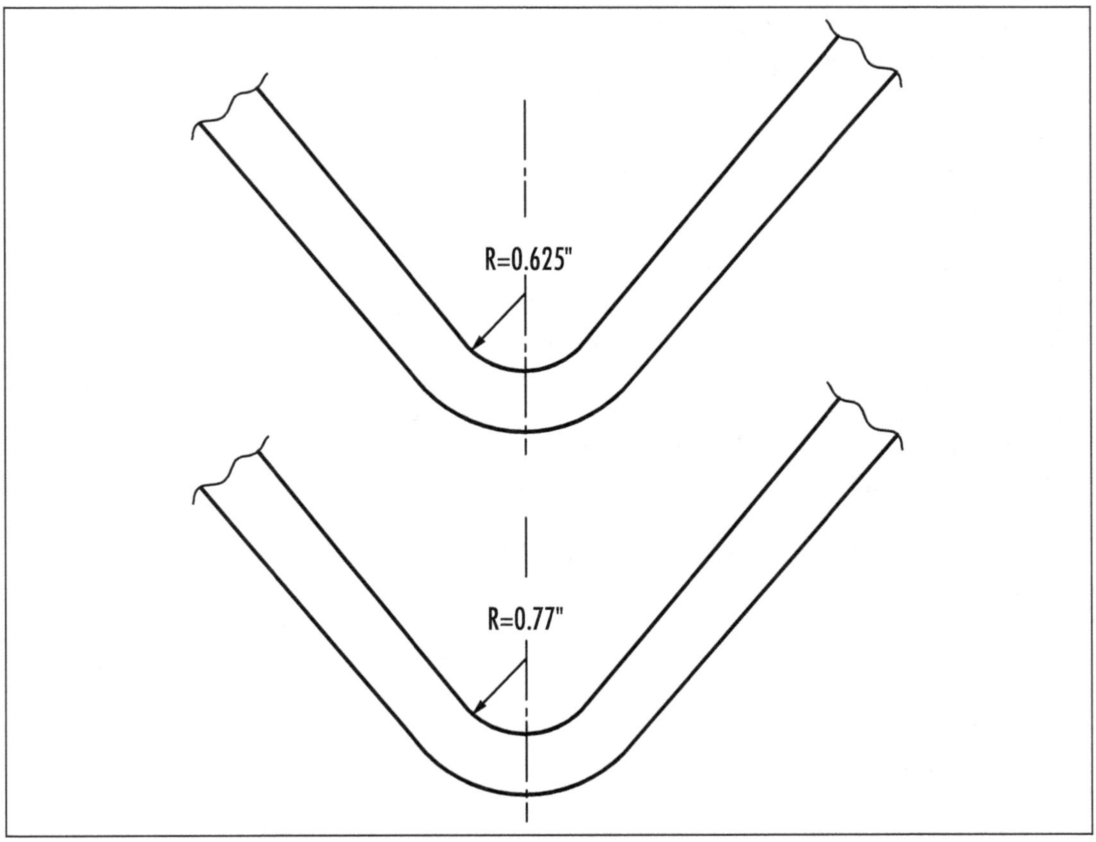

Fig. 20-2. Increase in the inside radius in $^1/_2$" material when the vee opening is increased from 4" to 5".

A problem that has been mentioned in previous chapters is that the inside radius is often specified as "one metal thickness" on the engineering drawing. The only way that is going to happen is if the $^1/_2$" plate is formed using a 3" vee die opening, which will provide an inside radius close to $^1/_2$". Of course, the risk of cracking is extremely high: because the forming tonnage increases by almost 50%, and the potential angular accuracy can vary as much as ±3.5° because of material thickness variations. Most press brake operators request increases in the inside radius to minimize forming problems. If the engineering staff will not change the radius, it is highly probable that the operator will still use conventional vee dies in order to hold accuracy and keep the forming load within the available machine capacity. This can create problems with the blank size required to hold the part dimensions. However, a competent operator will allow the variations to be hidden in flanges that don't affect the part use. Unfortunately, this will create problems later on if the part is subcontracted out to another company who will question the drawing (that may have been used for years), or the operator who knew how to form the parts is replaced with a new operator. This very real problem will not be solved until management forces engineering to correct drawings to include inside radii that conform to forming standards that must be used by the shop.

C) Tooling and the Press Brake

Since the forming tonnages will be higher to successfully form plate, the upper die must be made wider to allow the die shoulders to carry the additional force to ensure that the die will not "sink" into the ram of the press brake (reference Chapter 5-A). In addition, since the vee opening will be larger and the upper die will have to move further down into the die cavity to make the bend, consideration must be given to the closed height of the machine. It is important to check that the upper and lower die can come together to form the part. After the part is formed, sufficient stroke is necessary to remove the part from the press brake. Larger hydraulic press brakes have higher closed heights than smaller press brakes. As a result, some operators have found that their die selections were not high enough to allow the dies to come together at the bottom of the stroke. If special die adapters have to be acquired at the last minute to add to the die space, production schedules may be disrupted.

The larger opening of the vee die means that the lower die may be wider than the filler block (die holder) that is available. Consideration must be given to removing the filler block and mounting the lower die on the bed top. This creates other possible concerns. Is the bed top wide enough to provide adequate support? How will the lower die be fastened to the bed top? Is the upper die high enough to allow the dies to form the material at the bottom of the stroke?

The 0.5" wide tongue located at the bottom of lower press brake dies is used to locate the die in the filler block. Set screws hold the tongue against the opposite side of the filler block tongue slot. Although many press brake bed tops have a slot lengthwise on the bed top, the slot only provides for the tongue on the lower die to fit into the slot so that the shoulders of the lower die can sit on the bed top surface. There is no provision for front to back die alignment or clamping the die to the bed. Some manufacturers provide the cross tee slots that are used to bolt the filler block to the bed top. If those slots are available, the lower die might be drilled and counterbored through the bottom of the vee opening, allowing bolts to be used to fasten the vee die to the bed top.

The bed top must be wide enough to properly support the vee die body. For companies that know that plate forming will be done on their press brake, it is common practice to

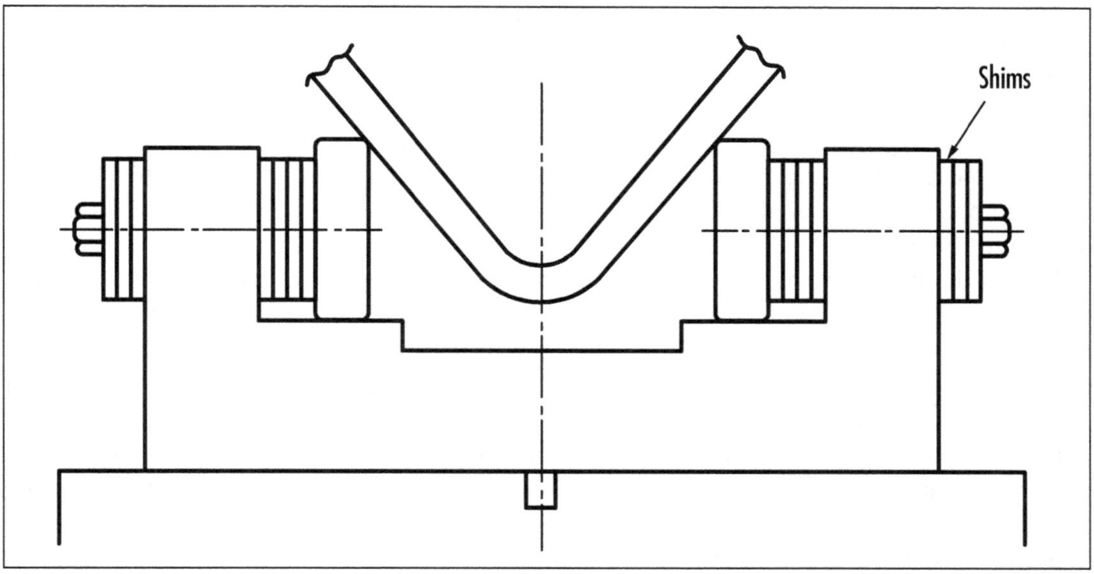

Fig. 20-3. Adjustable vee die.

purchase a wider than normal bed top. Often a wider adjustable lower die, that can have the vee die opening shimmed opened or closed to provide a range of vee openings for various thicknesses of plate, will be purchased at the time the press brake is acquired (Fig. 20-3). Most flanges in plate do not have reverse bends to create offset type shapes where a wider width lower die could cause an interference problem. If only a standard width bed top is available, the bed top might have been machined to allow angle brackets to be used. These 90° angle brackets (reference Chapter 1-C-4-c) are bolted to the front and rear of the bed, allowing the top surface of the bed to support wider tooling. If angle brackets cannot be easily mounted, contact the original press brake builder and request their assistance or suggestions.

D) What is Mild Steel?

Mild steel, as described in Chapter 4-B, is common steel having a tensile strength of 60,000 PSI and yield strength of no more than 40,000 PSI. When purchasing plate, usually described as material one-quarter inch or thicker, the steel service might not supply ordinary mild steel but rather structural grade mild steel. The ASTM designation for this steel is A-36. This material has a tensile strength of anywhere from 58,000 PSI to 80,000 PSI. The yield strength is listed as 36,000 PSI (minimum). In most parts of the United States, that will mean that the maximum yield strength will not exceed an increase of 15,000 PSI over the nominal, or a maximum of 51,000 PSI. Several steel companies on the West Coast do not seem to follow that maximum "limit," and actually allow their yield points to increase to as much as 65,000 PSI. As a result, the tonnage needed to form most A-36 mild steel will increase as much as 30% over the forming tonnage shown in a standard air bend tonnage chart if the yield strength stays under the 51,000 PSI figure. It can increase to as much as 60% over standard if the yield strength is as high as 65,000 PSI. To review the concept of bending factors, see Chapter 6-A. Remember that the forming tonnage calculated from charts or bending factors must never exceed 85% of the maximum available tonnage of the press brake (reference Chapter 4-D and E).

E) Making a Good 90° Bend in Plate

Material thickness variations in plate will be much greater than those found in thinner material. If the operator attempts to form a second bend at the same ram setting as the previous bend, there is a good possibility that the material could be thicker at that the new location, which will cause the angle to overbend. Plate that is overbent presents a major problem, since trying to open the bend angle can be very difficult. The main rule in forming plate is **do not overbend the angle.**

Unless the press brake is equipped with an NC plate gage, most gage locations for the heavier materials are done by using scribed lines. The main reason is that most plate forming is done with very small lot sizes. Setting simple gages often takes too much production time compared to simply laying out the bend lines with scribed line marks. (For more information on scribed lines, refer to Chapter 7-H-4.) Often, a crane assist is available to help load and unload the part (Fig. 20-4). At this time, the operator should raise the ram setting several thousandths above the ram setting needed to bend a 90° angle (or the angle that is to be formed). The operator then locates the sheet on the top of the lower die so that the scribed lines are close to the center of the die opening. The ram is lowered, allowing the upper die to stop just above the top of the material, with sufficient space that the operator can move the plate to align the scribe marks with the center of the upper die. The initial forming stroke can be made after the operator is satisfied that the location is very close to the proper bend line. It is recommended that, after a bend has been made, the ram motion be stopped just past the bottom of the stroke to allow the crane assist to catch up with the long leg of material sticking out toward the operator (Fig. 20-4). This will keep the part in place after the ram returns to the top of the stroke.

After the bend has been completed, the operator must check the angle of the bend using a gage that will show the angle that is needed. However, the gage must be modified so that the inside radius of the part will not touch the gage. Both ends of the formed part must be checked.

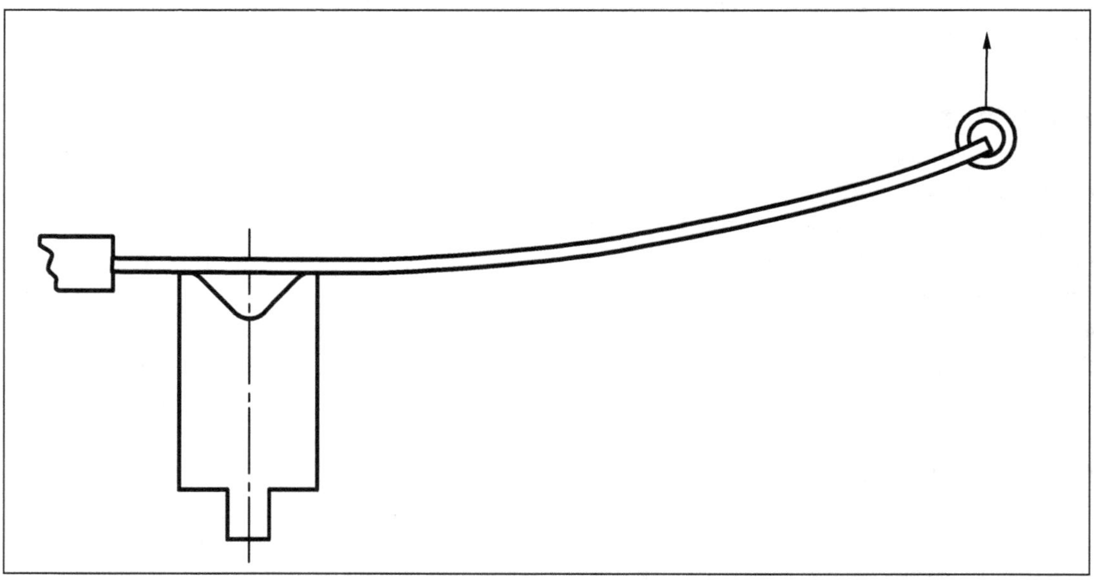

Fig. 20-4. A crane assist can be used for loading or unloading. Care must be taken, if a gage is to be used, that the gage assembly is strong enough to withstand any impact from the material while it is being located.

Hopefully, the angle will be open. The operator, based on his skill and experience, lowers the ram setting so that the next stroke will reform the bend angle very close to the desired angle. The part will still be supported by the crane after the second stroke. The part is again checked at both ends for the proper angle. There is a possibility that one end of the part will be correct but the other end will still be open. A skilled operator will take a piece of cardboard or a thin sheet metal that has probably been used with other parts. He will lay the cardboard or metal on top of the end of the part that is open, and then lower the ram carefully, ensuring that the shim material is properly located, and make a third cycle. Note that the ram setting has not been changed. The top shim simply forces the end of the angle that is open farther down into the vee die, creating additional bend angle in that area of the bend line. When the ram returns the top of the stroke, the part is again inspected. At this time, the operator may find that the proper angle has formed along the entire bend line and the part can be removed. In some cases, several more strokes may be needed to bring the bend up to the proper angle.

Many plates are designed with larger radii and require bump forming to create an acceptable shape. Some of the material may be classified as high tensile steel and will need an upper die with a special nose radius. Since scaly material and high tensile steels can wear the corners of the lower die rather quickly, consideration should be given to flame hardening the corners of the lower die. If the material has been flame cut, its edges may be very hard and will not form without causing cracking of some sort. Grinding the edges of the material where it enters the lower die may be required.

It is common for each end of the outside radius of the plate to peak upwards and often develop small cracks. This is similar to the eraser used as an example earlier (Chapter 8). When the eraser is looked at from the end, it can be seen that the inside of the eraser bulged out, but the outside edges peaked. In plate, a similar peak causes the tensile strength of the steel to be exceeded at that point only, and the material cracks. If the damaged area is simply ground off on the outside edge and the corner rounded slightly to ensure that any potential cracks are eliminated, there should be no problem in using the part. If hairline cracks are present along the full length of the outside radius, using a larger nose radius on the upper die may eliminate them. It is always important to check the material being formed to make sure it is the correct type and can be formed using the present tooling arrangement.

Chapter 21
European-Style Press Brake Tooling

In the 1960s, a company in France introduced an "up acting" press brake; that means that the moving member is at the bottom, which is the reverse of most press brakes. The machines are hydraulic with only one cylinder at the bottom lifting the "up moving" ram through the forming stroke. Therefore, the upper dies are mounted on the nonmoving bed. The lower moving member or ram holds the vee die. It is important to balance the forming load around the centerline since the press brake has limited ability to handle any off-center load.

The tooling made to work with this press brake does not mount in the conventional manner. The upper dies are designed with an "L" shaped top instead of the standard $^{1}/_{2}$" wide × $^{5}/_{8}$" high mounting tongue used on conventional press brakes. The ram has a matching "L" nose, so that when installed, special die clamps mounted to the outside of the bed plate held the upper die (reference Fig. 2-2). The ram clamps, each about 6" long, are bolted into the bed plate above the special die tongue. The die clamps have some space between each clamp. The lower die simply sits on a filler block that has a recess machined in the top that matched the width of the lower die. The dies are not always bolted down, although the filler block is bolted to the moving ram nose (Fig. 21-1).

The tooling is special. All of the die sections are made from a hardened type of tool steel—the entire die body is hardened to somewhere over 40 Rockwell C. The tooling is ground all over, and held to very close tolerances. The dies are made in short sections with a maximum length of about 12". Several sections can be lined up side by side, and appear to be one long length. The quality of each die shape from section to section is closely held, and the appearance is very impressive. The problem is that the tooling only works in a press brake arranged with the unique clamping system; also the tooling is very expensive. When special shapes or changes in the tooling design are requested, it is found that they are normally not available.

The European "up acting" press brake started to be sold in the United States around 1973. Although some interest was created, the press brake design and tooling limitations

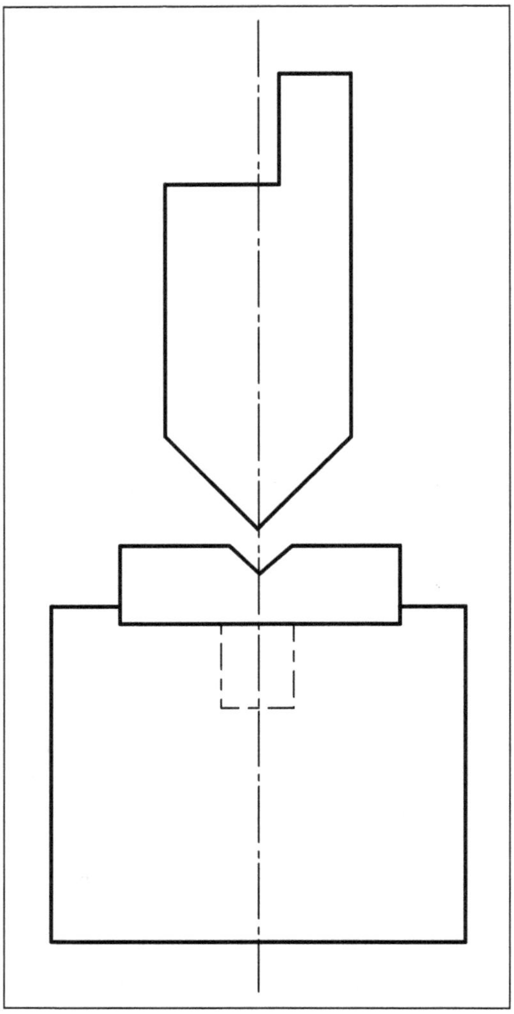

Fig. 21-1. Concept design of a European-style die setup. Note that a special tongue is required on the upper die. The lower die is often made to fit into a slotted filler block. Side clamping of the lower die into the filler block may or may not be required.

did not result in many machine sales. The up acting press brakes, however, became very popular in Europe because they were not too expensive, and many of the customers felt they only needed one set of dies. If a die section was damaged, a replacement section made to the exact dimensions of the original die was always available. The primary customers were small sheet metal shops requiring press brakes of 60 ton capacity or less.

In 1979, Amada, a Japanese company that had been selling the European machines under license, started selling the same design in the United States after their license agreement expired. After a great deal of marketing time and expense, a gradual acceptance of the tooling concept was gained in the sheet metal forming field. Amada has expanded the number of die cross sections available, but the selection is still very limited. The interesting point about the name "European-style" tooling is that most of the press brake manufacturers in Europe do not use this type of tooling. Its current use is with customers that have Amada-style press brakes or machines that have adapters that are made to accept the special mountings.

Since the late 1980s, companies in the United States—that would not spend money on any type of special press brake tooling—have been finding themselves in a bind. The number of qualified press brake operators, who actually know how to set up and align tooling, is dwindling. As a result, the expensive ground tooling with identical replaceable sections has started to become popular because the centerlines do not require realigning from set to set. A great deal of literature has been created explaining how longer part lengths can be made by placing identical sections side by side to make any needed die length. Keeping the short die sections in a drawer instead of on a die rack can minimize setup time, since the sections are easy to handle. Damaged die sections are easily replaced with identical sections.

Very little is said about the need to have a special adapter to install the tooling in a press brake that is not made with the European style ram nose. Very little is said about the limitation of die cross sections. The hardened sections do wear and, when replaced with new sections that do not have any wear, can cause angular variations to occur. At this time, many press brake users have purchased the European-style sectionalized tooling. As would be expected, the people who have spent the money to buy the expensive tooling have nothing but good things to say about it. However, the people in the shop, who have to use the tooling, have often encountered limitations and have adapted the dies to "what works well" and find creative ways to overcome its limitations.

A) Die Selection

The die cross sections are limited in number because each cross section requires special grinding wheels for their manufacturing process. As a result, only the grinding wheels that can form the angles and nose radii selected by the manufacturer are available. The nose radii of the upper dies tend to be rather sharp; they are often sold with sharp corners, such as a 0.020" radius, or sometimes as large as a 0.030" nose radius. The dies designed to form basic 90° angles are made 88° or 90°. This means they may be bottoming dies or coining type dies if truly bottomed out. How they are used depends on the available tonnage of the press brake using the 90° dies (review Chapter 3-C). Since many of the press brakes do not have the available tonnage to bottom a full-length part, much less coin a sharper inside radius, the concept of bottoming with springback is used to form many of the parts. This method requires some ability by the press brake operator to minimize the possibility of the part overbending if it is struck too hard. Many of the dies available are made to "metric" vee die openings, which can make estimating forming tonnages somewhat difficult. What also makes it more difficult is that the suppliers of this type of tooling often publish standard air bend tonnage charts in their tooling catalog, and then offer no air bend dies in their tooling selection. With the exception of one or two 30° dies, all of the other dies are often shown with 88° or 90° angles.

Most of the newer "high tech" press brakes offer the ability to input the desired angle that needs to be formed into the machine controls. If the ram has been properly set up, the stroke will reverse at the bottom allowing the desired angle to be made. All of the computer calculations for this type of forming are based on air bending. With very few exceptions, air bend dies are not available from the suppliers of European-style tooling. As a result, the "high-tech" computerized press brake controls may not work as expected.

B) Quality of the Dies

The ground tooling quality is excellent; each section supplied matches the cross section of the previously ordered cross section. The height and centerline tolerances are closely held, and practically eliminate any need to adjust for centerline variation. The surfaces of the dies are hardened to over 40 Rockwell C, which provides longer die life under normal use. Some of the die suppliers provide the entire die section heat treated to over 40 Rockwell C. This may cause a splitting or cracking problem if the die section is struck too hard. Hard sections will not deflect or move under load. Since many of the dies invite high forming loads by their bottoming type design, there is some risk that die breakage could occur.

The sharper the nose radius of an upper die, the greater the possibility that it will wear. Even the knives in our kitchen that are made from a good grade of hardened tool steel, and are only used to cut softer materials such as fruits and vegetables, have to be regularly sharpened. The small radius noses of upper dies that sink into sheet metal to start a simple 90° bend will be subject to wear. Hardened dies cannot be reworked. If a replacement section is purchased to take the place of a worn section, it will have the same radius as the original die section. As a result, the new section may not match any of the slightly worn sections that were not replaced. Since some sections will be used much more than other lengths, a great deal of planning should be made prior to the initial die purchase as to what will be the most economical method of maintaining all of the upper dies without encountering the problem of slightly different nose radii along the forming surface.

Sectional tooling has its strength in fast setup, great quality, and minimal storage space. Sectional tooling, however, can be a big problem if marking is an issue on the outside of the formed part. All press brakes deflect under load. In some cases, the lower die sections can separate after several bends due to the bed deflection. The greater the forming load and the taller the lower die, the greater the likelihood of the sections moving apart. If this does occur, the corners of the lower die sections will cause either "marks," or, in some parts, what appears to be a "lack" of a mark. This would be in the locations where the lower dies are apart and there is no die contact with the outside surface. This distance may only be 0.020", or it can become greater if no attention is paid to the separating problem. It is always wise to minimize the number of lower die sections if a full-length lower die is not an option (review Chapter 10).

C) Use of European-Style Tooling in Conventional American Press Brakes

Since the die mounting is completely different with the European style tools, a special ram adapter plus a special filler block will be needed if the tooling is to be used in a conventional press brake. Care must be taken to ensure that the added height needed for the ram adapter, plus the height of the new tooling, allows sufficient die space and stroke to form all of the parts scheduled to be made on the new setup. The replacement filler block may have to be lower in height in order to recapture some of the lost available stroke taken up by the ram adapter. If a new press brake is purchased with the buyer expecting to have the ability to use not only conventional press brake dies but also the European type dies, adding additional die space or more stroke length to the press brake is a wise decision. The special ram adapter will be expected to clamp short lengths of upper dies, and the clamping bolts in the die clamps must be no further than 6" apart—otherwise there will be the possibility that the short sections could fall out of the ram adapter while they are being used. This is also required for any press brake that uses short die sections to form such products as multiple box lengths.

Chapter 22
Punching

Press brakes have been used for punching holes and notches in parts since the 1930s. With the introduction of "punching machines" in the 1960s, most of the press brake punching work was off loaded to the newer equipment. Since the 1980s, lasers have taken over much of the work of making holes in a blank. Since a number of small "job shops" use press brakes as their only fabricating piece of equipment, it is still important to include an explanation of punching in any discussion of press brake tooling.

Normally, when punching is discussed, the subject refers to punching a round hole in a piece of metal. Punching works on the same principle that splitting a block of ice does. Some of you "old timers" may remember a block of ice being brought to a picnic. Crushed ice, as we know it today, was not as easily obtained (or was more expensive). It was easier simply to buy a large block of ice, wrap it in paper so that it did not melt, and split it into small chunks or pieces when the final destination was reached. Splitting a block of ice was rather straightforward. The tool used was an ice pick. This tool had a nice handle similar to a screwdriver; but, instead of a blade, the pick was a $^1/_8$" diameter round piece of relatively hardened steel which stuck out from the handle 4" to 6". The end of the pick tapered to a sharp point. If the person took the ice pick and plunged it into the middle of the block of ice, the sharp point of the ice pick would penetrate into the ice about $^1/_4$" to $^1/_2$" deep. At that point, a crack would start downward in the block of ice and quickly propagate throughout the block so that the large block would crack into two pieces. It was easy to continue to split all of the larger sections into smaller pieces of ice using the same ice pick by driving it into the various pieces of ice.

A big problem occurred when somebody forgot to bring the ice pick. Most of the time, someone would find a screwdriver and try to split the block using its blunt nose. When the screwdriver was brought down on the block of ice, the edge of the blade was not sharp enough to penetrate the hard ice. Therefore, on each stroke, the person attempting to split the block of ice would shower himself and anyone standing close by with an assortment of ice chips.

Punching holes into steel works similarly to using an ice pick. As the round, flat surface of the punch is lowered down onto the top of the part, the punch pushes the material down against the "die" located underneath the material. The stroke continues down, forcing the punch to penetrate through the metal. At the same time, the bottom of the hole being punched is bulging down into the hole opening in the die. The sharp corners around the circumference of the die create a small notch in the bottom of what will soon be a punched hole. When the shear strength of the material is exceeded, fracture of the material takes place (Fig. 22-1). A crack around the entire perimeter of the hole moves quickly down through the thickness of the material. At the same instant, a similar fracture starts at the bottom and moves upward. If everything works as planned, the two fractures meet each other somewhere inside the part thickness. The material that makes up the "hole" is no longer connected to the original part, and will be pushed down through the die and out of the punching unit as the punch continues toward the bottom of the stroke. The punch penetration is about 25% to 33% of the material thickness if the metal is mild steel. The penetration is only about 10% of the part thickness if the material is higher tensile steel.

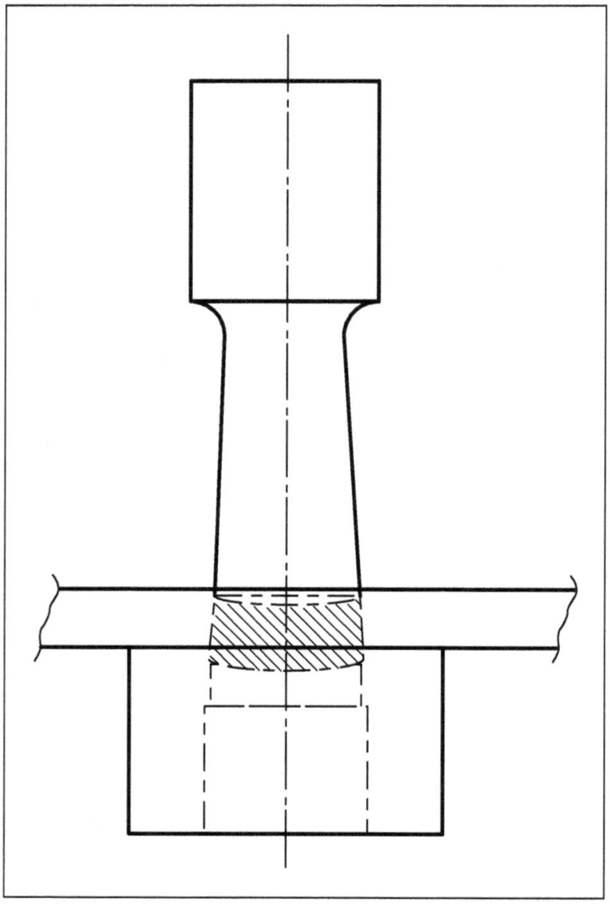

Fig. 22-1. Punch shown pushing a punch "slug" through the hole into the lower die. The punched material will pass through the die and lower holder and must be removed from the bed top. It is important the slugs do not accumulate under the lower die creating the possibility of the punched material stacking up and causing a problem.

A) Punch and Die Clearance

The tool that is used to create a hole is called a "punch." The punching surface is made the exact diameter or shape of the hole to be punched. The shank of most commercial punches slants inward slightly. The reason is that the taper helps maintain a sharp edge when the punch is sharpened. Some punches are sold with what is a small "cone" shaped extension to the punching surface (Fig. 22-2). The extension has nothing to do with reducing the punching load—a popular sales misconception. The cone shaped extension is what is left from a bar cutoff operation. It would raise the price of the punch if it had to be initially removed. As soon as the punch is used and needs to be sharpened, the cone shape is ground off and the punch surface is left flat. All punches and dies are made from a hardened tool steel.

The tool that is under the punch and has a mating diameter is the die button (commonly called the die). The diameter of the hole in the die is the punch diameter plus clearance. The amount of clearance needed has been a subject of much debate over the years.

The range varies from 5% to 20% of the material thickness being punched. If the punches and dies are standard commercial products, there is a good possibility that the clearance will be on the higher side. The best range of clearance when punching material 10 gauge or thinner is a clearance range of 5% to 7% of the material thickness. If the hole is round, the clearance must be all around the circumference of the hole. This means that the diameter of the die is the punch diameter plus an additional 10% to 14% times the material thickness (Fig. 22-3).

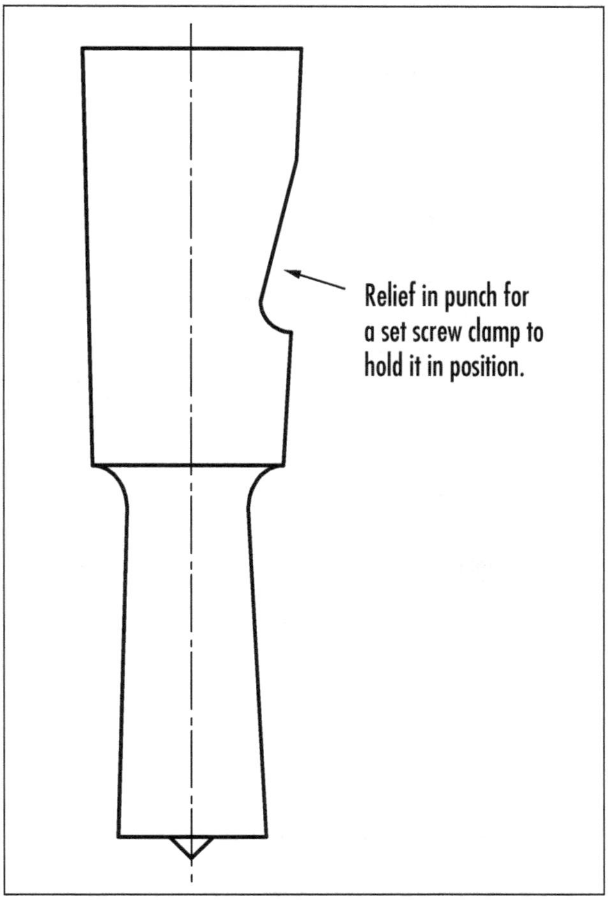

Fig. 22-2. A typical punch shown with the cuttoff cone shape still left on the punching surface. It will not cause a problem while it is there, but it will be ground off as soon as the punch is sharpened.

The "punch diameter plus clearance dimension" depth of the hole in the die is normally about one metal thickness. At that point, the diameter of the hole inside the die is made somewhat larger to allow the "slug" (the piece of metal that has been punched) to travel down through the die without becoming stuck or hung up in the tightly dimensioned die opening. The total height of the die varies from 0.5" to about 1" thick. The diameter of the die will vary depending on the size of the hole being punched. Its diameter must be large enough to provide sufficient area at the bottom to ensure that the support in the die holder that contains the die is adequate and will not allow the die to sink into the holder.

If the clearance is incorrect, the two fractures (one from the top and the one from the bottom of the material) may not meet during the punching stroke. As a result, each fracture will not extend completely through the material thickness. When the hole is inspected after punching, it will have what looks like a secondary fracture in the surface of the hole thickness. This imperfection is called a **double shear** since the punch actually recuts part of the hole during the downward stroke. If the slug that was punched from the part is inspected, it is probable that it will also have a matching double shear. If this is an issue, the clearance of the bottom die must be adjusted.

B) Guide Pins/Bushings

The smaller the clearance between the punch and die, the more important the punch and die alignment becomes. It is imperative to provide some method to keep the dies in alignment when the dies are removed from the press brake. For most punching die sets, guide pins and bushings are the most popular method used to maintain alignment. Guide pins are round,

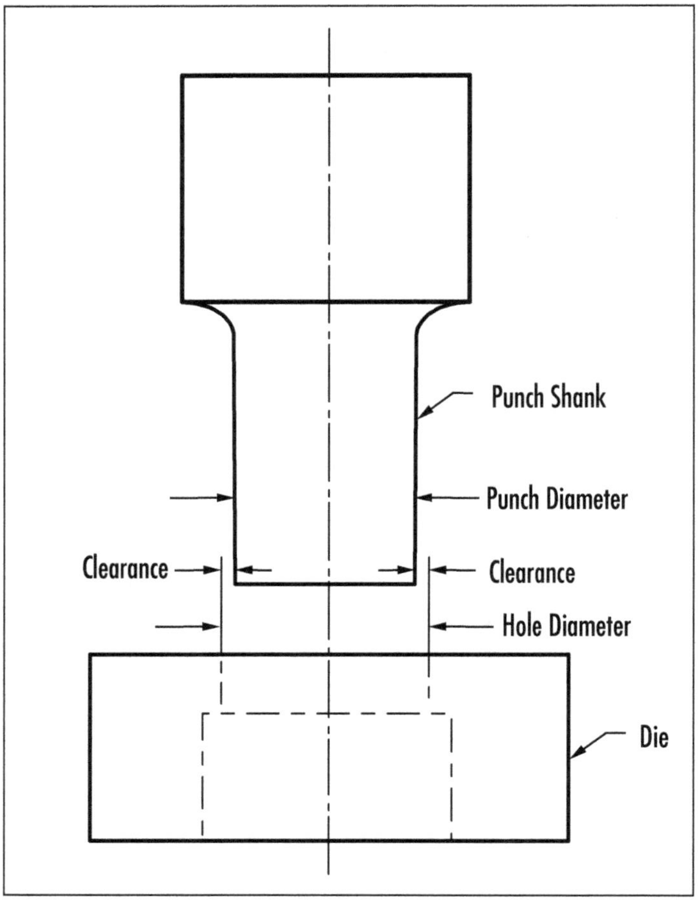

Fig. 22-3. Punch and die illustrating clearance.

ground pins that are long enough to be pressed into either the bottom base plate (the most common method) or occasionally into the upper mounting plate. If the guide pin is mounted on the bottom plate, a matching bushing is pressed into the upper plate. When the press brake cycles and reaches a point at least two inches from the bottom of the stroke, the guide pin slides into the bushing and helps to guide the upper die as it completes the punching stroke (Fig. 22-4). Each die needs at least two guide pins/bushings to ensure alignment. Although guide pins do theoretically help during the forming stroke, if the press brake is not well maintained and there is excessive clearance between the machine slides and guides and any end guides, the heavy weight of the ram can easily move, causing the relatively small guide pins in the die set to breathe. In that case, if the punch and die clearances are minimal, the punch could **clip** the edge of a die, causing serious damage to the tooling.

The main purpose of the guide pins and bushings is to ensure that the punch and die alignment is maintained whenever the entire die set is removed from the press brake. It is very time consuming to align multiple punches and dies mounted in a large die set if the top and bottom plates that make up the complete die set are installed separately into the press brake. Guide pins, which have been provided by the tooling builder to help locate all of the punches and dies, allow fast installation and setup of the press brake. Most users keep the upper and lower dies together during storage to minimize any possibility of damage to the punches.

CAUTION: The punching die set must be designed so that, when the punches reach the bottom of the stroke, the guide pins do not stick through the top of the upper die plate and contact any part of the press brake or angle brackets needed to support the die set.

C) Punching Load

The tonnage required to punch a hole is determined by the shear strength of the material being used. Engineering books specify that the shear strength of various steels is three-quarters of the tensile strength of that metal. Extensive punching tests conducted by the W.A.Whitney Company, and reported at various seminars, have pointed out that the shear strength was actually a range of from 72% to as much as 82% of the tensile strength. For many years, the published shear strength of mild steel was 50,000 PSI. For the engineer, this could present a problem in attempting a tonnage calculation.

If the specified maximum tensile strength of mild steel is 60,000 PSI, the formulated engineering shear strength found in handbooks would be 0.75 times 60,000 PSI, or equal to 45,000 PSI. If the maximum range found in testing by W.A.Whitney is to be believed, then the 60,000 PSI figure should be multiplied by about 0.8 which would result in a 48,000 PSI shear strength calculation. The figure that has been published since the 1920s is the 50,000 PSI figure.

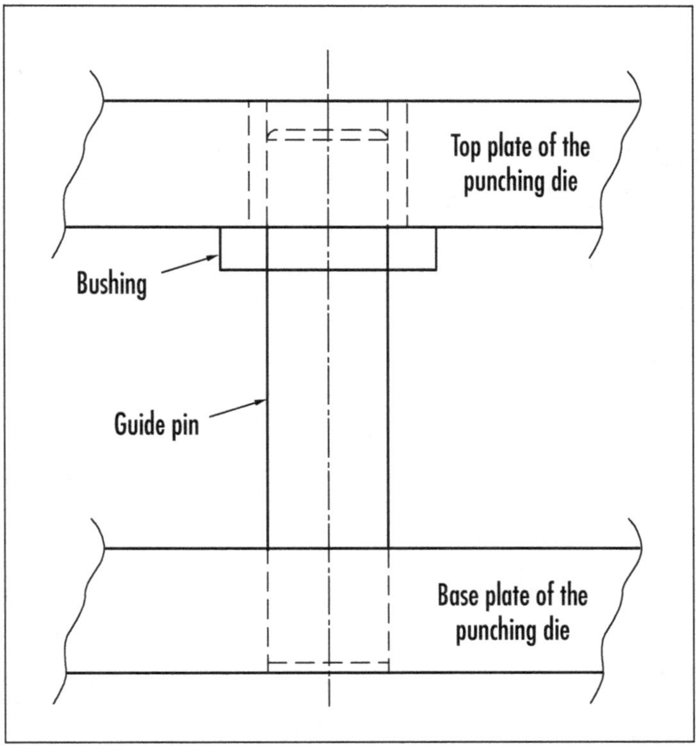

Fig. 22-4. Guide pin and bushing.

Keep in mind that calculators had not yet been invented in the 1920s. If the above figures are converted into tons (since that is the easiest unit of measurement that relates to machine capacity), the 45,000 PSI becomes 22.5 tons/in.^{2}; the 48,000 PSI figure is 24 tons/in.^{2}; the 50,000 PSI figure converts to a simple 25 tons/in.^{2}. Obviously, the 25 tons/in.^{2} is simpler to remember, easier to multiply under shop conditions, and compensates for any material thickness variations that will occur in the material. In addition, thin cold rolled steel will have surface hardening from the rolling process. This can increase the top and bottom surface to a tensile strength to 70,000 PSI. By using the simple 25 tons/in.^{2} figure, the shop personnel do not have to worry about any safety factors in their calculations, and will always arrive at a tonnage figure that works.

The formula used to calculate the tonnage to punch a hole in mild steel is to multiply the area being sheared (that is the perimeter of the circle times the material thickness) × the 25 tons/in.^{2} shear strength. For example, to calculate the load required to punch a 2" diameter hole in $^1/_4$" (0.250") mild steel plate (not A-36), multiply the area being sheared (2" × π × .25") × the shear strength (25 tons/in.^{2}) = 39.25 tons. A tonnage chart showing the force required to punch various size holes in mild steel is found in Fig. 22-5. For other types of material, it is best to use 80% of the tensile strength for the shear strength figure. If the material is high strength steel and is listed as having a minimum tensile strength, add 15,000 to the listed tensile strength (the normal range expected in most areas of the country) and then take 80% of the higher figure.

D) Stripping Loads

When the punch passes through the sheared section of the punched hole into the thickness of the part where fracture has occurred, the punch must not have its direction reversed until

Tons Pressure Required to Punch Mild Steel

In order to obtain the best life from a press brake, punching loads should be not more than the limits suggested by the press brake builders.

THICKNESS of METAL		HOLE DIAMETER														
Gauge	Inches	1/8"	3/16"	1/4"	5/16"	3/8"	7/16"	1/2"	9/16"	5/8"	11/16"	3/4"	13/16"	7/8"	15/16"	1"
20	.036	.35	.53	.71	.88	1.1	1.2	1.4	1.6	1.8	1.9	2.1	2.3	2.5	2.7	2.8
18	.048	.47	.71	.94	1.2	1.4	1.7	1.9	2.1	2.4	2.6	2.8	3.1	3.3	3.5	3.8
16	.060	.59	.89	1.2	1.5	1.8	2.1	2.4	2.7	2.9	3.2	3.5	3.8	4.1	4.4	4.7
14	.075	.74	1.1	1.5	1.9	2.2	2.6	2.9	3.3	3.7	4.1	4.4	4.8	5.2	5.5	5.9
12	.105	1.0	1.6	2.1	2.6	3.1	3.6	4.1	4.7	5.2	5.7	6.2	6.7	7.2	7.7	8.3
11	.120	1.2	1.8	2.4	3.0	3.5	4.1	4.7	5.3	5.9	6.5	7.1	7.7	8.3	8.8	9.4
10	.135	...	2.0	2.7	3.3	4.0	4.6	5.3	6.0	6.6	7.3	8.0	8.6	9.3	10.0	10.6
3/16"	.187	...	2.8	3.7	4.6	5.5	6.5	7.4	8.3	9.2	10.2	11.1	12.0	12.9	13.8	14.8
1/4"	.250	...	...	4.9	6.2	7.4	8.6	9.8	11.0	12.3	13.5	14.8	16.0	17.2	18.5	19.7
3/8"	.375	...	...	...	...	11.1	13.0	14.8	16.6	18.5	20.3	22.1	24.0	25.8	27.7	29.5
1/2"	.500	...	...	...	...	...	17.2	19.7	22.1	24.6	27.1	29.5	32.0	34.4	36.9	39.4
5/8"	.625	...	...	...	...	...	...	...	...	30.8	33.8	36.9	40.0	43.0	46.1	49.2
3/4"	.750	...	...	...	...	...	...	...	...	...	...	44.3	48.0	51.7	55.4	59.0

Courtesy of Cincinnati Incorporated

Punching/ Force in Metric Tons

Punching mild steel plate / shear strength 35 kg/mm²

HOLE DIAMETER mm.		3	4	5	6	7	8	9	10	11	12	13	14	15	16	17	18	19	20	21	22	23	24	25
METAL THICKNESS - mm.	1	.33	.44	.55	.66	.77	.88	1.0	1.1	1.2	1.3	1.4	1.5	1.7	1.8	1.9	2.0	2.1	2.2	2.3	2.4	2.5	2.7	2.8
	1.5	.50	.66	.83	1.0	1.2	1.3	1.5	1.7	1.8	2.0	2.2	2.3	2.5	2.6	2.8	3.0	3.2	3.3	3.5	3.7	3.8	4.0	4.1
	2	.63	.88	1.1	1.3	1.5	1.8	2.0	2.2	2.4	2.7	2.9	3.1	3.3	3.5	3.8	4.0	4.2	4.4	4.6	4.9	5.1	5.3	5.5
	2.5	.83	1.1	1.4	1.7	2.0	2.2	2.5	2.8	3.0	3.3	3.6	3.9	4.1	4.4	4.7	5.0	5.2	5.5	5.8	6.1	6.3	6.6	6.9
	3	1.0	1.3	1.7	2.0	2.3	2.7	3.0	3.3	3.7	4.0	4.3	4.6	5.0	5.3	5.7	6.0	6.3	6.6	7.0	7.3	7.6	8.0	8.3
	3.5	...	1.6	1.9	2.3	2.7	3.1	3.5	3.9	4.3	4.7	5.0	5.4	5.8	6.2	6.6	7.0	7.4	7.8	8.1	8.5	8.9	9.3	9.7
	4	...	1.8	2.2	2.7	3.1	3.5	4.0	4.4	4.9	5.3	5.8	6.2	6.6	7.1	7.5	8.0	8.4	8.9	9.3	9.8	10.4	10.6	11.1
	4.5	...	...	2.5	3.0	3.5	4.0	4.5	5.0	5.5	6.0	6.5	7.0	7.5	8.0	8.5	9.0	9.5	10.0	10.5	11.0	11.5	12.0	12.5
	5	...	...	2.8	3.3	3.9	4.4	5.0	5.5	6.1	6.6	7.2	7.8	8.3	8.8	9.4	10.0	10.5	11.0	11.6	12.2	12.7	13.3	13.8
	6	...	...	...	4.0	4.7	5.3	6.0	6.7	7.3	8.0	8.6	9.3	10.0	10.6	11.3	12.0	12.6	13.3	14.0	14.6	15.3	16.0	16.6
	8	...	...	...	...	...	7.1	8.0	8.8	9.7	10.6	11.5	12.4	13.3	14.2	15.0	15.9	16.8	17.7	18.6	19.5	20.3	21.2	22.1
	10	...	...	...	...	...	...	...	11.1	12.2	13.3	14.4	15.5	16.6	17.7	18.8	19.9	21.0	22.1	23.2	24.3	25.4	26.5	27.6
	12	...	...	...	...	...	...	...	...	...	15.9	17.3	18.6	19.9	21.2	22.5	23.9	25.2	26.5	27.8	29.2	30.5	31.8	33.2
	16	...	...	...	...	...	...	...	...	...	...	...	...	...	28.3	30.1	31.8	33.6	35.4	37.1	38.9	40.7	42.5	44.2
	20	...	...	...	...	...	...	...	...	...	...	...	...	...	...	...	...	...	44.2	46.4	48.6	50.9	53.1	55.2
	25	...			...	...	...	...	...	...	...	...	...	...	...	...	...	...	...	...	...	...	...	69.0

Courtesy of Cincinnati Incorporated

One metric ton equals 1000 kilograms or 2205 pounds.

To obtain tonnage for other steels, multiply by direct ratio of shearing strengths.

Fig. 22-5. Tonnage charts showing the force required to punch a hole in ordinary mild steel.

the flat surface of the punch passes into the die "hole" by about 0.020". This is to make sure that the slug is pushed out of the part and has not been trapped in the hole because of burrs or a tight fit in the hole.

When the hole is punched, the material being punched acts the same as if we took a pencil with a sharp point and tried to poke it through a marshmallow. As the pencil pushes down on the marshmallow, the marshmallow pushes down and becomes much wider. After the pencil has been completely pushed through and the pressure is released, the marshmallow tries to return to its original shape. It will stick tightly to the pencil and will be very difficult to remove. To remove a punch from a punched hole, the punch must be pulled out of the hole since the material will be holding on to it. The force required to remove the punch is called the stripping load. For those engineers who like to calculate this type of load, it is technically called a "normal" load. (In 45 years of working in this industry, the author has never met anyone who has attempted to calculate this problem!)

There are three common methods used to strip the part from the punches. If the hole or holes are near the edge of the part, an upside down "L" shaped metal **hook** is fastened to the rear of the lower die. The punch (or punches) passes around the hook on the down stroke. After the hole (or holes) has been punched, the punch is gripped by the hole in the part. As the punch returns to the top of the stroke, the part is lifted up until it touches the bottom of the stripper. The punch continues to the top of its stroke while the part is stripped or pulled off of the punch (Fig. 22-6). A second method is to slip a rubber or polyurethane tube around the shank of each punch, allowing about $^1/_8$" to $^1/_4$" of the material to extend beyond the length of the stem. When the punching stroke is made, the tube will contact the part and start to compress, building up pressure. As the stroke is completed, the tube has been compressed enough so that on the return stroke the compressed tube of polyurethane will push the part off the punch; these are often called rubber strippers. The maximum thickness of material that is recommended to be stripped using rubber strippers is 12 gauge.

The third method of stripping is used when combinations of punch or hole locations do not lend themselves to the other two methods. This third way requires a special pressure pad to provide sufficient spring pressure so that when compressed at the bottom of the stroke, the

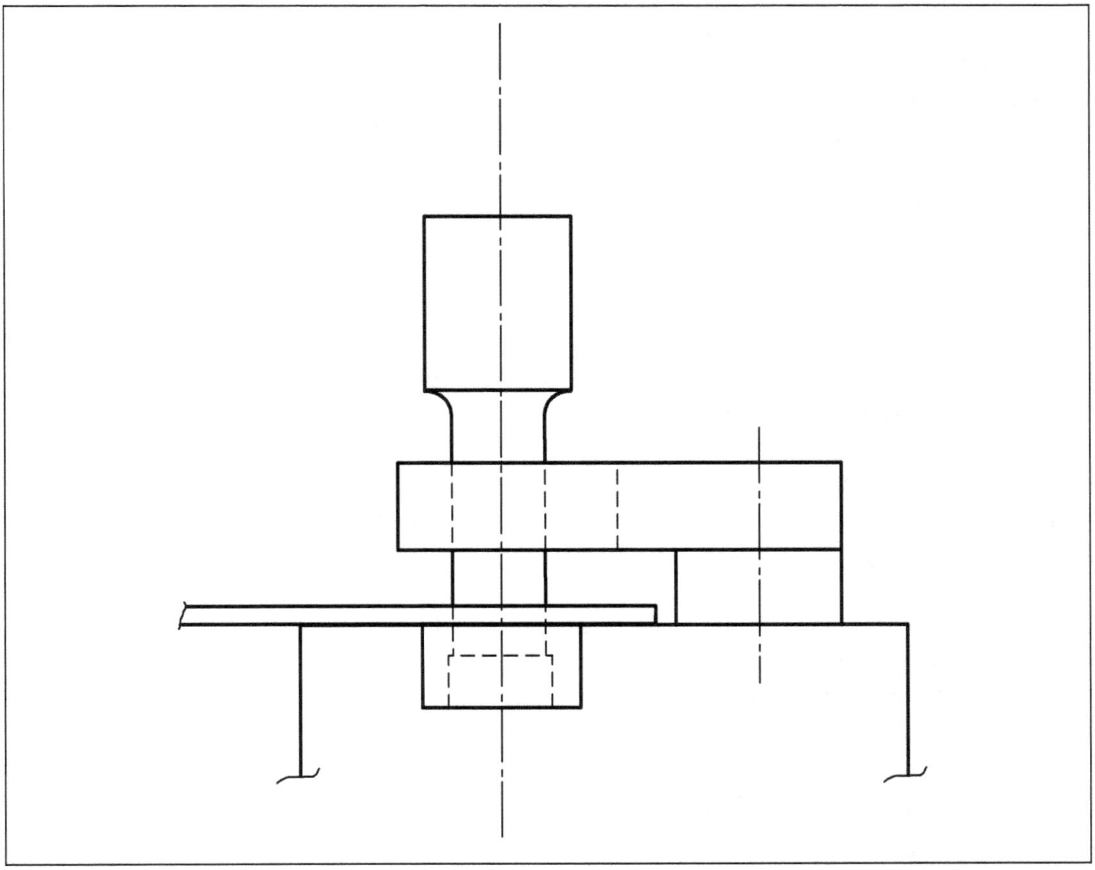

Fig. 22-6. A simple hook stripper needed to keep the punched material from being lifted upwards after the holes are punched. The punches will be held in the hole after the punching stroke, and must be pulled out on the upstroke while the part is restrained by the stripper.

flat surface of the pad will allow the part to be smoothly stripped from all of the punches. The normal stripping load is 2% to 5% of the total punching load. The spring pressure, or the pressure generated by polyurethane strippers, must be able to provide sufficient pressure to overcome the maximum stripping load for the material being punched. If, for some reason, the stripping pressure seems to increase and the parts are not being consistently pushed off the punches, the polyurethane is probably starting to wear out, the springs have broken, or the punches are dull. Dull punches will not only increase the punching load (the same as trying to split the block of ice with a screwdriver) but will also increase the stripping load.

E) Self Contained Punching Units

In the late 1930s, the first self-contained punching die set was invented. The unit looked like an elongated, flat "C." The thickness of the unit was about 1.25" wide. The gap of the "C" provided space for the material edge to be located. By placing a number of the "C" shaped units alongside each other, a number of holes could be punched along the edge of a part. The throat of the "C" limited the distance from the edge to the location of the punch. The punch and die button were mounted in the front of the "C" frame. On the top of the "C," a long punch was located in a bored hole. The bottom of the "C" was machined to hold a conventional die button. The punch and die were aligned by the "C" frame which allowed excellent guiding to be obtained. The top of the punch was made larger than the remainder of the stem. This allowed a spring to be located around the punch stem and to be contained between the top of the "C" frame and the larger diameter top of the punch itself. Underneath the die button, a hole was made to allow the punched slug to drop through the frame and fall out in the front of the unit. In the back of the "C" frame, a drilled and tapped hole allowed a gage rod to be installed to locate the part at the correct place in the "C" frame (this is an option in most units).

The "C" frame units are mounted using a tee slotted mounting plate. By placing bolts through slots in the bottom leg of the "C" frame, all of the units can be fastened in their proper location. Each unit has a dowel pin in the bottom that can be used for locating the unit. Templates are often made to locate each C frame unit at the proper spacing between holes. A simple flattening die mounted in the press brake ram is used to push the punches down and punch the holes. Care is taken to keep each frame and punch at a specific height. When the flattening die touches the top of the punch, it touches all of the units at almost the same time. As the stroke is made, the spring on the top of the frame is compressed. When the upper die returns to the top, the spring return lifts the punched part up against the inside surface of the C frame. As the spring returns to its normal length, the punch is lifted, stripping the material from the punch (Fig. 22-7).

This type of unit is still available today from a variety of manufacturers. If it is currently used in your plant, it may be wise to look at the setup while it is actually punching parts. The only items not available or supplied with the punching equipment is the flattening die, as well as a recommended strip of lexan or plexiglass that might be mounted to the ram to guard the unprotected space between the bottom of the flattening die and the top of each punch.

F) Press Brake Limitations

Many rules concerning punching load limitations for press brakes have been developed over the years. There are some manufacturers of press brakes today that discourage any punching being done on their machines. Prior to purchasing a press brake, a review should be made of all the potential work that may be done on the press brake, including punching

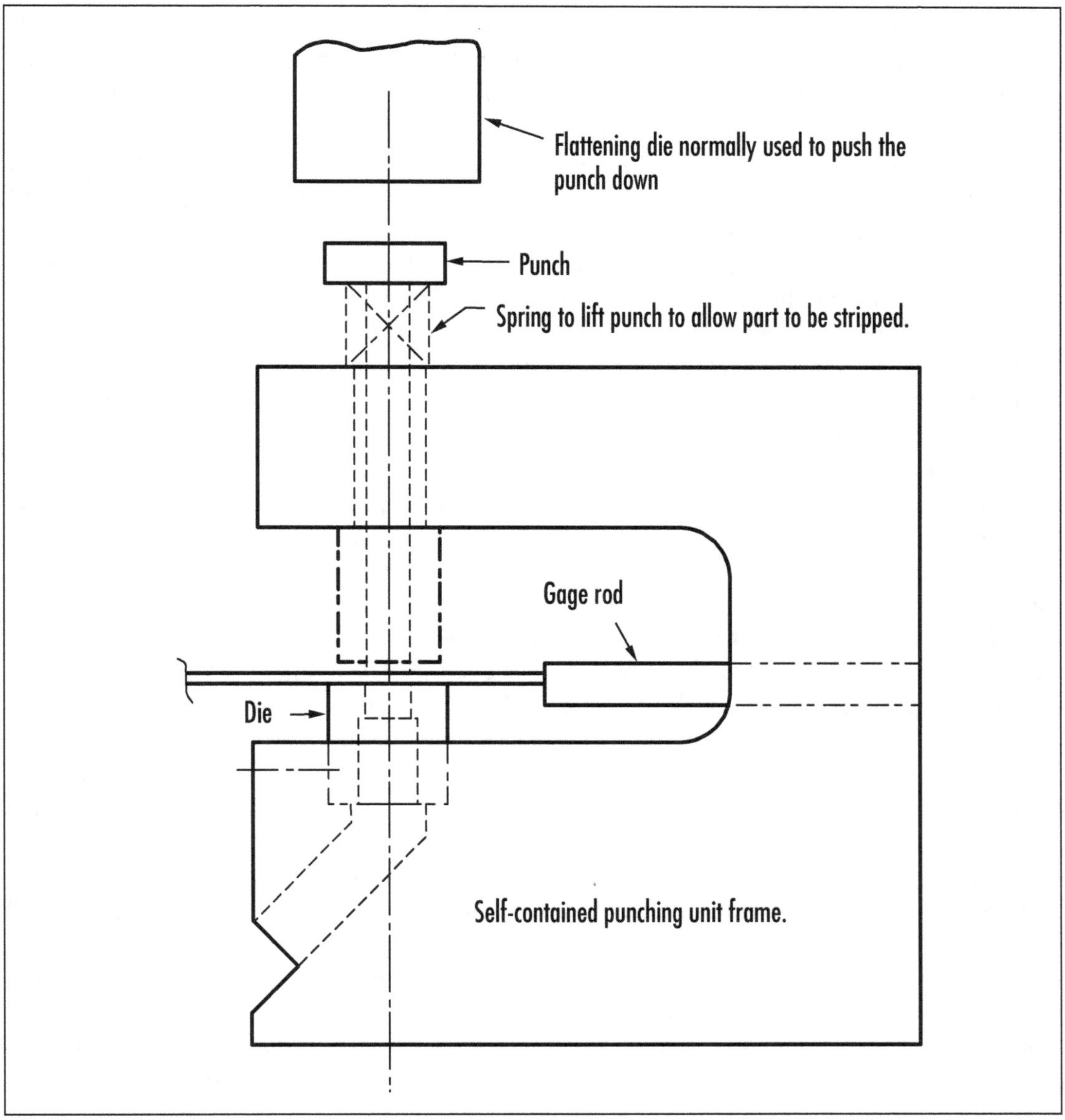

Fig. 22-7. Self-contained punching unit. Various designs are available from a number of manufacturers.

work. The goal is to ensure that any machine limitations do not restrict the type of future work that might be available for the equipment. The fact is: a truly versatile press brake should be adequate for some punching work.

Prior to 1958, all mechanical press brakes were sold with a mid stroke tonnage rating. To use those machines for punching, the load had to be balanced around the press brake centerline both left to right and front-to-back. In addition, the maximum punching tonnage was limited to two-thirds of the rated tonnage capacity of the machine. A 150 ton press brake had a punching limitation of 100 tons.

In the 1940s and 1950s, there was a need to mount larger punching die sets into press brakes. The larger die sets had many patterns of holes that were not always symmetrical in design. A great deal of engineering was spent developing charts to allow some front to back off center loading to be done on the press brakes without damaging the machine slides and guides (review Chapter 5-G). At least one press brake manufacturer provides this information with all of their new machines as well as their older models.

In most punching work, where special die sets are being used, all of the punches will strike the metal at the same time. This means that the load on the machine occurs almost at the same time. The machine deflection can cause some variation in the actual load on the press brake. It is very important to balance the load front to back and right to left the same way a playground teeter-totter must be balanced to be used effectively. The maximum off-center load allowed on the press brake is shown in **inch tons**. If there is a 10 ton punching load located 10" in the rear of the machine centerline, that is considered an eccentric load of 100 inch tons (the load times the distance involved). If a punching load was required on the front side of the machine centerline opposite the rear load, but this load was 40 tons and only 4" to the front of the centerline, the effective load would be 4 × 40, or 160 inch tons to the front. If both loads happened at the same time while holes were being punched, the press brake would only have to accommodate the total punching load plus the difference in the eccentric load. The added load would be 160 inch tons minus the 100 inch ton load, or a 60 inch ton eccentric load. For specific information, it is best to work with the press brake supplier.

In 1958, a double rating was used on the mechanical press brake to allow users to work with a higher machine load near the bottom of the stroke. Users immediately jumped to the conclusion that, since punching loads were at the bottom of the stroke, the punching load could also be increased. This worked in a few cases where the machines that were used for punching selected dies that were similar to dedicated punching units. It was soon found that large increases in punching loads did not work if "die set" punching was required. About the same time, large hydraulic press brakes were being introduced to the market.

In the early 1960s, a great deal of discussion was conducted to decide if hydraulic machines "damped" the punching load sufficiently that vibrations in the machines were reduced. If so, the punching load could be increased to the full tonnage of the hydraulic press brake. The discussion lasted until the first full-tonnage punching job was attempted on a new press brake and the machine had instant problems. The problems with vibrations and break-through immediately became obvious. The punching tonnage was reduced to two-thirds of the maximum machine tonnage and that worked most of the time. The conditions were that the punching dies did not "bottom out" after the holes were punched, that material being punched was mild steel, and that the normal symmetrical machine load was held.

The machine problems seemed to occur if die sets were being used, if the material was 16 gauge or thinner, or if the material was "high tensile" steel. Punching a hole is not like bending a piece of metal where the load gradually builds up during the first 30° of the bend cycle. When the punch strikes the metal, it is like running into a wall. The metal gives immediate resistance to the punch. As the stroke continues down, the punch will penetrate into the metal about one-third of the material thickness before the fracture takes place. The instantaneous load causes the throat area of the housings to stretch or gap.

If an ordinary "C" clamp is used to tighten two pieces of wood together, as the screw is tightened, the frame of the clamp gaps or opens up. This can easily be seen each time the screw is turned to develop more clamping pressure. The press brake throat area acts the same way, but it is often harder to see since it is located behind the bed and ram, and the operator is not in a position to watch its movement. When sufficient pressure has been built up in the housings to provide the required force, the holes are punched. Since there is no resistance from the material after the slugs are completely sheared, all of the deflections built up in the housings immediately cause the gapped throat to snap back. The snapback seems to act like releasing a stretched spring. The spring does not just return to its original position but starts to compress, extend, and repeat the vibration cycles until finally returning to its original position. In the case of the press brake housing, this vibration may take less than a second, depending on the load.

The snapback of the housings can cause major machine problems under certain conditions. Punching dies that include pressure pads, and/or stop blocks that are used for setup purposes, can interfere with the snapback and cause increased shock to the machine. Some housings, especially a press brake that has been made with a deeper-than-standard throat cutout, can have the housing gap over $^1/_8$" while the punching load is being built up in the housings. When the load is released, the housing will close up almost as much as it has gapped. This means that the original ram setting of the machine will be reduced momentarily by the amount of the gap. Since the penetration into the metal, prior to the actual punching taking place, is about one-third of the material thickness, the distance of the housing snapback is not only going to allow the punch to move through the remainder of the fractured hole but also cause the punch to enter into the die some amount. If the die allows this to happen, this may not be a problem. If a die set with pressure pad is being used, there is a high probability that the pressure pad will be bottomed out, causing a high secondary load to occur. This load will create a great deal more vibration in the press brake. The same problem happens if the die set is equipped with stop blocks (sometimes referred to as setup blocks). These blocks are often set to stop the pressure pad or the upper plate of the die itself when the punch enters the die button more than $^1/_{32}$" to $^1/_{16}$". The ram snapback might cause the bottoming out to take place on these blocks. If the punches are reground in order to sharpen the punching corners, the initial ram setting is automatically lower. If the stop blocks are not lowered by the amount of the regrind, a die that had worked previously may now bottom out.

A series of tests conducted in the late 1960s and early 1970s by Cincinnati Incorporated found that high tensile steels require much less punch penetration into the material. The higher tensile steel not only requires a higher calculated punching load as well as less penetration, but also causes the machine vibration to double during the punching cycle. It has been suggested that with normal mild steel penetration of one-third of the material thickness, there may be some "damping" effect from the punch being held by the punched hole itself while the machine vibration is happening. Since high tensile steel has only about 10% penetration prior to the punch breakthrough, the punch has little damping available from the hole initially and this may add to the machine vibration.

To make sure that the majority of punching work that may be done on a press brake will not cause vibration problems or machine accuracy concerns, all die set punching or high tensile steel punching should be limited to no more than 50% of the press brake capacity. It is always best to check with the press brake manufacturer if there are any doubts concerning a specific punching job. If punching work is common on your press brake, always check the machine mounting bolts holding the press brake to the floor to make sure the vibration created by the punching has not caused them to become loose. A press brake is, by design, front heavy. If the mounting bolts are not secure, there may be a risk of the machine tipping over toward the operator side. Before any such major disasters could occur, there would be many other possible problems showing up. On a mechanical press brake, the motor for the ram elevating adjustment will fall off; or the electrical control box on the side of the right hand housing may come loose. On hydraulic press brakes, the seals of the hydraulic cylinders may start to leak;

or the operator will be showered with oil. The operator controls or the encoder will come loose or be damaged. Abnormal vibrations can cause many strange problems to surface.

Another limitation of the press brake is that the maximum stripping load is limited to 10% of its punching capacity. This is especially true if a hydraulic press brake is being used for the punching work. The machine is not designed for an upward pulling load; it is designed to lift the weight of the ram plus the upper die. Rubber strippers and spring strippers actually help lift the ram during the up stroke. There should be concern if hook type strippers are being used.

G) Stepping of the Punches

When press brakes were initially considered for punching jobs, it became apparent that the punching load builds up and is released within a very short stroke distance. If the punches were made different heights, about one metal difference in actual length, and if the punches were arranged so that the each level of punches would contact the part in a balanced pattern, it was found that the total punching load on the machine could be doubled. This is called **stepping** the punches.

Stepping can be accomplished in several ways. The punches can be made different lengths (not recommended since this option can cause problems when the punches are sharpened). The other common methods are placing shims under heads of some punches to provide the pattern of stepping needed, or the die set may have different depth punch mountings machined in the punch holder. Two levels of punches or symmetrically stepped punches are the most common. Some special punching applications have used up to three steps. Additional stepping causes stripping problems with spring pad travel, or unequal loading on hook strippers (Fig. 22-8). Although each level of punching only uses the tonnage required for that level, all of the holes are stripped at the same time during the upstroke of the ram. This means that the stripping load will be 2% to 5% of the total

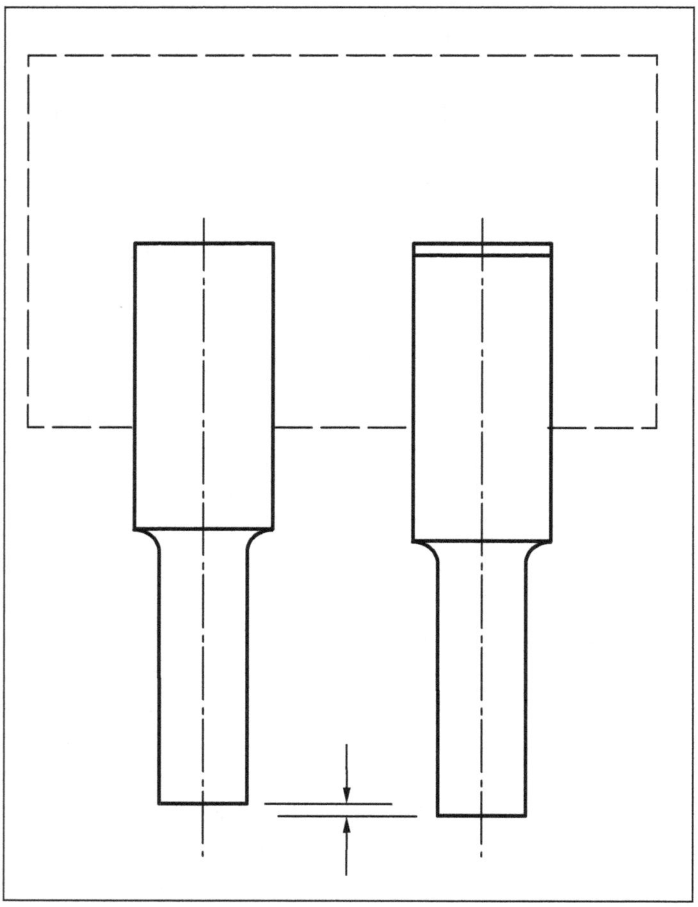

Fig. 22-8. Punches stepped one-half of the metal thickness. If done in a symmetrical pattern, the machine loading can be reduced. This may allow more holes to be punched each stroke, or a smaller tonnage press brake to be selected.

punching load. Care must be taken to ensure that the 10 % machine stripping limitation is not exceeded.

H) Rake on the Punches

In order to reduce the punching load if large holes or rectangular notches must be punched, it has been found that grinding a slope on the face of the punch will solve the problem. This slope is called **rake**, so named after a similar angle used on squaring shears to reduce the shearing load. The rake can also cause distortion as discussed in Chapter 7-B.

The formula for calculating the punching load (punching load = area being punched or cut × the shear stress of the material) is easy for punching a hole, since the entire area is the circumference of the circle × the material thickness. The fact that the only actual cutting being done on the metal is about one-third of the material thickness does not enter into the calculations. When a rake is ground on the upper punch, as the punch moves down through the punching stroke, only the "distance × metal thickness" that is cutting the one-third of the

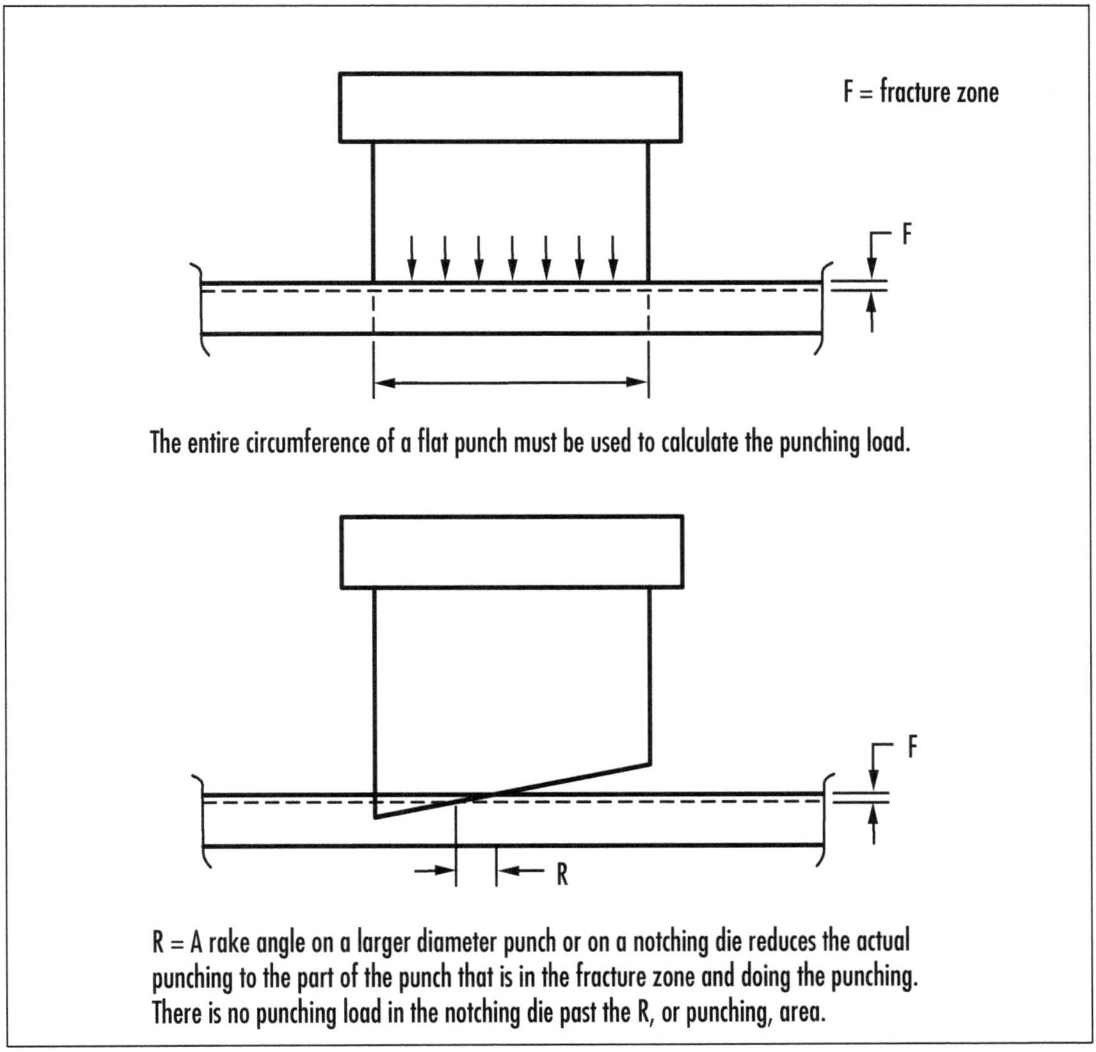

Fig. 22-9. Illustration showing how a "rake" angle on larger punches or notching dies can reduce the punching load.

material thickness makes up the area to be included in the formula. The greater the rake angle, the greater the distortion in the part that is being punched or cut. Too much of a rake angle also causes concerns with pressure pad travel and slug removal (Fig. 22-9).

When discussing rake with people who work around squaring shears, it is referenced as a specific angle, or how much the rake will cause the knife to change per foot of travel. For example, the rake angle may be referred to as a "1/4" per foot of rake angle." If a company calculates the load for punching based on a specific angle in order to reduce the punching area to conform to the press brake punching capacity, a communication problem can occur with the punching die supplier. If the buyer specifies a 1/4" rake, they may assume that the rake will be 1/4" per foot. The builder of the die may understand the rake will only be 1/4" from one end of the part to the other end of the part, or perhaps only a small percentage of the needed rake angle. As a result, the die will require a much higher punching load than the initial calculation. Correcting the problem will become very expensive since the stroke of the punch or knife having the incorrect rake angle is not close to the stroke required if the proper rake had been designed into the tooling. The spring travel that may be needed for the pressure pad may cause the guide pins not to engage, and in some cases the proper die would not even fit into the press brake because of the increased height of the die set. Prior to building any special tooling, all parties must agree on the needed tonnage, and must be aware of the press brake limitations.

Chapter 23
Multiple Dies in the Press Brake

There are some situations where it is advantageous to use multiple die sets running side by side in the press brake. This allows a part to be processed through the forming stations in one handling at significant cost savings. The easiest setup is when one full length of die can be sectioned in order to make different lengths of bend in the part. This was discussed in Chapter 17 concerning the forming of boxes. Often, a small section of die is used to make a small flange, while the remaining length of die forms the longer bends. Different types of tooling running in the press brake at the same time (reference making hems in Chapter 13-A) are not uncommon. Punching dies can be adapted to a side by side arrangement. The parameters that must be kept in mind are as follows.

1) Each of the forming dies must be made to work on the same machine centerline. This is very important with vee dies and special dies that normally require front to back alignment. Unless each die has a special length filler block that can be individually mounted on the bed top, multiple dies in the standard full-length filler block must all have a common centerline. Holding the centerlines of punching dies and flattening dies around the press brake centerline are not as critical.
2) If the tooling is to run in a mechanical press brake, each set of dies must be tested while making parts in the center of the machine so that the ram setting of each die, when it is forming material, is within a 0.020" range. The same will be true of a hydraulic press brake if all of the dies are going to be used at the same time. When the dies are located in their forming position on the press brake, each die will have to be shimmed in order to compensate for the differences in deflections that will occur since each die set will need a different tonnage to form the various shapes. This will cause the bed and ram to have unique deflections that a simple ram adjustment won't correct.
3) If the multiple die sets are going to be run in a programmable hydraulic press brake, but only one die set will be used at any one time, each of the upper dies should be

machined to barely touch the lower die before the ram is reversed. All of the die sets must be within a 0.020" ram setting at the bottom of the stroke. When the operator moves from station to station to form the part, the ram will have to be raised for the material clearance. This arrangement will keep any of the other die sets in the press brake from bottoming out or touching each other. It will still be necessary to shim each set to compensate for the machine deflections unique to that location on the bed.

4) Once the dies have been set up, and good parts have been made, all shimming information and ram settings should be recorded. The next setup will require some fine-tuning since the material will vary.
5) A thorough review of the operation with regard to operator safety as well as operator controls needed to cycle the press brake must be completed prior to the first production run. Multiple palm button stations, sequential foot switches, or multiple types of guarding should be considered.

It is noted that special types of filler blocks that adjust the amount of crown available in the center of the bed and ram do not work well with multiple die setups. Some suppliers claim that individual deflection compensation along the length of the bed is possible. All things might be possible if you want to spend enough time to work out the problem. However, if the operator wants to obtain good parts quickly within a minimum of time, the simple methods used to shim a variety of other dies appear to be the best decision

Chapter 24

Other Important Information Needed for Making Good Tooling Decisions

A) Calculating the Weight of Steel

In order not to overload a forklift truck or even a crane, it is often important to be able to calculate the weight of a stack of steel blanks or perhaps a single piece of plate. Engineers are insistent upon calculating weights down to the last half-pound. Under shop conditions, it is often important to have an immediate answer so that the material can be moved. Most engineers will calculate the steel weight by multiplying the number of cubic inches of steel in the stack by a figure like 0.283 (the weight of one cubic foot of steel weighing 489.6 pounds divided by 1728, which is the number of cubic inches in one cubic foot) to arrive at what they would give as an exact weight. In the meantime, the stack of steel is waiting to be moved.

Many years ago, a steel handbook published by a large steel service center provided this easy method of calculating the weight of the stack of steel. Using a standard tape measure, you can measure the height and width of the stack. Multiply the height and width dimensions to obtain the area in square inches of the side that has just been measured. Take the answer and add a zero to the calculated area. That figure will be very close to the weight of a 36" length of steel of that cross section.

For example, if the stack is 10" high by 27" wide by 48" long, the calculation could be done as follows. Multiply 10 × 27 to obtain 270. Add a zero, which gives a figure of 2700. This is the weight of 36" of steel in the stack. Since the parts are 48" long, the estimated weight would be 1 $^{1}/_{3}$ times the answer, or 3600 pounds. Measuring the other way, if the 10" by 48" side was selected, the answer would be 10 × 48 = 480; adding a zero gives a figure of 4800. The 27" length is $^{3}/_{4}$ (or 0.75) of the 36" length. Multiplying 4800 × 0.75 provides the same 3600 pounds answer.

If the number of cubic inches of steel in the stack is calculated, the answer is 10 × 27 × 48, or 12,960. Multiplying that figure by 0.283 gives the engineering answer of 3668 pounds. If the 12,960 figure is multiplied by a 0.28 figure that is used in some manuals, the answer is 3629 pounds. If the estimated 3600 pound calculated in the simple and quick shop method is used, the calculations can quickly be completed on the shop floor (often mentally), and the concern about whether the forklift or crane will be able to move the stack is quickly alleviated.

B) Alignment of the Press Brake Dies

It is very important that the upper and lower press brake dies are properly aligned. Forming a part can take tons of pressure, and it is imperative that the load on the tooling be equally balanced. Since the upper die is in a fixed location, clamped in the ram by the die clamps, it cannot be moved. The lower die is normally mounted on a filler block (or die holder) which will require that both bars of steel must be moved either to the front or to the rear to obtain a metal thickness clearance between the nose of the upper and the sloping sides of the vee in the lower die. The reason for a possible misalignment is that the commercial tolerances for the nose of the upper die and the vee opening around the theoretical forming centerline is about ±0.020" for each bar. Trying to manually push the filler block back and forth will often cause the bar to "spin" on a high point on the bed top. This causes one end of the bar to never align with the other.

The concept of jamming the upper die into the lower die (as described in Chapter 2-B) will not work, although it has been tried by many press brake operators. The downward force locks the filler block in place, and there is no "wedge" action that will move the lower die and filler block sideways. In many setups, the upper die is made with a relatively sharp nose radius ($^1/_{32}$" or even less). The radius in the bottom of most vee die openings $^3/_8$" or larger will be made $^1/_{16}$" or even greater in order to ensure that the vee opening will not split open under the high forming load. It is noted that the radius in the bottom of the vee opening must be smaller than the outside radius of the material being formed in the vee opening.

If the operator attempts to align the lower die with the upper die by forcing a $^1/_{32}$" nose radius into a $^1/_{16}$" radius in the bottom of the vee die, there will not be enough slope on the sides of the $^1/_{16}$" radius to start moving anything sideways, much less moving the lower die and filler block. If the operator assumes this method has actually accomplished some alignment, and proceeds to tighten the lower die in place, every time a bend is made the misaligned lower die will rock either to the front or back, thus stretching the hold-down bolts and failing to produce the quality of bend angle that is expected.

The bottom vee die should be aligned with the nose of the upper die so that the distance between the sloping sides of the upper and lower die surfaces are within 0.003" of each other for vee openings $^3/_4$" and smaller. Larger vee openings may accept proportionately greater misalignment.

C) How to Obtain Good Die Alignment

Some manufacturers of press brakes provide a full length dovetail or other type of slot in the front and rear of the bed of their machines that allows the mounting of gage and support arms. If this feature is available on your press brake, it is very easy to mount die aligners (Fig. 24-1). A die aligner is a piece of steel about $^3/_4$" thick, 2" wide, and about 5" or 6" high. About 1" down from the top of the bar, a $^1/_2$" or $^5/_8$" hole is drilled and tapped in the center of the 2" side. On the other end of the bar, a clearance hole is drilled that will allow the bar to be clamped to the bed into the dovetail slot on the side of the bed. A standard $^1/_2$" (or $^5/_8$") hex head bolt about 4" long is screwed into the tapped hole in the bar. This description is for one aligner of a set of six needed to allow the operator to obtain good die alignment.

If a set of die aligners are purchased from a manufacturer, the 2" wide bar may have been machined to provide a $^1/_8$" stepped tongue that fits into the dovetail slot on one side of the bed to ensure that the bar does not turn while being used. The hex head would be replaced by a $^3/_{16}$" bar about 4" long that is mounted at 90° to the threaded bolt to make a handle that allows simple hand turning of the screw.

Die aligners are made in a set of six. Two are mounted at each end of the bed with the bars bolted to the side of the bed. The threaded screws are then hand tightened against the front and back of the filler block. The remaining two aligners are used in the center of the bed if it is determined that the filler block has bowed slightly during its use over a period of time. In this manner, the operator can go to one end of the bed, with the nose of the upper die located slightly above the bottom of the radius of the lower die, and visually see how good the initial die alignment is. The end of the filler block can be easily moved back and forth by hand turning the screws in the die aligners. The dies can be aligned within 0.010" visually. To obtain the closer alignment needed, it will be necessary to use feeler gages.

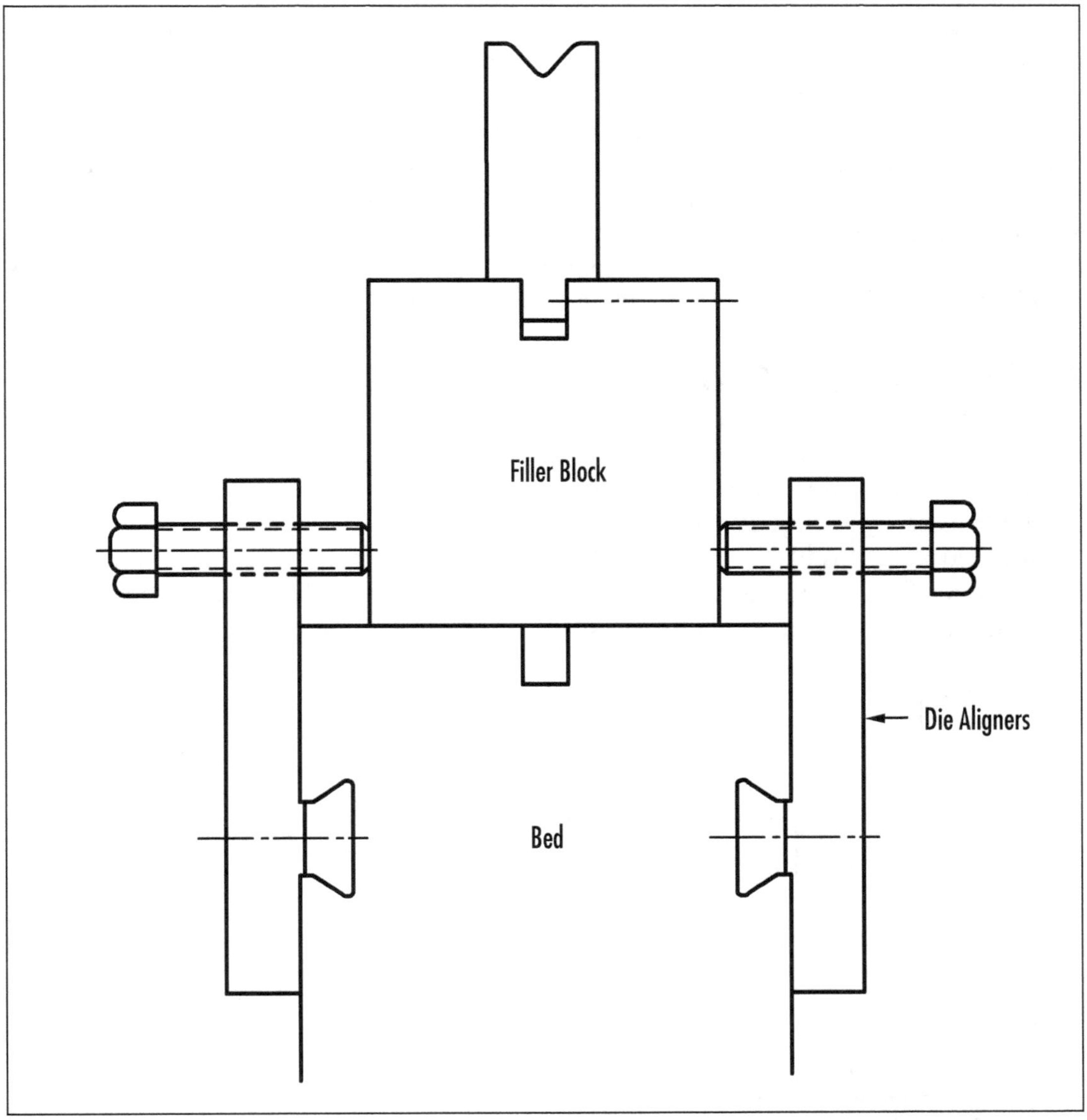

Fig. 24-1. Die aligners.

For those who are used to checking clearances in various machines, feeler gages are common tools. For those not familiar with them, a feeler gage is a piece of stainless steel approximately $^1/_2$" × 3" to 4" long, made to a specific decimal thickness. The thinnest gage would be 0.001" thick. The gage increments are normally from 0.0015, 0.002, 0.003, 0.004, 0.005 inch, and then in various steps up to about 0.025" thick. Most of the time, the gages are mounted together, with a screw through one end that allows the other end to swing out to be used to check a specific clearance. Many times, in order to obtain a certain thickness, several feeler gages of various thicknesses are stacked together to obtain a decimal thickness not found in an individual gage. A 0.001" feeler gage is seldom used. If the gage is pushed against any type of resistance, it will simply bow or bend like a thin piece of paper. If the feeler gage is 0.0015" or thicker, there is enough stiffness in the metal that it will stay reasonably straight while it is being used to measure clearance between two dies.

It is common practice for the press brake operator to use an arbitrary group of feeler gages that may approximate the thickness of the material that is going to be formed, and use those to check the forming surfaces between the upper and lower dies. After the first end has been aligned to the satisfaction of the operator (or setup person), he/she must walk to the other end of the die set and complete the die alignment procedure. The die aligners, since they are still hand tightened at the original alignment, will help hold the filler block in place, while the second end is being moved to obtain proper clearance. When the second alignment is complete, the operator must go back to the first end to make sure it has remained in position. When both ends are acceptably aligned, the filler block can be clamped down. At this point, the operator can start setting gages and support arms, plus begin making any needed test parts in order to determine the proper ram setting to complete the setup.

D) Selecting the Correct Filler Block (or Die Holder)

In Chapter 1-B-4, and in other chapters, reference is made to the use of filler blocks. Filler blocks are used to fill up some of the available die space in the press brake in order to reduce the height and expense of standard vee dies, to provide a simple method of allowing the tooling to be easily aligned, and to provide support for the lower die.

Filler blocks are available in various heights and widths from the original press brake manufacturer (Fig. 24-2). Each bed top design has a unique method of mounting its filler block. If the one that is currently being used on your press brake needs to be replaced, note

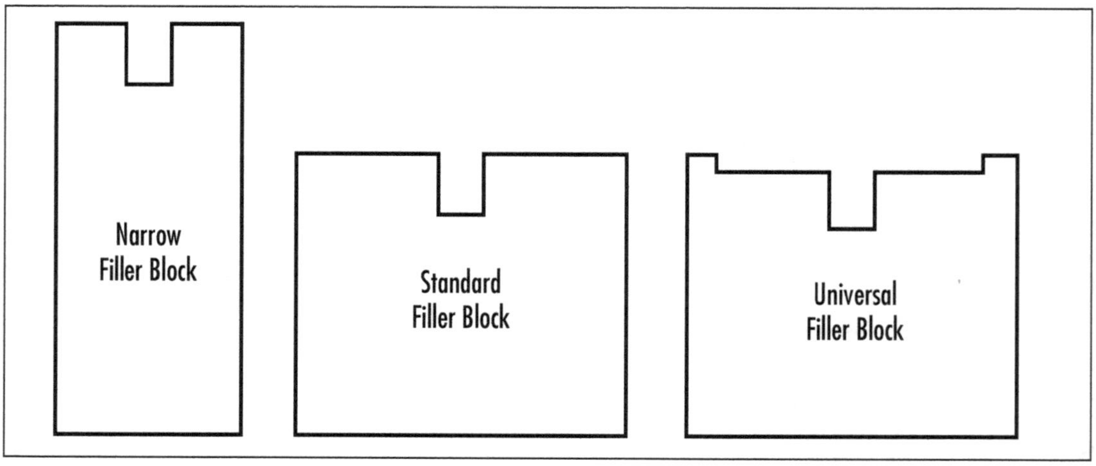

Fig. 24-1. Die aligners.

the following: the method by which the filler block is held to the bed; the dimensions of the bar that is being used; and the actual location of the hold-down spacing. If contact with the press brake builder is made, they will want to know the serial number of the machine to ensure correct replacement.

The filler block has acquired many names over the years. Many of the names are only used in certain parts of the country, while others seem to be used in a specific industry. The other common names of a filler block include: die holder, bolster, riser block, and rail. The term **bolster** is used by companies that utilize presses as well as press brakes. The bolster on a press is used to hold dies just as a filler block holds the lower die on the press brake. The term **riser block** seems to be found in the industry that manufactures tractor trailers. The **rail** name came from a press brake manufacturer named Dries & Krump who supplied some of their filler blocks made from the bars used for railroad tracks. Rail material is not available in ample quantities at this time and has not been supplied as a common filler block for over 40 years. Currently, the available filler blocks fall into the following four categories.

1) The most popular style is a bar that is made from 2" to 2.5" wide to 4" to 4.5" steel having a slightly higher carbon content than mild steel. The additional carbon minimizes galling of the top surface of the filler block as the lower die is slid into position. Narrow filler block bars offer clearance for 20 gauge to 10 gauge parts that may need special room in the rear of the lower die for the flanges of previous bends. The bar width is wide enough for die support, but narrow and high enough for good general purpose sheet metal bending. The top and bottom of the bar are machined flat and parallel to each other.
2) Wider bars, also flat on the top, are available to hold larger die bars and provide the support needed to carry the higher forming loads. Most filler blocks of this style are made from square bars of steel and are 4" or 5" wide. The working height is reduced about $^1/_2$" to allow the top and bottom to be properly machined.
3) An optional style of filler block, often called a universal filler block, has a special $^1/_4$" deep groove machined to a decimal opening on the top surface of the filler block to allow a square die bar with a different vee opening on each side (called a four-way die) to be located. The standard $^1/_2$" tongue slot needed to mount standard lower vee dies is also machined in the center of the filler block. Most manufacturers offer three universal sizes of filler blocks, each with a different groove size, to accommodate the three standard size square four-way dies that are common in the industry.
4) Special shape filler blocks that are designed to hold the European style press brake dies are available from a limited number of press brake manufacturers as well as press brake tooling companies. Care must be taken if "off of the shelf" filler blocks are purchased, since mounting locations for the filler blocks are not standardized throughout the industry.
5) A variety of "crownable" filler blocks have been designed over the years. The purpose of this design is to eliminate the need for the operator shimming under the lower die to compensate for any bed and ram deflection that occurs during the forming cycle. The designs normally have a series of wedges that can be adjusted front to back that can create a "crown" in the top surface of the filler block, allowing a straight surface to be produced at the bottom of the forming cycle. The problem is adjusting the wedges properly prior to making the bend. The original designs required the operator to adjust screws in the front of the filler block to achieve the needed crown. This was very time consuming, but also became a "guess" on the part of the operator, and often

required several test bends to determine if the adjustment was even close to correct. Newer designs offer motorized movement of the wedges, but still leave the final position to the operator's discretion. The ultimate method at this time is to tie the filler block wedge adjustment to the computer control of the press brake, which is programmed for the forming tonnage and theoretical deflection. Prior to investing a great deal of money in this option, it is recommended that an extensive test program be made using typical tooling setups that will be used with the new press brake to ensure that the programmable unit will actually save time and money.

Consideration should be given to the fact that many production runs on a press brake are 6 pieces or less. Test bends may not be practical. The new press brake must have a proven tonnage control adjustment and a compatible computer linkup. The operator must be able to input the required forming tonnage. The load must be centered left to right and not have multiple die setups in the machine at the same time. Can bends be made in a sequence that requires different forming tonnages as is often needed when forming a rectangular part? What type of maintenance should be expected?

6) Many special filler blocks are supplied to accommodate the need for higher dies to fit into the press brake die space or to provide added clearance for part flanges or part removal. Remember, if a special filler block is needed for some particular work, it may not provide enough flexibility for the remaining jobs. Some companies may need two filler blocks to do all of their fabricating work.

E) Four-way Dies

By machining a square bar into a specific square decimal dimension, and then locating a different vee opening on each side of the square, the operator will have a vee die that can be rotated to allow different thicknesses of material to be formed. This will minimize the normal setup time needed to change the lower die, plus do any needed die alignment. There are some obvious limitations. The four-way die bar must be strong enough to allow the vee openings to be used to form the recommended thickness of material without the bar breaking (Fig. 24-3). The vee openings machined into the sides of the bar will limit the forming capability of the bar compared to a die having solid sides. If NC back gaging is available on the press brake that is using a four-way die, the gage fingers could have some interference if a short flange is needed when the smallest vee opening in the four-way bar is being used. Most of the vee openings in the four-way bars are selected to form material 11 gauge and thicker. The main reason is that the thicker materials do not have to be turned over very often, which would cause flange interference with a larger lower die bar. If an offset type bend is needed, the distance between the bends must be spaced far enough apart to clear the four-way die block.

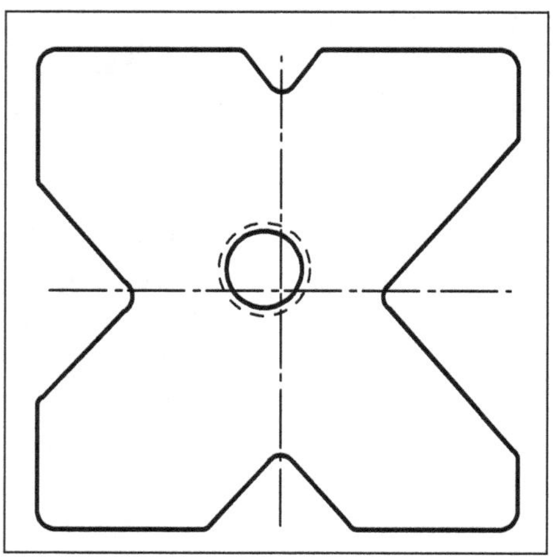

Fig. 24-3. Four-way die.

Four-way dies cannot be flame hardened and kept straight. The different openings on the side of the die may also cause twist in the bar to occur if abnormal heat is applied in flame hardening. As a result, four-way dies should not be considered for the forming of higher tensile steels due to the higher forming loads in a limited strength lower die, plus the high probability of excessive die wear. Each end of most four-way dies are drilled and tapped with a 1" hole. The holes allow a special stud to be used in order to permit the four-way die to be lifted off the universal filler block and rotated so that the different vee openings can be used. Each supplier of four-way dies has a special design of lifting unit, normally called lifting links, which they make available to the user at added cost, in order to allow the operator to easily rotate the die. Most of the lifting devices work at the ends of the bed or ram. This requires that the four-way die be made the full length of the press brake. The method of turning the die must require that the operator does **not** place his fingers into the die area. The turning should be accomplished by the operator standing at the end of the bed and using a standard wrench, or a special tool supplied with the lifting units, rotate the entire die length without any additional assistance. Special lifting links are needed if a shorter four-way die is being used.

If a four-way die is to be purchased, prior to ordering the die make sure it can be rotated in the die space that is available in your press brake. The height of the filler block and the upper die take up some of the available die space. When the four-way is lifted and turned, the lifting device must have positioned the square die so that the diagonals of the square will miss the nose of the upper die and the top surface of the filler block. Interference can occur if the maximum die space (when the ram stroke is up) will not accommodate an upper die, the filler block, and the diagonal of the fourway die. The fact that this concern is even mentioned should point out that many users have found interference after a great deal of money was spent for tooling that did not work as expected.

The straightness of a four-way die will not be as good as a simple vee die. The finished bar must be straight enough to be mounted easily into the groove of a universal filler block (normally machined to an opening about 0.015" larger than the outside dimensions of the four-way die). The die must be able to be pushed in from the end of the press brake by the operator, or dropped into the groove while being held by lifting links. Often, just handling the die by lifting it at each end will cause a slight bow to remain in the bar. The result may be that on two opposite sides the die will be flat when lying in the filler block. When the die is rotated 90°, either the ends or the center will be lifted away from the filler block about $^1/_{32}$" if the four-way die is 10' to 12' long. The bow that often creeps into the four-way die will vary depending on the length of the die.

F) Adjustable Vee Dies

Large hydraulic press brakes are normally used for bending heavier material. The thicker plates will require larger vee die openings. The size of individual vee dies, their cost, plus the problem of handling and die storage makes the purchase of an adjustable vee die opening very attractive. The most common size is a die that will have die openings that range from 5" up to a maximum of 10" (Fig. 20-3). A groove, similar to the decimal groove found in the top of a universal filler block, is often provided to allow a 5 $^1/_4$" four-way die to be used in the setup. The reason that this size four-way is selected is that the normal vee die openings in the bar are 1, 2, 3, and 4 inches. This provides the operator a range of vee openings of from 1" up to 10", or the ability to form many parts from 11 gauge up to 1" thick plate.

An adjustable vee die requires support under the entire die body, which is wider than the width of the normal press brake bed top. If an adjustable vee die is going to be used with a new press brake, the top of the bed should be modified. The bed should accommodate the width of the die body plus any shims that will be hanging outside the die body. If the bed top is wide enough, support arms can be fastened to the front or rear of the bed to assist in the handling of the heavy material. An adjustable vee die is mounted on the bed top. Since the press brake will have the higher tonnage needed to form thicker plate, the manufacturer of the press brake will assume that larger dies are going to be required. To provide for the tooling and the additional space needed to remove heavier material after it has been formed, the press brake will be made with more stroke and die space than smaller tonnage machines.

When the upper die penetrates into the larger vee die opening, and added forming stroke is required, the conventional upper die height will not be sufficient to reach down far enough into the lower die, even when the ram reaches the bottom of its stroke. It is very possible that the upper die will need to be made much higher, or that a ram adapter (a spacer on the top that fills up the die space like a filler block does under the lower die) will have to be used to be able to form simple 90° bends.

Shims are used to locate the two full-length forming inserts that are in the middle of the lower die body. The shims are normally made about 48" long and slotted at each of the 12" centerline clamping bolts; that way, the shims can be lifted up from behind the forming insert if the vee is being opened, or placed behind the insert if the vee is being closed. If the inserts must be moved outwards in the rectangular die cavity to obtain a larger vee opening, after the shims are removed, the forming insert is pulled out toward the setup person. The bolts that are used to secure the insert will be moved out through the hole in the side of the die body. They will stick out from the die body the same distance as the thickness of the shims that have been removed. The removed shims are then put over the extended clamp bolts. In this manner, only one length of clamping bolt is required. The shims should be made in thicknesses that can easily be lifted and moved by the setup person. The most common shim thickness is $^1/_4$", and the maximum thickness is 1". The die manufacturer may have a set of standard thicknesses that they supply to minimize the number of shims that have to be moved, yet provide a good incremental range of vee openings.

The rectangular forming inserts often have different radii on each of the four corners. These can be rotated, as required, to obtain good die life and minimize wear. If the inserts are made from standard die steel, all of the radii can be reworked when worn and be put back in "like new" condition. If abrasive materials or high tensile steel is being formed, two of the forming radii on one of the narrow ends of each insert can be flame hardened to extend the die life of the inserts. Only two sides of an insert can be flame hardened if acceptable straightness is to be maintained.

G) Polyurethane Die Pads

Polyurethane is an elastomer plastic; it is not rubber. Its use in press brake tooling is found in square or rectangular blocks that have been used as the lower die. Its main advantages are as follows.

1) It eliminates objectionable marking that sometimes occurs when using steel dies to form stainless steel, aluminum, or painted stock.
2) It eliminates of a great number of lower dies, since the pad will try to push the material around the upper die regardless of the thickness of the pad.
3) A polyurethane pad may be used as a pressure pad in some circumstances.
4) The cross sectional shape of the polyurethane pad can be molded to a unique shape that can help form some special shapes.
5) A simple 90° bend can be formed with a good, sharp, inside radius at about the same tonnage that is required to make a conventional bottoming bend using steel dies.

If it has some advantages, why isn't the pad commonly used in more applications? As is true with any product, there are disadvantages. It is always up to the user to decide if the advantages outweigh the disadvantages. The main limitations of polyurethane are as follows.

1) The relationship of cost per hit to the life of the pad is very high. Metal dies provide the possibility of obtaining hundreds of thousands of bends from a simple vee die. With a polyurethane pad, depending on the application, the life may be from several hundred bends to as many as 40,000.
2) The pad needs a holder not only to keep it from moving, but to ensure the pressure generated is kept in the right direction (Fig.10-8).
3) Gage fingers extending over the polyurethane are not too practical unless a flip-up gage is used. This means more expensive gaging is required. During the forming stroke, the pad moves up and above the top surface of the holder. Standard fixed gaging would be deformed.
4) Although small flanges have been made, production runs with flanges smaller than 25% of the pad width will considerably shorten the pad life, especially if the material is 18 gauge or heavier.
5) Any dirt on the metal will tend to imbed itself into the polyurethane pad. If the pad is to be used to prevent marking, the material and pad must be kept very clean.
6) The forming pressure needed to make bends using a polyurethane pad is similar to bottoming. This high tonnage may be a concern.
7) If production rates are high or the bend requires deep penetration into the pad to make the bend, heat can be generated in the pad. If the temperature surpasses 250°, some areas of the pad seem to decompose or become hard, and large chips may fall out.
8) After repeated use, the pad will take a permanent set. This can require frequent adjustment of the ram setting, or the operator may hit the part at a higher tonnage but only use one ram setting.
9) Because of the size of the pad and holder, prior bends can make it impractical to properly locate subsequent bends because of lack of clearance.
10) Making multiple bends such as offsets is not possible since the pad cannot penetrate into the upper die in order to obtain the part definition if one hit offset forming is attempted. If the offset is to be made in two hits, the pad may cause interference with the first flange.
11) If the upper die is too narrow and the penetration of the upper die too deep, the pad may wrap the material around the entire upper die nose, making unwanted, additional bends in the part.
12) It is difficult to determine if the resulting bends made using the polyurethane pads are made by air bending, bottoming, or bottoming with springback. The material must be formed against the upper die in order to obtain consistency, but some springback seems to take place.

It is normal to do some development work prior to using a polyurethane die in production. The normal hardness for the polyurethane pad should be 90A to 95A durometer. The ends of the holder should have plates to keep the pad from moving out the end of the pad holder when under pressure. The ends of the upper die should be rounded off to keep any cutting of the pad by the upper die to a minimum. The life of the top surface of the base pad can be increased by laying a separate $^{1}/_{8}$" thick piece of polyurethane on the top of the larger pad, which will allow all of the actual wear caused by the forming of parts to be absorbed by the thin pad.

H) Ram Extensions and Horns

A simple way to extend the forming surface of a press brake is to make the bed and the ram longer than the conventionally designed machine. A standard 10' press brake means that the distance between the housings is a little longer than 10'. (Remember that some builders make their machines 10' overall, not the distance between the housings.) The overall length of the bed and ram for the normal 10' press brake is 12' total length. If a company that was purchasing a new press brake needed to form some 14' long parts, assuming that the press brake tonnage is sufficient to make the bends, and if the throat allows enough clearance for the forming of the part, the builder could add 12" to each side of the bed and ram at nominal cost. This type of addition to the bed and ram is called a **bed and ram extension**. The ram and bed can be extended on one side or both sides, but it is always best to keep the loads balanced if possible.

When the bed and ram are extended, the ram appears to have the same design as a standard ram. The end of the extension is normally made straight out from the slide and guide mounting or, if it is sloped somewhat on either side, the design simply continues out from the additional distance (Fig. 24-4). Some press brakes are limited to a maximum extension of 12" on either side. Larger machines can have the bed and ram extended as much as 24" on each side, and the maximum extension is 36" on each side. The maximum can only be achieved on larger tonnage press brakes. The main reason for the limitation is that, if an accidental nonbalanced load was placed on one end of the ram, the load could cause the ram to go out of level in such a manner that the slides and guides of the press brake could be broken. The repair cost would be very high, and it is unlikely the damage would be covered by any machine warranty.

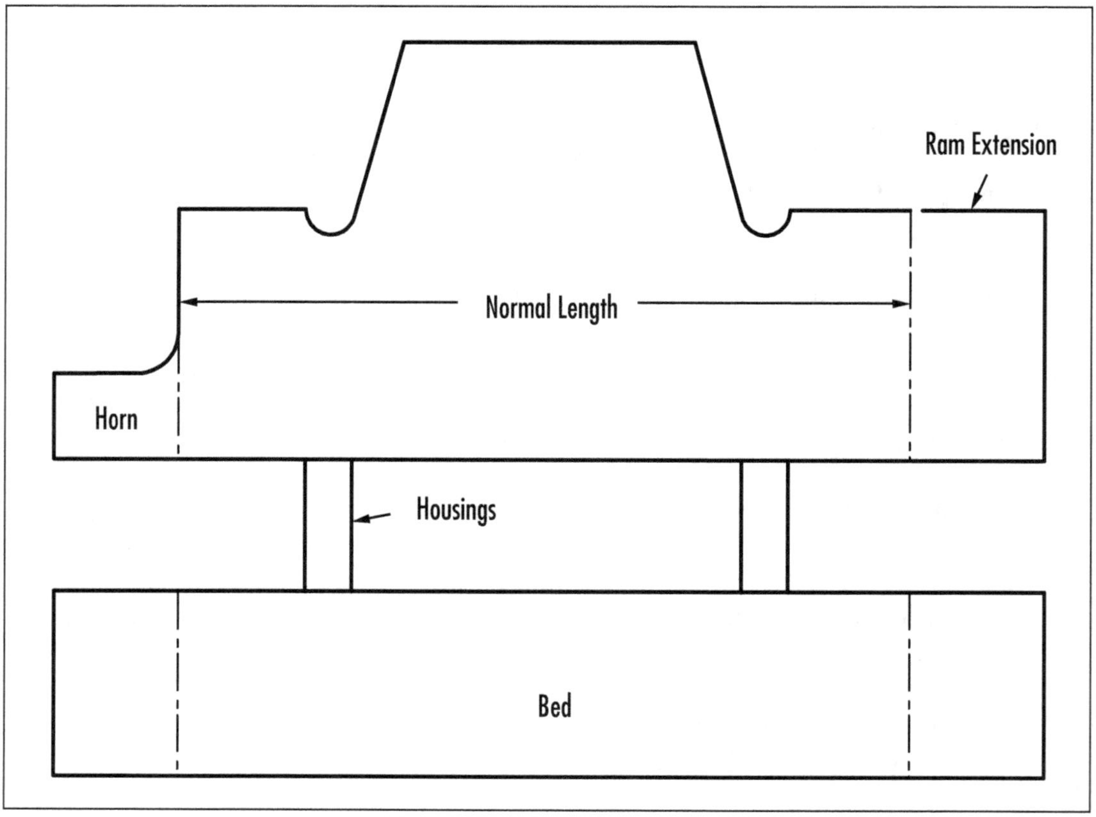

Fig. 24-4. Bed and ram extensions.

Some users have special parts that are too large to form in the conventional die space of their press brake. Often, the parts are large round tubes, or square or rectangular wrappers only 12" to 20" long. If a press brake is sized to have sufficient forming tonnage available at the end of a ram extension, a special shaped ram extension called a **horn** can be supplied. The horn would appear to have the top surface of the extension only 12" to 18" above the ram nose. It is normally parallel with the ram nose from the end of the machine until it resumes the original ram height near the normal length of the ram, to allow the ram slide to be properly mounted. The forming tonnage available on the horn will be specified by the press brake manufacturer and will be very low compared to the total machine tonnage. Some unusual ram deflection and die shimming should also be expected.

I) Rocker Dies

Occasionally, a part must be formed that requires one or two legs of the part being radically overbent. The die is then designed with a moving part called a **rocker**. Rockers come in a variety of designs in order to overbend parts with greater than normal springback. If the part is a channel shape, symmetrical rockers are located opposite each other and form the channel, complete with overbend, in one stroke (Fig. 24-5). When the upper die returns to the top of the stroke, the rocker or rockers return to their open position, allowing the pressure pad to move back to the top surface of the lower die. The part, which has its overbent flange(s) opened slightly by normal springback, may drop off the upper die or may be lifted up with the die if

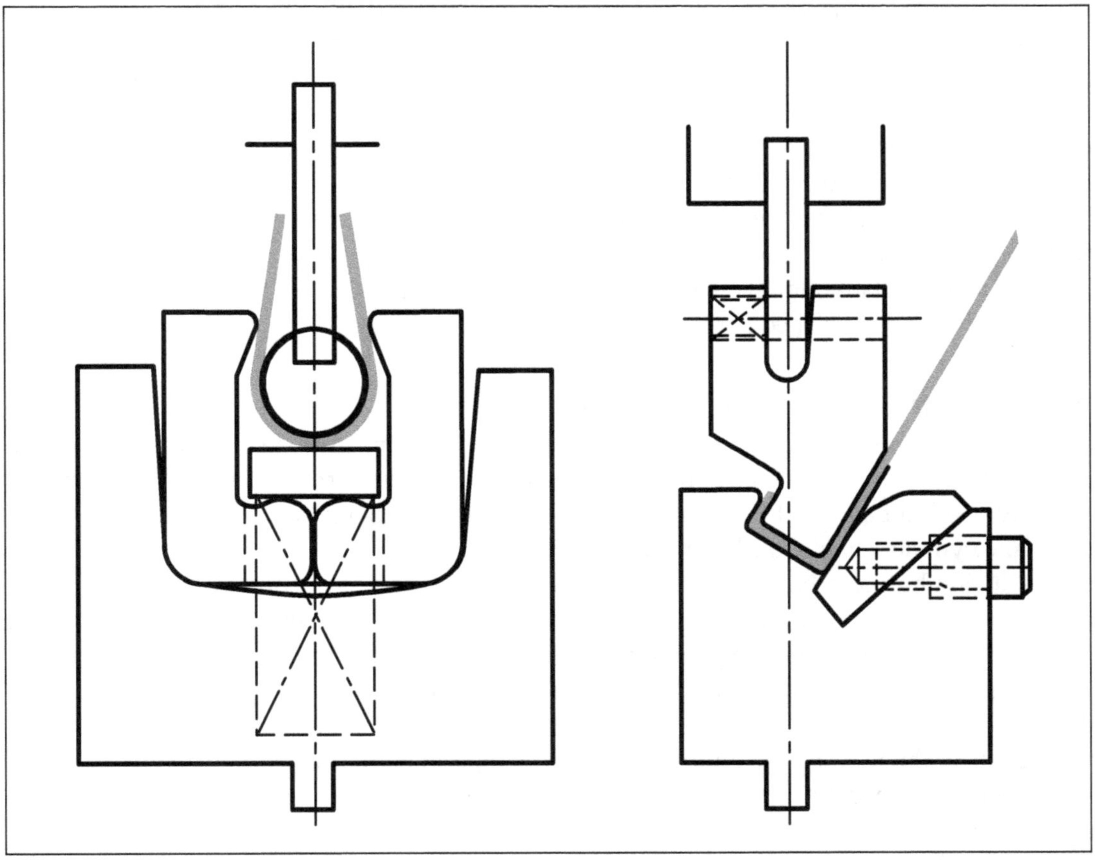

Fig. 24-5. Rocker dies.

the springback did not allow the flanges to clear the upper die width. It may be necessary to provide the upper die with a wedge release to allow for easy part removal.

Rocker designs are found most frequently in channel die and wiping die applications. Not shown in most illustrations are the end plates generally needed to keep the rocker sections in their proper die location. Make sure this type of design is tested prior to putting the die in production. Die setup and alignment should be done only by experienced personnel, since the setting of clearances can be somewhat difficult.

J) Roller Dies

Since the 1940s, a special lower die called a **roller die** has been made available (Fig. 10-11). The die consists of a wide holder that has two semicircular rollers mounted side by side. The dies are designed to form only simple vee type bends. End plates may be added to the ends of the die to provide assistance in holding the rollers in place, and to provide spring locations to pull the rollers back to their starting position after each stroke. Some designs have located the springs and restraints inside the die body.

The purpose of the design is to allow one die to form a number of different material thicknesses with one lower die, minimize marking on polished material, and control distortion caused when holes or slots are formed near the bend line. The design also can reduce the upward forming speed when the front flange is long and subject to backbend distortion.

One may question why this type of die isn't very popular. The reason is that the inside radii produced by the die are larger than most designs will accommodate. Each of the rollers has a centerline about which it rotates. That centerline will approximate the corners of a conventional vee opening of the lower die. When the rule of finding the estimated inside radius (assuming mild steel is being formed) by multiplying the vee opening times 0.156, the centerline created by the roller centers far exceeds almost all of the recommended vee die openings shown on the standard air bend tonnage chart. In some cases, aluminum parts may stick close enough to the nose radius of the upper die that the inside radius may be accepted, but the probability is that the radius will be distinctly elliptical if the radius on the upper die was close to one metal thickness. Prior to purchasing this type lower die, it may be best to inspect some similar materials that have been made to ensure that the shape and consistency will work with your finished part requirements.

K) Excessive Spring Pressure

On occasion, a special lower die will need a great deal of pressure pad tonnage to hold the material that is being formed against the upper die during the forming stroke. The excessive pressure is often generated by the use of polyurethane pads being compressed, or the use of hydraulic cylinders being used in place of springs.

If a hydraulic press brake is used, when the forming stroke is completed at the bottom of the stroke, the downward pressure stops when the valves are switched. During the down stroke, the ram normally not only provides sufficient pressure to form the part, but also pushes against a counterbalance pressure that will lift the ram back up to the top of the stroke when the ram direction is reversed. The counterbalance pressure must be sufficient not only to lift the weight of the ram, but also the weight of the upper die plus accommodate any stripping loads required in part removal. The counterbalance pressure is designed to raise the ram at a specific speed depending on the valve, pump, and pipe size used in the press brake.

If the tooling has generated an abnormal amount of pressure pad tonnage at the bottom of the stroke, when the ram reaches the bottom of the stroke and the hydraulic valves switch, the compressed tonnage in the die is released. If the pressure from the die exceeds the weight of the ram and upper die, the ram will be lifted upward at whatever speed the pressure pad can move, and will lift the ram up at a speed much faster than it was designed to travel. The distance that it will move at this faster speed will be the length of travel of the pressure pad. The shock wave through the hydraulic system of the press brake may cause a loud noise, high vibrations throughout the machine frame, and possible damage to the hydraulic system. If this is the case, it will be immediately evident when new dies are loaded into the machine for tryout and the first cycle is made.

The only answer is either to eliminate or reduce the pressure from the die or to hold the pressure pad down until the upper die has cleared the lower die cavity. If the pressure pad tonnage is generated by hydraulic cylinders (instead of normal spring pressure), installing a flow control valve on the return hydraulic line will often slow up the pressure pad speed sufficiently to eliminate the problem.

Chapter 25
Determining How to Make a Good Part

The job is not easy for the person who determines how a part is to be processed in order to make a usable product. In many companies, the responsibility is given to the press brake operator, who is supposed to magically debug the gaging methods, determine the sequence of bends, hold all of the tolerances, and produce good parts without any scrap. All this is often accomplished without the opportunity to have any input into how the part should be made prior to it arriving at the press brake. In order to minimize the time that can be wasted on the initial forming of a new part, it is always recommended that the entire process be reviewed and discussed prior to any part blanks being made. A simple review of the part drawing should identify the following items.

A) Press Brake

Is a press brake available that has sufficient tonnage to make the bends called for on the drawing? For example, the inside radii called for on the drawing may be shown as one metal thickness or even less than one metal thickness. This means that standard air bend forming cannot be used. Bottoming or coining dies (if the inside radius is less than one metal thickness) will be required, which will increase the needed tonnage by a substantial amount. Is the inside radius really needed, or can it be modified to stay within the tonnage limitations of the press brake that has been selected?

B) Tooling

Is the tooling that is currently in the fabricating area the proper type needed to form the part? If new tooling is required, will the cost of the new dies fit into the current tooling budget? Can new dies be obtained fast enough to allow the delivery schedule to be met? If the present tooling must be used, who is going to approve any changes in the part design to ensure that the parts can be used?

C) Blank Size

The size of the "flat blank" must be estimated. Who calculates the blank size? Do they understand that the bend allowance formula will not give a perfect answer? It is good practice to make a sample blank and form it on the dies that will be used. Do this prior to making a large number of blanks that may be punched or notched; otherwise, it may be found that the blank size is incorrect. Remember that the inside radius is a function of the vee die opening that is to be used. If the radius that is shown on the drawing is used for calculating the blank size, and the actual radius formed in the selected vee die is different, the calculated blank size will be wrong. Someone must communicate to the press brake operator which vee die opening should be used to insure that the results will be consistent. If the inside radius is to be smaller than one metal thickness, any formula obtained from an engineering handbook will not work. Some companies have developed unique bend allowances for their dies that work for them, but they are rarely published.

D) Bend Sequence

What will be the sequence of bends if more than one bend is needed in the part? Each bend must be made with the majority of the weight of the part in the front of the machine to ensure that the part can be easily supported by the operator during each forming cycle. This eliminates the risk that the part will fall to the rear of the machine, thus tempting the operator to reach through the die opening and putting his safety at risk. After the final bend is made, can the finished part be easily removed toward the operator side of the press brake?

The sequence must also ensure that any tolerances critical to the part usage can be held. The variations expected from angular tolerances, as well as gage location and blank tolerances must all be anticipated before forming the first part. If proper planning is not done prior to bending the first part, good parts will not be formed.

E) Gaging

How is the part going to be gaged? If an NC gage is available on the press brake, most of the time and effort will be made to use that gage. Obviously, it will take less time to set up, thereby keeping the total setup time and overall cost of the part to a minimum. Can the tolerances of the part as shown on the drawing be held? It is reasonable to assume that the blank size may vary in width and length $\pm^{1}/_{32}$". If the part is back gaged, any error will be on the operator side. If a second bend is required, where will the blank error be located? If the part is pushed forward to form the second bend, the blank error will stay on the operator side. If the part is turned around, and a bend made on the other side of the part, it is likely that the error will be between the two bends.

If the inside dimensions of the part must be held, it may be desirable to front gage the first bend, thereby locating the blank error into the height of the back flange. The remaining bends can be gaged, and any gaging errors will cause only linear variation.

F) Angular Variations

The potential error caused by angular variations must now be addressed. If a sheet metal part is to be air bent, the resultant angle in sheet metal parts could vary ±1.5° for part lengths up to 10'. Bottoming dies cut the potential angular variations in half. Any type of compromise die that involves bottoming the material in some manner, but anticipates some springback, will have a tolerance somewhere in between.

If a 90° formed angle made in 16 gauge mild steel actually has an angular variation of 1°, the variation will be easy to measure. Put the longer flat leg on a flat surface, allowing the vertical leg to stick up in the air. Place a simple 90° square next to the vertical leg. A 1° variation will result in the vertical leg being out of square by $^1/_{64}$" for each vertical inch of the flange. If the vertical leg is 4" high, 1° will cause the top of the leg to be $\pm^1/_{16}$" out of square with the bottom of the flange. If the part had been turned 180°, and a 4" high flange formed on the other side, the total variation at the top opening of the channel shaped part would be a potential plus or minus angular variation of $^1/_{16}$" for each vertical leg. The total variation at the top would be $\pm^1/_8$"; this is in addition to any potential linear variation in the based caused by a blank error or a gaging error. This does not address any angular variations caused by rough handling of the part from the time of forming to the point of the part being used. Knowing the angular variation possibilities, will the formed part be able to be used? A bigger question may be "are the drawing tolerances practical?"

G) Linear Tolerances

Tolerances seem to make it practically impossible to obtain a good part. The fact is that all tolerances follow a statistical bell shaped curve. Most of the bend angles are good with only a small variation. Many of the bend lines will be very consistent. Most of the blanks will be identical. Although blank tolerances of $\pm^1/_{32}$" have been mentioned, most of the production blanks may be actually held to a 0.010" to 0.015" length and width variation.

The initial effort must be to minimize as many of the obvious problems prior to going into production. The proper type of forming dies must be selected. The blank size must be modified to eliminate any fudging by the operator in an attempt to hold dimensions. The gaging process must locate the best position to minimize blank error. The operator must know how the final part is to be used so that he can try to control the angular variation to minimize any assembly concerns. Handling the completed parts should be reviewed to keep finished parts from being deformed.

If these initial steps are taken, the production problems that do arise should be minimal. Unless a major problem is encountered, the parts that vary may be salvageable.

H) Instructions

It is important to put all of the instructions in writing. They should not only be available to the supervisor, but also to all the operators who will work on the part. After production has started, update the changes or modifications so that the next time the parts are made, the production debugging is minimal. It is also important for the engineering drawings to be updated. Upper management should emphasize that making changes and updating drawings is not finger pointing, but simply a team approach to solving problems needed to make the best product at the lowest cost.

I) Material

Most problems in the press brake area are caused by material variations. Unfortunately, the purchasing department is often directed by management to buy the cheapest material possible. Material thickness variations; wide changes in tensile or yield strength from order to order; material that is not flat and that contains variations of bow, twist, or camber are all too common occurrences. It seems that certain areas of the country too frequently encounter the problem of the part not staying flat after the forming operation; this has been traced to steel service centers providing steel that was in tolerance but did not have consistent hardness.

The above problems reflect the saying "penny wise and pound foolish." A company may save pennies per pound purchasing the steel, but spend dollars per pound trying to form it. Successful press brake forming depends on all parts of the organization working together as a team in order to minimize product costs. Engineering must start the process by specifying the proper materials needed either on the drawing or in agreed upon standards for the fabricating area. Management must be wise enough to listen to the personnel making the parts on the press brake to ensure that all decisions are working. When a problem is solved, accurate record keeping assures that it doesn't return as soon as someone changes jobs or goes on vacation.

A classic example of forming problems waiting to happen is the tractor trailer industry. Many of their engineering drawings specify that their steel material be purchased at 50,000 PSI minimum yield strength. This means that the purchasing department can obtain a variety of high tensile steel materials, many of which may require that special minimum inside radii be used. The shop press brake operators only have certain dies available to make the parts which have one machined radius based on the specification of a specific type of material. If a different type of steel is supplied by the purchasing department, although it meets the engineering specifications, the forming of the different type of material may increase the forming tonnage to the point that the press brake does not have sufficient capacity to make the bend or that the proper angles cannot be held. This does not even consider the possibility that the inside radius may be formed too small, thereby creating the potential for the material to crack. The industry seems to make their method work, but in some cases it has caused steel companies, tooling suppliers, as well as the trailer company a great deal of time and money.

Chapter 26

Understanding Words Used in the Context of Press Brake Tooling

Most of the terms used when discussing press brake tooling—including the parts formed, the material, or the method of bending—are common words. However, they have specific meanings in the context of this subject. Therefore, those words will be reviewed below to minimize any misunderstanding.

Accuracy: The angular variation in a bend angle after it has been formed from a flat sheet of metal. It can also refer to the dimensional difference in the legs of a formed part that may have been caused by material variations or gaging error.

Air Bend: To form a bend angle by allowing the metal to contact only the nose radius of the upper die and the two corner radii of the lower die. This type of bending is also referred to as **three points of contact**. All estimated forming tonnages to bend metal using a press brake are based on air bending the part.

Alignment: To inspect the nonsymmetrical variations that may be present between the upper die and the lower die. Also, to move the lower die as required to maintain the same distance between the upper and lower die forming surfaces. When punching dies are being used, alignment is the adjustment of the round die opening to provide the same clearance between the circumference of the punch and the die.

Back Bend: A distortion caused by the corners of the lower die in the long flange of a part that can occur during the movement of the material from a horizontal plane to the formed angle. (Often the flange moves up in the front or rear to a 45° angle.) The weight of the material, if not supported, will cause the flange to bend back over the front or rear corner of the vee opening. This results in a flange that has the proper angle formed only in the material that actually was pushed down into the vee opening. The remaining flange length will have a second apparent bend line located where the material contacted the corner of the vee opening. This very visual reverse bend can vary from several degrees to as much as 10° to 15°. Not only is the appearance of the part distorted, but the total angle will be unacceptable.

Bead: In press brake work, it is a common term for a shape that looks like a hinge. Some areas of the country call the rounded and closed shape a **curl**. In press work, a bead is often a rounded rib shape pressed into a flat sheet, but it usually does not extend the full length of the blank. The material remains flat at each end.

Bed: The bottom, nonmoving plate in the front of a press brake. The top surface holds the filler block and the lower dies. It provides support and stiffness to allow a straight bend to be made.

Belly: A distortion that can occur in a long bend having a small flange being formed with the part width in the front of the press brake exceeding 24". If the material in the front is supported at each end by the operator and a helper, the weight of the center of the material will not move upward during the forming cycle at the same rate as the supported ends. As a result, a conical shape will appear in the wide area of the part affecting the formed angle and the appearance of the entire part.

Bend Allowance: An estimate of the length of the neutral axis in the area of each of the corner radius that is to be formed. The distance of this arc is often estimated by using one of the many formulas available in engineering handbooks.

Blade: A term used in parts of the country for the upper press brake die. In the fabricating area of some companies, it may also refer to the upper knife used in a squaring shear.

Bolster: When used in conjunction with press brakes, it is another name for a filler block.

Bottoming: A method of press brake forming where the part being formed is forced into the lower die far enough for the outside surface of the material to contact the sloping surfaces of the vee die. If any of the different methods of bottom forming a part is used, the forming tonnage will be noticeably greater than the charted tonnage to air bend the same thickness of mild steel.

Bow: The distortion in an unformed sheet where the material will touch the floor at each end but be raised up in the center a measurable distance. If the part is already formed, in some cases the bend line may not be straight, but may have an arc (or **bow)** that has caused problems with part dimensions or assembly with mating pieces.

Brinell Hardness Number: A commercial scale that measures the hardness of different materials by measuring the distance a ball shaped instrument can be pushed into a material's surface. A special testing machine is needed. The amount of penetration is measured using instruments supplied with the test equipment. The numbering system was developed by comparing the different hardnesses of natural materials to the material being tested. The softest natural material selected is chalk; the hardest is diamond.

Buildup: An accumulation of dimensions that can affect total tolerances or the actual fit of tooling or several parts into a certain limited space.

Bushing : A hardened round cylinder made to a tight tolerance outside diameter and having a tightly held round inside diameter. The bushing is pressed into a bored hole in the top of a die set to locate a matching guide pin, which is mounted into the lower plate of the die set. See **Guide Pin**.

Calculation: The mathematical process for determining the length of the blank size prior to forming the part, the method of ensuring that the tooling will fit into the available die space, or the forming tonnage needed to bend a specific part.

Centerline: A theoretical line establishing the symmetrical distance to the front or back of a specific part. The alignment of tooling is based on certain parts of the press brake tooling being located on the theoretical centerline of the top and bottom dies.

Channel: The word used to describe a part having a flat bottom with flanges at each end formed to 90° angles. A hat shaped channel has two flanges at the top of the two vertical legs formed at 90° away from the flat base of the part.

Clearance: The recommended distance between a punch and die to ensure that a hole will be cleanly punched. Clearance is also the distance between two dies to ensure that the upper and lower dies do not interfere with each other when the material is being formed. Some dies actually have surfaces that move next to, or even touch, each other. Specific clearance (or distance between the surfaces) is required to minimize any chance of the parts causing damage to each other, such as "galling" or "pick up."

Cold Rolled Steel: Steel that is processed at the mill through a specialized series of rolling techniques after it has cooled sufficiently. The process reduces the steel to a specific gauge thickness. Cold rolled steel (abbreviated **c.r.s.**) is available only in thicknesses up to 10 gauge (0.135").

Compressive Strength of Steel : The maximum amount of pressure that can be placed on a piece or section of steel before some type of distortion occurs. If exceeded, the steel could be reduced in thickness (or height), or some part of the material could bulge or expand like a marshmallow when it is pressed down. Most engineering handbooks suggest that the compressive strength of steel is the same as the material's tensile strength.

Compromise Method: A method of forming using bottoming type dies, where the user does not have sufficient press brake tonnage available to truly bottom and set the material. The operator uses press brake dies that do not allow for normal material springback, but that have an allowance of only 1° or 2°. The part is formed into the vee die until the outside of the material touches the lower die and a minimum amount of forming pressure is built up. The ram is then reversed, allowing the formed part to spring back to the desired angle. This method depends on the experience and skill of the press brake operator.

Consistency: The ability of the operator to produce formed parts that fall within an acceptable tolerance range, time after time; it is important for all press brake work.

Corner Radius: Normally, the radius measured is the shape of the inside radius produced when a bend is made in the part. If the radius being discussed is the outside radius, it should be referred to as the outside radius The reason that the inside radius is considered the main radius is that most engineering drawings call for a specific inside radius; also most inspection measurements using radius gages are made using the inside radius of the part.

Corrugation: A part having a repeated shape that, when formed, has the same dimensions if the part is turned over.

Cracking: The separation or breaking that occurs on the outside of a bend radius where the material has exceeded its tensile strength. The outside radius may develop small hairline openings along the bend line. In severe cases, the two legs of the formed part may simply come apart. If high tensile steel is formed using too sharp a top die radius, and the plate breaks, it may actually **fly** some distance to the front or rear of the press brake. An operator or helper should always stand to the side of the plate, never in the front, during the forming cycle.

Crease: An indention in the center of the inside radius where the upper die contacted the part, and the resistance of the material to move downward caused the upper die nose radius to sink into the surface of the part prior to the bend being completed. When that happens, the yield point of the top surface of the material is exceeded. The indention depth provides sufficient area for the upper die to continue the forming stroke without further abnormal distortion in the material. The sharper the nose radius machined on the upper die, the deeper the crease in the center of the inside radius. The thickness of the material, and the forming pressure, also have a direct effect on the crease depth.

Curl: See **Bead**.

Degree(s): One degree is 1/360 of a circle. When used to describe a formed angle, it is the number of degrees measured using a simple combination square. The angle measured is normally the inside angle as measured in degrees. The angular tolerance variation measured in the formed angle is given in the number of degrees that the actual formed angle varies from the desired angle.

Die: The upper or lower die is the designation for the tooling used in a press brake. Each forming tool or bar is designated as a die. A die set can be used to refer to both the upper and lower dies when they are used together, even though they are not made to be used exclusively with each other. The press brake upper die may be used with several different lower dies. This use of a "die set" is not the same as a specialized design of tools that are specifically used to form only one shape. Nor is it the type of tooling used in presses. A die is also the tool that is placed under a punch and material when a hole of some sort is to be punched into a part. The die will have a shape matched to the punch with appropriate clearance to prevent any clipping or interference when the punch pushes through the material and enters the top opening of the die.

Die Aligners: Screw adjustment tools mounted in the front and rear at each end of the press brake bed. They are used to move the filler block back and forth in order to achieve the proper lower die alignment with the upper die. A third pair is always available to mount on the front and rear near the center of the bed to assist in straightening any slight bow that may be present in the filler block.

Die holder: See **Filler Block**.

Distortion: Any change in the original flat blank that is caused by the press brake bending process. This may include such concerns as: the flaring of a prepunched hole that occurs if the edge of the hole is too near the bend line; back bending of the blank near the bend line (see back bend); or "bow" along the bend line discovered after the part has been formed. Occasionally, the problems that may have occurred in the blank prior to the unformed parts arriving at the press brake may be called "distortion." The problems in not having a flat blank are normally bow, twist, or camber concerns. Poor sheared or punched edges can be classified as distortion problems in extreme cases.

Draw: The process whereby a flat blank is formed into a shape with the perimeter of the part held flat. A high clamping pressure is used to hold the outside flanges of the metal while the interior of the part is formed, normally by stretching (and possibly causing thinning) the material into the proper shape while ensuring that no material tensile failure occurs. This type of forming is normally done using presses, not press brakes.

Elliptical: Most engineering drawings specify that an exact inside radius be formed in each inside radius of the part being made. Most radius bends made through the air bend forming method, or that result when a specific radius die is used, are not true radii. They must be considered some type of elliptical shape. Most inspectors, who measure the inside radius of a part, use a simple radius gage to check the part quality. If some daylight appears under the radius gage, indicating that the inside radius is not exact but elliptical, but still seems to be fairly close to the radius gage for most of the bend, the part is judged acceptable.

Engineer: A person, who may not be a graduate engineer, who is responsible for making the part drawings that are used to produce a company's product.

Female Die: A common name, used prior to the 1980s, for the lower press brake die.

Filler Block: The rectangular steel block mounted to the top of the press brake bed that holds the lower press brake die in place. It helps fill up excess die space provided in most press brakes, and it provides a convenient method for the operator to clamp the lower die to the bed after it has been properly aligned with the upper die.

Flatten: A term used primarily when a short flange has been bent to an acute angle and is pushed down using flat top upper and lower dies to close the short leg into a parallel position with the longer unformed leg. The resulting part shape is called a hem or a seam. The dies that do the actual closing of the leg into this shape are called flattening dies.

Foot Switch: An electrical operating switch used to cycle the ram. It rests on the floor in the front of the press brake and is fastened by a flexible cable to the electrical controls. When the electrical circuit is turned on, the ram will start to move if the operator steps on top of the switch. Most forming strokes are started with the ram located at the top of its stroke. The operator can take his foot off the switch on the downward stroke at any time, and the ram movement will stop. Depending on what control options are available, the ram may automatically return to the top of the stroke once it has reached the bottom point of the stroke, or the operator may retain control of the ram motion throughout the entire cycle. The upper portion of the foot pedal should have some type of guard or cover to ensure the pedal is not accidentally activated.

Forming: This is the type of bending associated with a press brake. The thickness of the material does not appreciably change in the area of the bend angle.

Friction: The amount of resistance present when one surface moves over another. Friction is present between the corners of the vee die and the outside of the part being formed as the metal moves down into the vee die cavity. Friction occurs when the operator pushes the nose of the upper die along the vee opening in order to position both dies during the setup of a job. Since the resistance is very minimal in most cases, it is often ignored as a potential problem.

Gage: The fixed (or adjustable) stop; or the method used to properly locate the bend line prior to making the bend in the part.

Galling: A rough surface produced on a piece of metal unintentionally by one surface moving over another surface. Often, this is caused when both pieces of steel are the same type. For example, rubbing two pieces of the same hardness of die steel together may gall one or both pieces. Often, cold rolled steel will not accept a rubbing or moving contact with another steel without a tearing or roughing up its smooth surface. Inserting a copper or bronze liner in place of the die surfaces that have encountered galling will solve the problem. Another name for galling is **pick-up**.

Gauge: The thickness of steel indicated by the number placed prior to the word "gauge" when specifying thin sheet metal.

Guide Pin: A round, hardened and ground, solid steel cylinder (of varying lengths) that slides into a matching bushing. At least two guide pins and bushings are required for each die set in which they are used. They are mounted into die sets to align the upper and lower tool holders in order to maintain close clearances when the dies are removed from the machine.

Handbook: An engineering book that contains many charts, formulas, and answers to typical questions that arise during part design and process selection. It should be pointed out that all books do not provide the same answers but, in many cases, will provide the reader with sufficient information to make a good decision.

Hot Rolled Steel: Steel produced by rolling hot steel into a flat sheet of various thicknesses. When the sheet or plate cools, the outside surfaces will have a scale or crust that is not smooth. When the material is handled or formed, the scale may flake off, causing the thickness of the material to change slightly. Thin hot rolled steel may be further processed by a "pickling and oiling" process that removes the scale and leaves a fairly smooth top and bottom surface.

Knife: In some parts of the country, the term is used as a description of the upper press brake die. The normal definition used in the fabricating area is the tool steel blade used to cut metal in a squaring shear, or the round tool steel cutting tool used in a slitting line.

Laser: A machine that uses concentrated light technology to cut flat material into various patterns, usually limited to a maximum of 1" thickness or less. Some lasers are designed to weld parts.

Loopy: A slang expression to describe a radius that is not true but that has an undefined appearance. The radius often looks like it is made up of several different radii or is a constantly changing arc.

Male Die: A common name, used prior to the 1980s, for the upper press brake die.

Microinch Finish: A method of comparing the surface finish (or roughness) of a piece of metal with other surfaces to determine product acceptability. Originally developed by General Electric engineers, the initial intent was for a company to purchase a series of small templates. Each of the templates would have a number of small surfaces machined by various methods using different speeds and feeds (i.e., planers, milling machine, shapers, grinders). A "visual" comparison was made between the surface of the material being inspected and a "microinch" template that seemed closest to that surface finish, which then became a standard for future parts. The "microinch" is, from a technical viewpoint, the "root-mean-square average of the measured height deviations taken within the evaluation length or area, and measured from the mean linear surface." The microinch finish that is chosen is specified by using a square root symbol with the microinch roughness desired placed in the "v" section of the square root symbol. Common numbers used to identify the surface roughness are 16, 32, 64, 90, 125, and 250. Special machines are now available to measure the surface roughness and identify the actual microinch surface of the part in question.

Neutral Axis: The theoretical length of a line somewhere in the thickness of a piece of metal that, after it is formed, is the same length as it was when it was in the flat condition.

Offset: A formed part shape where two reverse bends result in one leg of the part being parallel to the original surface, but a specific distance from the original plane.

Operator: The main person designated by management to operate the press brake. It may also be the person who operates the press brake without specific approval of management, but who does so at his own risk. An operator is expected to be a person trained to safely operate the press brake, and who knows the proper way to adjust the ram settings, set gages, and sequence the bends in order to form good parts.

Patent: The legal means whereby a person may own exclusive rights to the use of their idea. A patent must be obtained through the United States Patent Office.

Peaked: In the center of a formed arc of a radius bend, it is possible to have an area which appears to stick out from the rest of the radiused shape and look like a smaller radius than the rest of the arc.

Pick up: See **Galling**.

Pipe: A round, closed tube that can be formed on a press brake. The press brake forming operation will not bend the material out to the very edge of the part; as a result, the pipe may have a peak where it is closed, unless a separate operation is performed to flatten the edges after the pipe has been welded at the seam.

Plate: Flat, hot rolled steel material rolled to a thickness of $^{3}/_{16}$" or thicker. Often the material can range from mild steel to various types of high tensile steel that have severe limitations if it is formed into angles or radiused bends using a press brake.

Polish: To rub (by hand or by power tools) a piece of metal in order to remove any evidence of the forming or machining process, and to obtain a high quality appearance with a very low microinch finish.

PSI: Abbreviation for **pounds per square inch** commonly used to identify the strength of a material.

Punch: The (tool steel) tool that is used to produce a hole in a piece of sheet metal to a specific dimension. The tool directly under the punch and material is called the die or die button. The word "punch" is also used in some parts of the country as another name for the upper press brake die.

Rockwell Hardness: A commercial scale used to measure the hardness of different materials. Instead of a ball shaped tool used in a Brinell Hardness Tester, the Rockwell equipment uses a diamond type tool to penetrate into the surface of the material being tested. The diamond tool is more consistent at penetrating harder surfaces, and is usually used when the test material surface would exceed a Brinell hardness of 350. For metals less than 350 Brinell in hardness, inspectors often select the Brinell hardness method since it uses smaller numbering increments. The Rockwell method is more reliable when the material is heat treated and has a higher hardness.

Seam: A term used to describe a small flange that has been formed in the same manner as a hem. However, the short, parallel flange is kept open or does not touch the longer leg of the part, allowing a clearance of at least one metal thickness.

Seat: The action required to ensure that the die is properly located in the press brake without any clearance between the die shoulders and the ram nose or the filler block.

Set Screw: A headless screw requiring an Allen wrench to turn it. It is used to clamp a part into a location in order to keep it from moving while the part is in use.

Springs: Those used in press brake dies are made from specialized heavy coiled steel. They are color coded to identify the amount of pressure they can generate, how much travel can be safely expected, and to ensure all of the springs being used are the same type. They are often referred to as **die springs**.

Stripper: The method used to pull, push, or remove a completed part that is clamped to the upper die or punch. The design of strippers varies from polyurethane cylinders or spring loaded pressure pads, to solid hook designs used to pull the part off the upper die while the press brake ram returns to its top position.

Stroke: The distance that the press brake ram travels either up or down.

Tongue: The narrow stem machined on the top or on the bottom of a press brake die to provide a method of clamping the die to the ram or onto the filler block. Several styles of tongues are available depending on the manufacturer of the press brake being used.

Wire Draw: When metal is pulled over a sharper than metal thickness radius for some distance, an unwanted distortion, looking like a large arc, can appear in the length of the leg that has contacted the radius. It will change the formed angle and the flange appearance, and it normally causes the part to be discarded.

www.ingramcontent.com/pod-product-compliance
Lightning Source LLC
LaVergne TN
LVHW081937200926
848484LV00009B/2452